GEOPHYSICS AND ASTROPHYSICS MONOGRAPHS

Editor

B. M. McCORMAC, *Lockheed Palo Alto Research Laboratory, Palo Alto, Calif., U.S.A.*

Editorial Board

R. GRANT ATHAY, *High Altitude Observatory, Boulder, Colo., U.S.A.*
W. S. BROECKER, *Lamont-Doherty Geological Observatory, Palisades, New York, U.S.A.*
P. J. COLEMAN, JR., *University of California, Los Angeles, Calif., U.S.A.*
G. T. CSANADY, *Woods Hole Oceanographic Institution, Woods Hole, Mass., U.S.A.*
D. M. HUNTEN, *University of Arizona, Tucson, Ariz., U.S.A.*
C. DE JAGER, *The Astronomical Institute, Utrecht, The Netherlands*
J. KLECZEK, *Czechoslovak Academy of Science, Ondřejov, Czechoslovakia*
R. LÜST, *President Max-Planck Gesellschaft für Förderung der Wissenschaften, München, F.R.G.*
R. E. MUNN, *University of Toronto, Toronto, Ont., Canada*
Z. ŠVESTKA, *The Astronomical Institute, Utrecht, The Netherlands*
G. WEILL, *Service d'Aéronomie, Verrières-le-Buisson, France*

QUANTITATIVE ASPECTS OF MAGNETOSPHERIC PHYSICS

L. R. LYONS

Space Environment Laboratory, National Oceanic and Atmospheric Administration, Boulder, Colorado, U.S.A.

and

D. J. WILLIAMS

Applied Physics Laboratory, Johns Hopkins University, Laurel, Maryland, U.S.A.

D. REIDEL PUBLISHING COMPANY

A MEMBER OF THE KLUWER ACADEMIC PUBLISHERS GROUP

DORDRECHT / BOSTON / LANCASTER

Library of Congress Cataloging in Publication Data

Lyons, L. R.
  Quantitative aspects of magnetospheric physics.

  (Geophysics and astrophysics monographs)
  Includes bibliographical references and index.
    1. Magnetosphere.   I. Williams, D. J., 1933–
  II. Title. III. Series.
  QC809.M35L96   1984   538'.766   83–26886
  ISBN 90–277–1663–3

Published by D. Reidel Publishing Company,
P.O. Box 17, 3300 AA Dordrecht, Holland.

Sold and distributed in the U.S.A. and Canada
by Kluwer Academic Publishers
190 Old Derby Street, Hingham, MA 02043, U.S.A.

In all other countries, sold and distributed
by Kluwer Academic Publishers Group,
P.O. Box 322, 3300 AH Dordrecht, Holland.

All Rights Reserved
© 1984 by D. Reidel Publishing Company, Dordrecht, Holland
and copyright holders as specified on appropriate pages within.
No part of the material protected by this copyright notice may be reproduced or
utilized in any form or by any means, electronic or mechanical,
including photocopying, recording or by any information storage and
retrieval system, without written permission from the copyright owner.

Printed in The Netherlands

*To Robin Lyons and Priscilla Williams,
for their patience, understanding, and, most of all,
for being our wives.*

TABLE OF CONTENTS

PREFACE	ix
LIST OF SYMBOLS	xi
CHAPTER 1: INTRODUCTION	1
CHAPTER 2: CHARGED-PARTICLE MOTION IN MAGNETIC AND ELECTRIC FIELDS	6
2.1. Guiding Center	6
2.2. Dipole Magnetic Field	7
2.3. Gyration	8
2.4. Bounce	11
2.5. Drift	14
2.5.1. Electric Fields	15
2.5.2. Magnetic Field Gradient	16
2.5.3. Field Line Curvature	16
2.5.4. Inertial Forces	17
2.5.5. Total Drift	18
2.6. Particle Distribution Functions	20
2.6.1. Phase Space Density and Differential Flux	20
2.6.2. Omnidirectional Flux	22
2.7. Summary	22
CHAPTER 3: TRAPPING REGION AND CURRENTS DUE TO TRAPPED PARTICLES	28
3.1. Verification of Geomagnetic Trapping Coordinate Systems	28
3.2. Trapping Regions	30
3.3. Trapped Particle Currents	40
3.3.1. Particle Distributions and Currents	40
3.3.2. The Ring Current	44
3.3.3. Ring Current Generation	48
3.3.4. Ring Current Decay	50
CHAPTER 4: ELECTRIC FIELDS	56
4.1. Introduction	56
4.2 The Convection Electric Field	56
4.2.1. Proposed Large-Scale Convection Electric Field	56
4.2.2. Mapping of the Convection Electric Field to the Ionosphere and Resulting Ionosphere Currents	59

	4.2.3.	Evidence for an Open, Polar Cap Magnetic Field	64
	4.2.4.	Cold Plasma Convection and the Plasmapause	75
	4.2.5.	Energetic Plasma Convection	79
4.3.	Current Sheet Energization in the Tail	86
	4.3.1.	Estimate of Total Particle Energization Rate	86
	4.3.2.	Particle Motion in a Current Sheet	87
	4.3.3.	Effects of Current Sheet Energization of Ions in the Geomagnetic Tail	91
4.4.	Auroras and Parallel Electric Fields	99
	4.4.1.	Evidence for the Acceleration of Auroral Electrons by Parallel Electric Fields	99
	4.4.2.	Association of Auroras with Field-Aligned Currents	104
	4.4.3.	Latitudinal Variations over Auroras	104
	4.4.4.	The Current-Voltage Relation along Auroral Field Lines	109
	4.4.5.	Generation of Large-Scale Inverted-V Precipitation Regions	112
	4.4.6.	Smaller Scale Discrete Auroral Structure	120

CHAPTER 5: WAVE-PARTICLE INTERACTIONS — 133

5.1.	General Relations for Wave Growth and Particle Diffusion		133
5.2.	General Results from Cold Plasma Theory		142
5.3.	$\|B_k\|^2$ in Terms of the Measurable Wave Intensity		143
5.4.	Auroral Kilometric Radiation (AKR)		145
5.5.	Whistler-Mode Waves and the Radiation Belts		156
	5.5.1.	Generation of Plasmaspheric Hiss	157
	5.5.2.	General Concepts on Pitch-Angle Diffusion in the Radiation Belts	163
	5.5.3.	Pitch-Angle Diffusion of Radiation Belt Electrons within the Plasmasphere	168
5.6.	Loss of Ring Current Ions by Ion-Cyclotron Waves		180
5.7.	Electrostatic Waves Outside the Plasmapause		190
5.8.	Balance Between Radial Diffusion and Radiation Belt Particle Losses		195
	5.8.1.	Equilibrium Structure of Radiation Belt Electrons within the Plasmasphere	199
	5.8.2.	Equilibrium Structure of Radiation Belt Ions within the Plasmasphere	206

INDEX — 229

PREFACE

The discovery of the earth's radiation belts in 1957 marked the beginning of what is now known as magnetospheric physics. The field has evolved normally from an early discovery phase through a period of exploration and into an era of quantitative studies of the dynamics of magnetized plasmas as they occur in nature. Such environments are common throughout the universe and have been studied in varying detail at the sun, the planets, pulsars, and certain radio galaxies.

The purpose of this book is to describe basic quantitative aspects of magnetospheric physics. We use selected examples from the earth's magnetosphere to show how theory and data together form a quantitative framework for magnetospheric research. We have tried to organize the material along the philosophy of starting simply and adding complexity only as necessary. We have avoided controversial and relatively new research topics and have tried to use as examples physical processes generally accepted as important within the earth's magnetospheric system. However, even in some of our examples, the question of whether the physical process applied to a particular problem is the dominant process, has yet to be answered.

MKS units are used in Chapters 1 through 4. Because of historical precedent and a desire to allow results in this book to be compared with the published literature, Gaussian units are used in Chapter 5, Wave-Particle Interactions. Chapter 1 presents a brief description of the earth's magnetosphere. Chapter 2 describes the motion of charged particles in magnetic and electric fields and introduces the useful approximation of guiding center particle motion in these fields. The earth's trapping regions and the currents established by particles trapped in these regions are presented in Chapter 3. Chapter 4 introduces the magnetospheric electric field and describes the major influence it has on magnetospheric dynamics. Chapter 5 describes wave-particle interactions and their role within the magnetospheric system.

Thanks are due to many people who have helped us in this effort. Specifically, we wish to thank Viola Hill and Gayle Snyder for typing, Jim Adams for drafting, and Lindsay Murdock for editing.

March 1983

L. R. LYONS
D. J. WILLIAMS

LIST OF SYMBOLS

A	area
A	vector potential
a	semi-minor axis of ellipse
B	magnetic field
B(ω)	wave magnetic field as a function of ω, normalized so that $B_{wave}^2 = \int B^2(\omega)\,d\omega$
B_e	equatorial magnetic field strength at earth's surface
B_i	ionospheric magnetic field strength
B$_\mathbf{k}$	magnetic field of a wave mode as a function of **k**
B_0	geomagnetic field strength in equatorial plane
B_D	magnetic field at $r = 0$ due to a particle's azimuthal drift
B_M	magnetic field at mirror point
B_s	magnetic field at earth's surface
$B_{V_\parallel}$	magnetic field strength at top of a potential variation along field lines
B_μ	magnetic field at $r = 0$ due to particle's spiral motion about **B**
B$_{wave}$	total wave magnetic field as a function of position and time
B$_{wave}(\omega, \theta)$	wave magnetic field strength as a function of ω and θ at a particular position and time
c	speed of light
c	as subscript, refers to cold plasma population
D	quasi-linear diffusion matrix in velocity space
$D(\omega^*, \mathbf{k})$	$D(\omega^*, \mathbf{k}) = 0$ is the electrostatic wave dispersion relation
D_{LL}	radial diffusion coefficient
D_{LL}^E	radial diffusion coefficient from electric potential field fluctuations
D_{LL}^M	radial diffusion coefficient from magnetic field fluctuations
D_{oe}	D_{LL}^E for $L = 1$ and $\mu = 0$
D_{om}	D_{LL}^M for $L = 1$
$D_{\alpha\alpha}$	pitch-angle diffusion coefficient, element of **D**
$D_{\alpha v}$	mixed diffusion coefficient, element of **D**
$D_{v\alpha}$	mixed diffusion coefficient, element of **D**
D_{vv}	speed diffusion coefficient, element of **D**
$\mathscr{D}$	cold plasma parameter as defined by Equation (5.13)
d	half-width of current sheet, also used for width of open magnetic field line region as mapped into the interplanetary medium
E	electric field
E$_c$	convection electric field in the equatorial plane
E$_\mathbf{k}$	electric field of wave mode as a function of **k**
E$_{\mathbf{k}, R}$	right-hand polarized component of the perpendicular wave electric field
E$_{\mathbf{k}, L}$	left-hand polarized component of the perpendicular wave electric field

LIST OF SYMBOLS

$\mathbf{E}_R$	corotation electric field
$\mathbf{E}_{\text{wave}}$	total wave electric field as a function of position and time
e	electronic charge, also, as subscript, refers to electrons
$\langle E^2 \rangle$	mean square strength of convection electric field fluctuations
$F(t)$	time-varying part of equatorial particle distribution function when it is separated into two parts, $f_0 = F(t)g(\alpha_0)$
$F(\omega/\Omega_i)$	the function $(\Omega_i/\omega)^2 (1 - \omega/\Omega_i)^3$
$\mathbf{F}$	force
f	particle distribution function in velocity space, normalized so that $\int f(\mathbf{x}, \mathbf{v}) \, d^3\mathbf{v} = N(\mathbf{x})$
f_c	critical value of f for net growth of whistler-mode waves in the radiation belts, based on Kennel and Petschek (1966) (see Chapter 5)
f_0	equatorial particle distribution function
f_p	particle distribution function in momentum space, normalized so that $\int f(\mathbf{x}, \mathbf{p}) \, d^3\mathbf{p} = N(\mathbf{x})$
$f(v, \alpha)$	number of particles having velocities between v and $v + dv$ and pitch angles between α and $\alpha + d\alpha$
$G(L)$	weighting function for energy loss from Coulomb collisions, defined by Equation (5.57)
$\hat{G}$	operator defined by $([\omega/k_\parallel] - v_\parallel)(\partial/\partial v_\perp) + v_\perp(\partial/\partial v_\parallel)$
$g(\alpha_0)$	time independent shape for the equatorial particle distribution function
$g_\omega(\theta)$	distribution of wave energy with wave normal angle at a particular ω
$[H]$	neutral hydrogen density
$\hbar$	Planck's constant divided by 2π
$h_\parallel$	$ck_\parallel/n\Omega$
I	integral invariant for particle bounce motion between mirror points
I_0	Bessel function of order zero and imaginary argument
I_p	height-integrated ionospheric Pedersen current
I_t	current per unit length
$I_\parallel$	total field-aligned current per unit distance
i	as subscript, refers to ions, initial value, or ionospheric value
$\mathbf{i}$	current
$\mathbf{i}_c$	current due to gyration effects
$\mathbf{i}_D$	current due to magnetic field gradient and curvature drifts
J	2nd adiabatic invariant
J_n	Bessel function of the first kind
J_λ	omnidirectional particle flux
$J_\parallel$	total electron flux per unit area precipitating into atmosphere
$\mathbf{j}$	current density
j	differential particle flux, i.e., particles/unit area-s-ster-unit energy
j	as subscript, plasma species
j_H	ionospheric Hall current density
j_p	ionospheric Pedersen current density
$j_\parallel$	magnetic field-aligned current density
K	kinetic energy
$\mathbf{K}_h$	Hermitian part of the dielectric tensor

LIST OF SYMBOLS

K_{th}	thermal energy
$K_\perp$	particle kinetic energy perpendicular to magnetic field
$K_{\parallel,\,min}$	minimum parallel kinetic energy $[(1/2)\,mv_\parallel^2]$ for cyclotron resonance
$K_{\parallel,\,res}$	parallel kinetic energy for resonance with a wave
k	wave vector
k	constant relating ϵ_p to $V_\parallel$
$k_{\parallel,\,res}$	value of $k_\parallel$ for resonance with a wave
L	McIlwain's L-parameter
L_{pp}	L-value of plasmapause
$\mathscr{L}$	cold plasma parameter as defined by Equation (5.13)
ℓ	distance along magnetic field lines
ℓ_c	distance along a single wave characteristic
ℓ_I	length of open magnetic field region as mapped into the interplanetary medium
ℓ_M	distance of mirror point from magnetic equator along field line
m	particle rest mass
m_p	proton rest mass
$\hat{n}$	unit vector along outward normal to a magnetic field line
M	earth's dipole moment
N	plasma density
$N(\omega)$	normalization factor given by Equation (5.22) in transformation from $\mathbf{B}_{wave}(\omega,\theta)$ to $\mathbf{B}_\mathbf{k}$
N_b	bound electron density
n	harmonic resonance number
P^E	power spectrum of electric field fluctuations
P^M	power spectrum of magnetic field fluctuations
$P_\perp$	particle pressure normal to **B**
$P_\parallel$	particle pressure parallel to **B**
p	integer $\geqslant 1$
p	momentum
$\mathscr{P}$	cold plasma parameter as defined by Equation (5.13)
$Q(\alpha_0)$	variation of bounce-averaged angular drift velocity with α_0 multiplied by $s(\alpha_0)$; $\sim (0.35 + 0.15\,\sin\alpha_0)s(\alpha_0)$
q	particle charge
$Q'(\alpha_0)$	variation of Ω_{DE} with α_0 multiplied by $s(\alpha_0)$
r	radial distance from origin
r_0	equatorial crossing distance of a magnetic field line
r_s	r at earth's surface $\equiv 1R_e \sim 6380$ km
R_c	radius of curvature of a magnetic field line
R_e	one earth radius
$\mathbf{R}_g$	guiding center location
$\mathscr{R}$	cold plasma parameter as defined by Equation (5.13)
S	surface separating closed and open geomagnetic field lines
$s(\alpha_0)$	variation of τ_b with pitch angle, $\sim 1.38 - 0.32(\sin\alpha_0 + \sin^{1/2}\alpha_0)$
S_D	disturbed polar ionospheric current system
S_q	quiet-time ionospheric current system

LIST OF SYMBOLS

S_q^p	quiet-time polar ionospheric current system
$\mathscr{S}$	cold plasma parameter as defined by Equation (5.13)
T	exponential decay time
T_c	temperature of cold plasma population
t	time
U^M	total energy of earth's dipole field above earth's surface
$V_\parallel$	total magnetic field-aligned electric potential difference
$V_{\parallel,0}$	maximum value of $V_\parallel$
$\mathbf{V}_D$	drift velocity of particle guiding center
$\mathbf{V}_{DB}$	total guiding center drift velocity due to magnetic field
$\mathbf{V}_{DC}$	guiding center drift velocity due to magnetic field line curvature
$\mathbf{V}_{DE}$	electric field drift velocity
$\mathbf{V}_{DI}$	guiding center drift velocity due to inertial forces
$\mathbf{V}_{DM}$	guiding center drift velocity due to magnetic field gradient
V_e	volume of the earth, $1.08(10)^{21}$ m^3
$\mathbf{V}_{sw}$	solar wind velocity
$\mathbf{v}$	velocity
v_g	group velocity
$\mathbf{v}_R$	corotation velocity
$\mathbf{v}_{\parallel,\min}$	minimum parallel particle velocity for cyclotron resonance
$v_{\parallel,0}$	center of relativistic resonance ellipse in $(v_\perp v_\parallel)$-plane
$v_{\parallel,\text{res}}$	parallel particle velocity for resonance with waves
$\mathscr{V}$	volume
W	sum of kinetic and potential energy
$W_\mathbf{k}$	total wave energy as a function of $\mathbf{k}$, electromagnetic and kinetic
$\mathbf{x}$	position in space
x	horizontal distance
x_i	horizontal distance in the ionosphere
x_w	half-width of large-scale auroral precipitation regions
Z	charge number of ions
α	pitch angle
α_{LC}	equatorial pitch angle of loss cone
α_0	equatorial pitch angle
β	ratio of particle speed to speed of light, v/c
$\Gamma_w(\alpha_0)$	pitch-angle dependence of D_{LL}^M
γ	wave growth rate
γ_r	relativistic mass correction factor $(1 - v^2/c^2)^{-1/2}$
$\mathscr{E}$	electron energy flux per unit energy
$\mathscr{E}_p$	total electron energy flux per unit area precipitating into atmosphere
ϵ	sign of charge
ζ	angle between the tangent to a single-wave characteristic and the $v_\perp$-axis
η	fraction of total number of particles on a flux tube that are within the loss cone, given by Equation (5.36)
θ	wave normal angle (the angle between $\mathbf{B}$ and $\mathbf{k}$)
$\Theta_{n\mathbf{k}}$	weighting factor in linear and quasi-linear wave theory given by Equation (5.4)

LIST OF SYMBOLS

λ	geomagnetic latitude
λ_e	magnetic invariant latitude of a magnetic field line at the earth's surface
λ_M	geomagnetic latitude of mirror point
μ	first adiabatic invariant (magnetic moment = μ/γ_r)
$\tilde{\mu}$	refractive index
μ_B	magnetic moment
μ_0	magnetic permeability of free space, $4\pi(10)^7$ henry m^{-1}
ξ	angle between lobe magnetic field and midplane of current sheet
ρ	gyroradius
Σ_p	height-integrated ionospheric Pedersen conductivity
σ	cross section for charge exchange with neutral hydrogen
σ_H	ionospheric Hall conductivity
σ_p	ionospheric Pedersen conductivity
τ_B	particle bounce period
τ_c	gyroperiod
τ_{CE}	ion lifetime from charge exchange
τ_{DB}	particle drift period around magnetic field axis due to magnetic field gradients and curvature drift
τ_{DE}	particle drift period around magnetic field axis due to electric field drift
τ_m	minimum lifetime for trapped particles under strong pitch-angle diffusion
τ_0	$4LR_e/v$
τ_p	lifetime of trapped particles from precipitation into the atmosphere
$\tau_s(\alpha_0)$	particle source time scale as a function of α_0
ϕ	electric potential
ϕ_i	ionospheric potential
ϕ_{pp}	electric potential of plasmapause
Φ	magnetic flux through particle's gyro-orbit
Φ_M	3rd adiabatic invariant
Φ_{pc}	potential differences across polar cap
Ψ	longitude angle measured eastward with respect to 2400 LT
Ω	particle gyrofrequency
Ω_B	frequency of particle bounce between mirror points
Ω_{DB}	frequency for particle drift around the earth due to magnetic field gradient and field line curvature
Ω_{DE}	frequency for particle drift around the earth due to electric fields
ω	real part of wave frequency
ω^*	complex wave frequency $\omega^* = \omega + i\gamma$
ω_{LH}	lower hybrid frequency
ω_p	plasma frequency
ω_{UH}	upper hybrid frequency
ω_x	cutoff frequency above Ω_e
$\parallel$	as subscript, component of vector quantity parallel to **B**
$\perp$	as subscript, component of vector quantity perpendicular to **B**

CHAPTER 1

INTRODUCTION

The earth's magnetosphere is that region of space defined by the interaction of the solar wind with the earth's dipole-like magnetic field. It extends from approximately 100 km above the earth's surface, where the proton-neutral atom collision frequency is equal to the proton gyrofrequency, to about ten earth radii ($\sim 63\,800$ km) in the sunward direction and to several hundred earth radii in the anti-sunward direction. It is shown schematically in Figure 1.1.

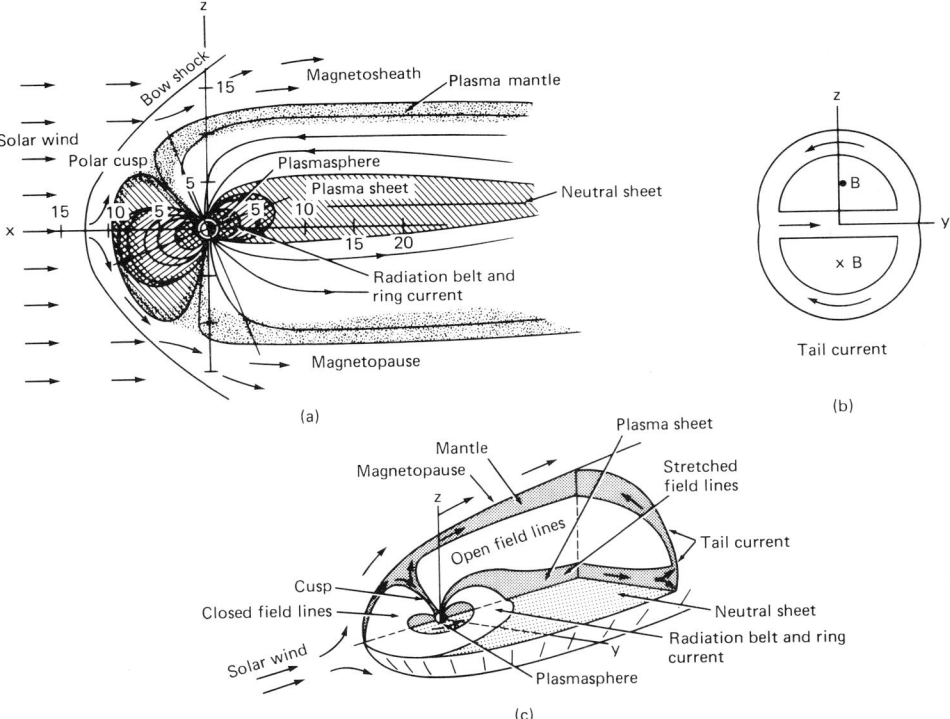

Fig. 1.1. Schematic of major regions of the earth's magnetosphere. Auroral ovals encircle earth at ~ 100 km and in the $65°-75°$ magnetic latitude band. (a) Noon-midnight meridian projection; (b) tail current system as viewed from sun; (c) three-dimensional perspective of (a).

The interaction of the supersonic solar wind with the intrinsic dipole magnetic field of the earth forms the magnetosphere whose boundary, the magnetopause (Figure 1.1), separates interplanetary and geophysical magnetic field and plasma environments.

Upstream of the magnetopause a collisionless bow shock is formed in the solar wind—magnetosphere interaction process. At the bow shock the solar wind becomes thermalized and subsonic and continues its flow around the magnetosphere as magnetosheath plasma, ultimately rejoining the undisturbed solar wind.

A rough estimate of the position of the dayside magnetopause is obtained by balancing the solar wind pressure against the geomagnetic field with the resistive pressure of the geomagnetic field itself:

$$\frac{1}{2} N m_p V_{sw}^2 = \frac{B^2}{2\mu_0}$$

where N = solar wind number density, m_p = proton mass, V_{sw} = solar wind velocity, $B = 0.311 \times 10^{-4}/R_0^3$ tesla is the earth's field at the magnetic equator, and R_0 = geocentric distance in units of earth radii.

Using quiescent solar wind values of $300 \leq V_{sw} \leq 400$ km s^{-1} and $4 \leq N \leq 6$ cm^{-3} gives a magnetopause position in the range 8.8–10.4 earth radii as compared with an average observed value of ~10.8 earth radii. That the observed value is somewhat larger than that predicted by the simple expression above indicates that all terms in the pressure balance equation should be considered, particularly the term representing the pressure of the particle population on the earthward side of the magnetopause.

In the anti-solar direction, observations show that the earth's magnetic field is stretched out in an elongated geomagnetic tail (in analogy with a cometary tail) to distances of several hundred earth radii. The geomagnetic tail field lines originate in high geomagnetic latitudes from the vicinity of the auroral ovals to the geomagnetic pole. Topologically the geomagnetic tail consists of roughly oppositely directed field lines separated by a 'neutral' sheet of nearly zero magnetic field. Surrounding the neutral sheet is a plasma of 'hot' particles, the plasma sheet, having a temperature of 1–10 keV, a density of ~0.01–1 particle cm^{-3}, and a bulk flow velocity of a few tens to a few hundreds of km s^{-1}. An overall tail current system in the shape of a figure eight when observed from the solar direction, is inferred to support the observed tail magnetic field topology (Figure 1.1b).

A further result of the solar wind—magnetosphere interaction is the establishment of a particle population made up of magnetosheath and magnetospheric particles just inside the magnetopause in the region where magnetic field lines have at least one connection to the ionosphere. This region is termed the low latitude boundary layer in the dayside equatorial region, the plasma mantle at high latitudes, and the magnetotail boundary layer in the tail where the particles exhibit strong anti-solar flow (Figure 1.1c).

Deeper in the magnetosphere is the plasmasphere, a population of cold (≤ 1 eV) ionospheric ions and electrons in diffusive equilibrium and corotating with the earth. The region of cold plasma corotation and thus the extent of the plasmasphere is determined in large part by the magnitude of the cross-tail electric field (see Chapter 4).

Throughout the inner magnetosphere there exists a region of trapped particles (the radiation belts and ring current) extending from altitudes where atmospheric losses prevent particle drift around the earth ($\leq$ a few hundred kilometers) to altitudes where magnetic field distortions prevent particle drift around the earth (~ 10 earth radii — see Chapter 3).

Radiation belt particles represent significant energy storage in the magnetosphere ($2(10)^{15}$–$2(10)^{18}$ J). Their gradient and curvature drifts establish a current encircling the earth, the ring current, which is responsible for worldwide depressions of the earth's surface magnetic field. During times of enhanced radiation belt intensities, particle energy densities significantly greater than the ambient magnetic field energy density are observed and can cause surface field variations up to several hundred nanoteslas (nT = 10^{-5} G). The bulk energy density of the ring current particles is contained within the energy range $\sim 1-200$ keV and has a mean of ~ 85 keV. This high-β plasma ($1/2\ \rho v^2 > B^2/2\mu_0$) decays primarily through charge exchange and ion-cyclotron wave generation.

Protons, helium, and oxygen together form the ring current but their relative contributions are unknown. Thus the ultimate source of radiation belt particles, the solar wind or the ionosphere, is still unknown. It is expected that strong energy and spatial dependencies will be evident in the source mixture for the bulk of these particles. The very high energy ($\geqslant$ several 10's of MeV) protons observed at low altitudes ($\leqslant 1.8$ earth radii) are supplied by neutrons, generated in the atmosphere by cosmic rays, which leave the earth's atmosphere and decay in the geomagnetic field.

Radiation belt particles are accelerated to their final energies by $\mathbf{E} \times \mathbf{B}$ convection across field lines and betatron and Fermi acceleration due to slow diffusion across magnetic field lines under conservation of the first two adiabatic invariants. The relative importance of each process has not yet been established. Radiation belt particles having a solar wind source obtain an initial heating in the geomagnetic tail where drift across electric field equipotentials can increase particle energies by amounts up to several tens of keV. Ions from the ionosphere are known to be accelerated to energies of $\sim 1-10$ keV by electric fields parallel to magnetic field lines emanating from the auroral zones. In either case a radiation belt source, most likely the plasma sheet, can be formed at altitudes $\gtrsim 6.5$ earth radii and then can be accelerated as discussed above to form the trapped particle population.

The second large energy storage region in the magnetosphere is the extended geomagnetic tail ($3(10)^{15}-3(10)^{18}$ J). The plasma sheet particles and the extended geomagnetic field lines contribute roughly equal parts to this energy storage. The relationship between the plasma sheet, the aurora, and the radiation belts is an intimate one but is not yet fully understood. As discussed above, the earthward portion of the plasma sheet is a likely source for the radiation belts but the relative contributions of the solar wind and ionosphere are unknown. It is also likely that diffuse auroral forms are due to plasma sheet electrons scattered into the loss cone by electrostatic waves. On the other hand discrete auroral forms are most probably caused by electric fields parallel to auroral magnetic field lines.

Associated with the parallel electric fields responsible for discrete auroral forms is an intense electromagnetic radiation in the 100–1000 kHz frequency band, called auroral kilometric radiation. This radiation, normalized to a planetary radius distance scale, is comparable in intensity with Jupiter's emissions. In fact, integration over respective radiating solid angles may make the earth a radio source of the same order as Jupiter in total power output.

If we consider the earth's magnetosphere in a general sense as a rotating magnetized plasma, we find that such objects are plentiful throughout our solar system and perhaps throughout the universe. This is not surprising since most of the universe is filled with plasma, and the basic interactions between plasmas, electric fields, and magnetic fields being revealed in the earth's magnetosphere are present in the development of cosmological regions from small interstellar clouds to entire galaxies.

Interplanetary spacecraft have identified magnetospheres around Mercury, Saturn, and Jupiter.[1] Astronomers have detected similar structures around rotating neutron stars (pulsars) and radiowave-emitting galaxies. Figure 1.2 illustrates the scale sizes observed for these various magnetosphere-like systems.

In all these cases it is evident that nature has been able to accelerate charged particles to very high energies. In the earth's magnetosphere there are at least four established methods of accelerating particles. The most general of these is charged particle drift across electric field equipotentials. More specific mechanisms are betatron acceleration, Fermi acceleration, and acceleration by electric fields parallel to magnetic field lines. It remains to be shown that field line interconnection can directly transfer energy from the magnetic field to charged particles (field line merging, reconnection) or that plasma turbulence effects are important as acceleration processes.

Magnetospheric systems, while similar, often have their own unique characteristics. For example, Jupiter and Saturn have moons in the heart of the charged particle populations which are effective absorbers creating distinctive features in their radiation belts. At Jupiter, the volcanic moon Io is a copious source of sulfur and oxygen, both of which have been detected at all energies throughout the Jovian magnetosphere. Jupiter's high spin rate (rotation period = 9 hr, 55 min, 29.7 s) can produce effects to accelerate particles in addition to those found in the earth's magnetosphere. For example, low energy plasma corotating with Jupiter's magnetic field will exceed the Alfvén speed and become supersonic well within the Jovian magnetosphere (30–40 Jovian radii). Even tiny Mercury, having neither atmosphere nor ionosphere, possesses a magnetosphere capable of accelerating large numbers of particles to high energies.

Much work remains to be done to understand how the laws of physics operate in interacting magnetized plasma systems that display the range of boundary conditions seen throughout the solar system and in the universe. In this book we describe certain quantitative aspects of magnetospheric physics as used at the earth which illustrate the complex and wondrous ways in which the basic laws of physics enable us to obtain an understanding of our surroundings.

Note

[1] For Mariner-10 Mercury results see *Science* **185**, 141, 1974.
 For Voyager-1 Jupiter results see *Science* **204**, 945, 1979.
 For Voyager-2 Jupiter results see *Science* **206**, 925, 1979.
 For Voyager-1 Saturn results see *Science* **212**, 159, 1981.
 For Voyager-2 Saturn results see *Science* **215**, 499, 1982.

INTRODUCTION

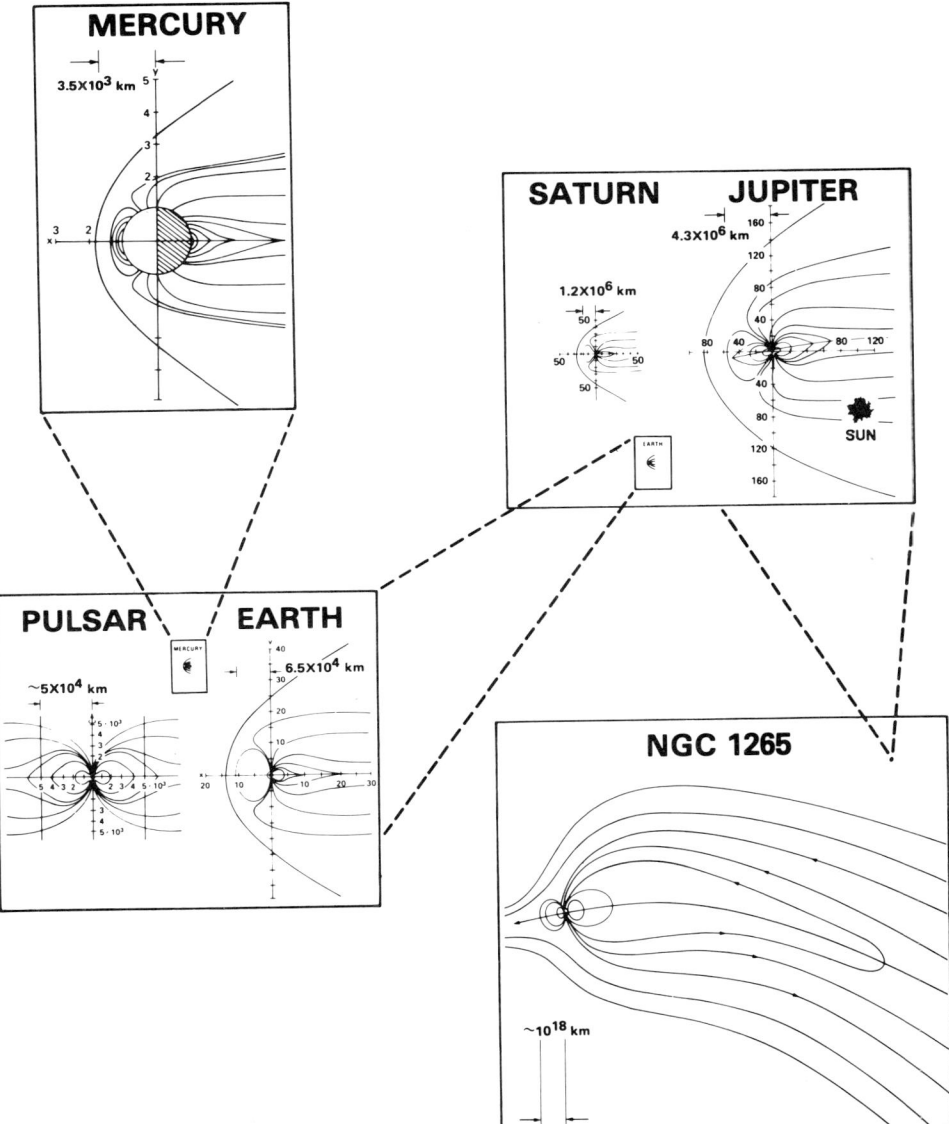

Fig. 1.2. The magnetic field configurations around several familiar objects. Magnetosphere-like systems, in a wide range of scales, are probably common throughout the universe. The subsolar magnetopause distance for Mercury is 3.5 (10^3) km; for the radio galaxy NGC 1265, the analogous distance is roughly 10^{18} km. The visible sun is shown to scale in Jupiter's magnetosphere.

CHAPTER 2

CHARGED-PARTICLE MOTION IN MAGNETIC AND ELECTRIC FIELDS

2.1. Guiding Center

The equation of motion for a charged particle in magnetic and electric fields is well known. In all but the simplest cases, however, this motion is difficult to conceptualize and formal solutions most often must be obtained by numerical integration techniques. For example the pioneering work of Störmer (1907, 1933, 1955) on the motion of a charged particle in a dipole magnetic field showed that there exists no general solution and that the particle trajectory must be obtained by a lengthy numerical integration process.

Fortunately, in many instances there exist approximate solutions to the equation of motion which allow a ready visualization and quick calculation of the particle's trajectory. The most useful approach in tracking charged particles in the earth's magnetosphere has been the guiding center approximation and subsequent development of adiabatic invariant concepts (Alfvén, 1950; Spitzer, 1956; Northrup, 1963). In the guiding center approximation, the instantaneous position of a charged particle moving in a magnetic field is broken down into its circular motion of radius ρ about the magnetic field and the displacement of the center of this circular motion, referred to as the guiding center $\mathbf{R}_g$ (Figure 2.1).

$$\mathbf{r} = \mathbf{R}_g + \rho. \tag{2.1}$$

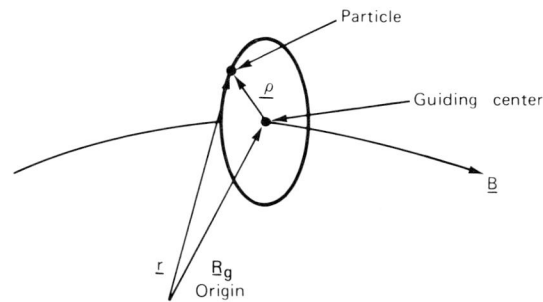

Fig. 2.1. Charged-particle motion in a magnetic field often can be treated by the guiding center approximation where the motion is broken down into the gyration of a particle around the field in a circle of radius ρ and the motion of the center of that circle along and across field lines.

A general expression for the motion of the guiding center can be obtained (Northrup, 1963) by substituting (2.1) into the particle's equation of motion

$$m \frac{d^2 \mathbf{r}}{dt^2} = q\mathbf{E} + q\left(\frac{d\mathbf{r}}{dt} \times \mathbf{B}\right) + m\mathbf{g} \tag{2.2}$$

where m = particle rest mass, q = particle charge, $\mathbf{g}$ = acceleration of gravity, $\mathbf{B}$ = magnetic field, and $\mathbf{E}$ = electric field. This is done by expressing $\boldsymbol{\rho}$ as $\boldsymbol{\rho} = \rho(\hat{j}\sin\theta + \hat{k}\cos\theta)$, time averaging over one gyroperiod so that $\langle\boldsymbol{\rho}\rangle = \langle\dot{\boldsymbol{\rho}}\rangle = \langle\ddot{\boldsymbol{\rho}}\rangle = 0$, and expanding $\mathbf{B}$ and $\mathbf{E}$ in a Taylor series about $\mathbf{R}_g$. The result is the nonrelativistic guiding center equation

$$m\frac{d^2\mathbf{R}_g}{dt^2} = q\mathbf{E} - \mu_B\nabla B + q\left(\frac{d\mathbf{R}_g}{dt}\times\mathbf{B}\right) + m\mathbf{g} + 0(\rho/x) \tag{2.3}$$

where μ_B = particle's magnetic moment due to gyration, x = scale length over which the magnetic field changes appreciably, and $0(\rho/x)$ are terms of order ρ/x. Note that except for the terms $0(\rho/x)$, the guiding center obeys the same equation of motion as a particle in a magnetic field $\mathbf{B}$ and an effective electric field of $\mathbf{E} - (\mu_B/q)\nabla B$.

In the guiding center approximation the particle's motion in the magnetosphere is broken down into three components: gyration about a field line, bounce back and forth along a field line between reflection (mirror) points, and a slow longitudinal drift around the earth. Although these motions are not strictly separate from one another, the vast differences in time scale associated with them makes a mathematical separation possible and leads directly to consideration of the adiabatic invariants. These motions are described in the following sections and are illustrated in Figure 2.2.

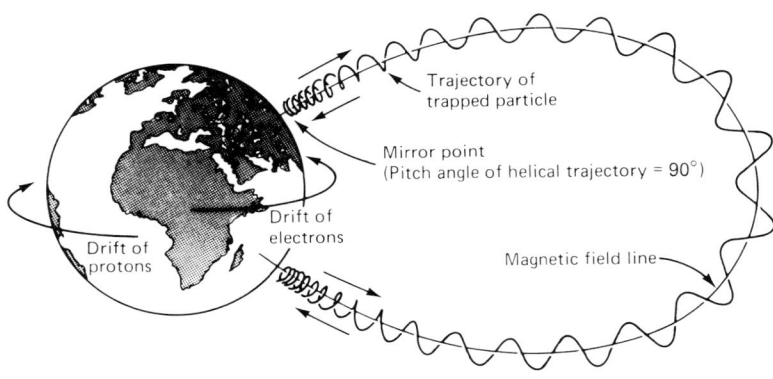

Fig. 2.2. The motion of a charged particle trapped in the earth's magnetic field.

2.2. Dipole Magnetic Field

Since the earth's magnetic field is nearly dipolar, good insight into the motion of charged particles in the earth's field is obtained by using a dipole magnetic field as an example. We summarize here several useful expressions for such a field.

The earth's dipole field can be expressed as

$$\mathbf{B} = \frac{M}{r^3}(-2\sin\lambda\hat{r} + \cos\lambda\hat{\lambda}) \tag{2.4}$$

$$B = \frac{M}{r^3}(1 + 3\sin^2\lambda)^{1/2}. \tag{2.5}$$

Here λ is magnetic latitude, $\hat{r}$ and $\hat{\lambda}$ are unit vectors in the r and λ directions, and M is the earth's dipole moment

$$M = 8.02(10)^{15} \, T$$
$$= 0.311 \, G - R_0^3.$$

The equation of a field line is

$$r = r_0 \cos^2 \lambda \tag{2.6}$$

where r_0 is the equatorial crossing distance of the field line. The element of arc length along a field line is given by

$$(dl^2) = (dr)^2 + r^2(d\lambda)^2. \tag{2.7}$$

Using (2.6), we get

$$\frac{dl}{d\lambda} = r_0 \cos \lambda \, (1 + 3 \sin^2 \lambda)^{1/2} \tag{2.8}$$

or

$$\frac{dl}{dr} = \frac{(1 + 3 \sin^2 \lambda)^{1/2}}{2 \sin \lambda}. \tag{2.9}$$

It is often useful to know the volume contained between a dipole field line and the surface of the earth. This is given by

$$4\pi \int_0^{\lambda_e} \cos \lambda \, d\lambda \int_{r_s}^{r_1} r^2 \, dr = V_e \left[L^3 \left(1 - \frac{1}{L}\right)^{1/2} \times \right.$$
$$\left. \times \left(\frac{0.43}{L^3} + \frac{0.171}{L^2} + \frac{0.229}{L} + 0.457 \right) - \left(1 - \frac{1}{L}\right)^{1/2} \right] \tag{2.10}$$

where r_s = the earth's surface $\equiv 1R_e$ = 6380 km, $r_1 = Lr_s \cos \lambda$, L is the equatorial crossing distance of a field in units of earth radii, $\lambda_e = \arccos(L)^{-1/2}$ and V_e, the earth's volume = $1.08(10)^{21}$ m^3. From (2.10) we obtain the volume between field lines separated by $\Delta L = 1$ (Figure 2.3).

2.3. Gyration

The motion of a charged particle in a magnetic field can be pictured generally as a spiral trajectory along the field line. This motion consists of a gyration about the field line plus a translation along the field line. If the magnetic field doesn't vary much in time and space, the motion can be treated as a circular motion around a guiding center (which to good approximation is the instantaneous center of gyration) that moves along the field line. Additional force terms (electric fields, gravity, magnetic field gradients) in general give the guiding center a drift velocity transverse to the magnetic field. It is

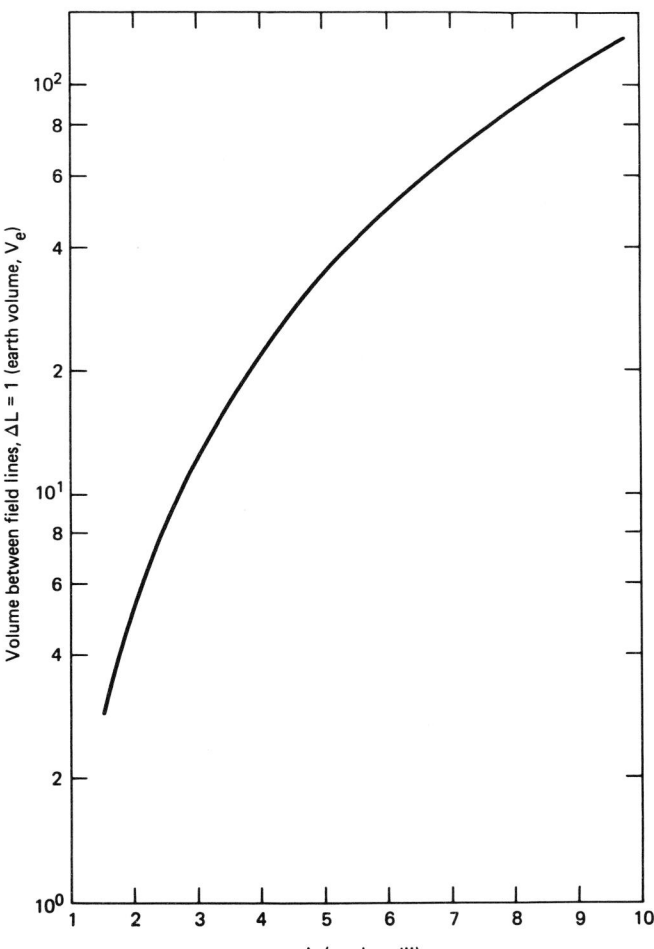

Fig. 2.3. Volume between dipole field lines separated by an equatorial crossing distance of 1 earth radius, in units of an earth volume = $1.08(10)^{21}$ m^3.

only in a reference frame moving with the transverse drift velocity that the particle has a periodic, circular motion about the line of force. Formally, calculations of the particle gyroradius and magnetic moment should be performed in the moving system. However, under most conditions in the magnetosphere, the drift velocities are much less than the particle velocity and negligible errors occur by using the particle velocity in the fixed reference system. Consequently we will use fixed reference system values in the following discussion.

The radius of gyration of a charged particle in a magnetic field B is given by

$$\rho = \frac{m\gamma_r v_\perp}{qB} = \frac{m\gamma_r v}{qB} \sin \alpha \quad \text{(m)}. \tag{2.11}$$

Here m is the rest mass, $\gamma_r = (1 - v^2/c^2)^{-1/2}$, c is the speed of light, $\mathbf{v}$ is the particle velocity, α is the angle between the particle velocity vector and the magnetic field (pitch angle), and q is the particle charge. We also have for the gyroperiod, τ_c, and gyrofrequency, Ω,

$$\tau_c = \frac{2\pi m \gamma_r}{qB} = \frac{1}{\Omega} \quad \text{(s)}. \tag{2.12}$$

If magnetic field changes are small over one gyroradius and gyroperiod, i.e., if

$$\left|\frac{\nabla B}{B}\right| \ll \frac{1}{\rho} \tag{2.13}$$

and

$$\frac{dB}{dt} \ll \frac{B}{\tau_c}, \tag{2.14}$$

then the particle can be considered to be in a nearly static magnetic field which does no work on the particle. By Faraday's Law this implies that the magnetic flux, Φ, through the particle's orbit is constant; i.e., $d\Phi/dt = 0$. We have

$$\Phi = B\pi\rho^2 = \pi \frac{m^2 \gamma_r^2 v_\perp^2}{q^2 B} = \frac{\pi}{q^2} \frac{p_\perp^2}{B} = \text{const}.$$

By multiplying by the constant $q^2/2\pi m$, we get the familiar expression for the first adiabatic invariant μ:

$$\mu = \frac{p_\perp^2}{2mB} = \text{const} \quad \text{1}^{\text{st}} \text{ Adiabatic Invariant} \tag{2.15}$$

Formal calculation of the effects on the particle trajectory of small temporal and spatial changes in B show that (2.15) is the correct relativistic adiabatic invariant associated with the particle's gyration about the field line (Alfvén, 1950; Alfvén and Fälthammar, 1963; Spitzer, 1956; Roederer, 1970).

Under the conditions of (2.13) and (2.14), a particle gyrating around a field line can be considered a circular current $i = (qv_\perp/2\pi\rho)$ generating a magnetic moment

$$\mu_B = iA = \frac{mv_\perp^2 \gamma_r}{2B}. \tag{2.16}$$

We have from (2.15)

$$\gamma_r \mu_B = \mu = \text{const} \tag{2.17}$$

which shows that the particle's magnetic moment, μ_B, is not an adiabatic invariant in the relativistic sense. However, non-relativistically μ_B is an invariant and historically has been identified as the first adiabatic invariant:

$$\mu_B = \frac{mv_\perp^2}{2B} = \frac{K_\perp}{B} = \text{const (non-relativistic)}. \tag{2.18}$$

CHARGED-PARTICLE MOTION IN MAGNETIC AND ELECTRIC FIELDS

Here $K_\perp$ is the particle kinetic energy perpendicular to the field line. It should be remembered that (2.15) or (2.17) must be used in the relativistic case. It is noted again that all quantities in (2.15) through (2.18) refer to the moving coordinate system in which the particle's transverse motion is a circular one about the line of force.

2.4. Bounce

Equation (2.18) leads directly to the mirror equation describing the reflection of a charged particle in a region of converging field lines. We have

$$\mu_B = \frac{K_\perp}{B} = \frac{K \sin^2 \alpha}{B} = \text{const.}$$

In a static field, K is constant and

$$\frac{\sin^2 \alpha_1}{B_1} = \frac{\sin^2 \alpha_2}{B_2} = \text{const.} \qquad (2.19)$$

As a particle moves into a converging field region, α increases to $\pi/2$ ($\sin \alpha = 1$), at which point the particle reverses its motion and travels back along the field line in the direction from which it came. Resolution of the $\mathbf{v} \times \mathbf{B}$ force acting on particles traveling toward higher B values in a converging field yields a component, $qvB_\perp$, which is always directed away from the higher field, resulting in mirroring as described by (2.19). When the earth's equatorial magnetic field, B_0, is used as a reference, the particle's pitch angle anywhere along a dipole field line is determined by the particle's pitch angle at the equator. From (2.19), (2.5), and (2.6), we have

$$\sin^2 \alpha = \sin^2 \alpha_0 \frac{B}{B_0} = \sin^2 \alpha_0 \frac{(1 + 3 \sin^2 \lambda)^{1/2}}{\cos^6 \lambda}. \qquad (2.20)$$

At the mirror point, $\sin^2 \alpha \equiv 1$, and we obtain for the relation between the equatorial pitch angle and mirror point B value, B_M, and latitude, λ_M,

$$B_M = \frac{B_0}{\sin^2 \alpha_0} \qquad (2.21a)$$

$$\sin^2 \alpha_0 = \frac{\cos^6 \lambda_M}{(1 + 3 \sin^2 \lambda_M)^{1/2}}. \qquad (2.21b)$$

To a good approximation, $\sin \alpha_0 \sim \cos^4 \lambda_M$.

We also have for the particle's perpendicular and parallel velocity components

$$v_\perp^2 = v^2 \sin^2 \alpha = v^2 \sin^2 \alpha_0 \frac{B}{B_0} \qquad (2.22a)$$

$$v_\parallel^2 = v^2 \cos^2 \alpha = v^2 \left(1 - \sin^2 \alpha_0 \frac{B}{B_0}\right). \qquad (2.22b)$$

In a dipole-like field configuration such as the earth's having a minimum value of B along field lines, particles simply will bounce back and forth between conjugate mirror points in northern and southern hemispheres as described by Equations (2.19)–(2.22) (see Figure 2.2). Figure 2.4 shows the variation of α_0 with λ_M, $v_\parallel/v$ with λ, and B/B_0 (an often-used measure of distance from the geomagnetic equator) with λ. For example a particle with equatorial pitch angle of $\sim 12°$ travels to a magnetic latitude of $50°$ and spends most of its latitudinal travel with $v_\parallel > v_\perp$.

A range of pitch angles (the loss cone) can be defined from (2.21a) in which trapping is not possible because particles with such pitch angles impact the atmosphere and are

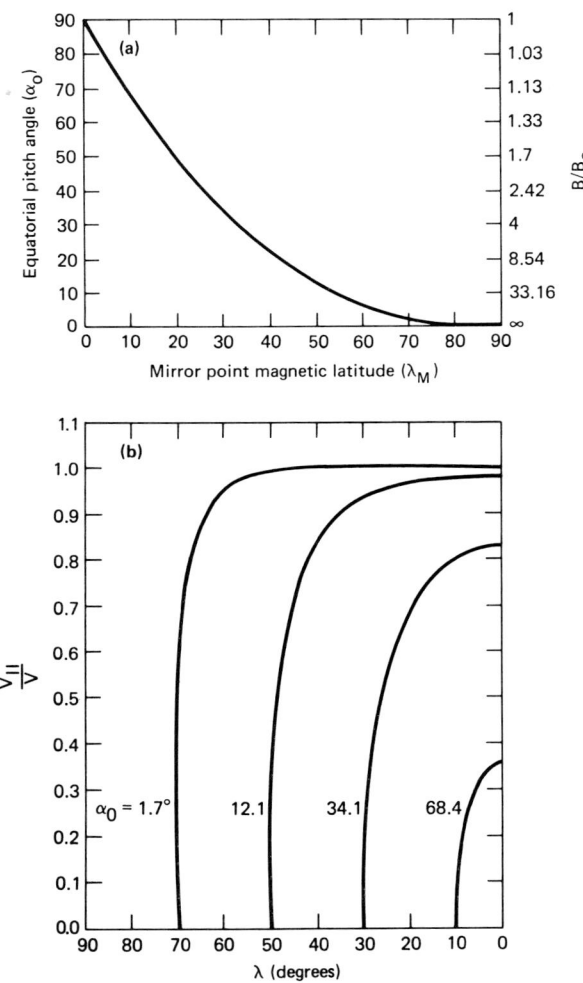

Fig. 2.4. Particle motion on dipole field lines: (a) equatorial pitch angle, α_0, and mirror point to equatorial magnetic field ratio, B/B_0, vs mirror point latitude, λ_M; (b) parallel component of particle velocity, $v_\parallel/v$, vs magnetic latitude, λ. At $v_\parallel = 0$, $\lambda = \lambda_M$.

absorbed. When we use an altitude of 100 km as the effective edge of the dense atmosphere, the atmospheric loss cone is given by

$$\alpha_0 = \arcsin (B_0/B_{100})^{1/2} \qquad (2.23)$$

where B_{100} is the field value at 100 km. All particles with equatorial pitch angles in the range 0 to α_0 will enter the atmosphere.

The total travel time from a mirror point to its opposite hemisphere conjugate and back is the bounce period. It is given by

$$\tau_B = 4 \int_0^{\lambda_M} \frac{dl}{v_\parallel} = 4 \int_0^{\lambda_M} \frac{dl}{d\lambda} \frac{d\lambda}{v_\parallel} = \frac{4r_0}{v} s(\alpha_0) \quad (s) \qquad (2.24a)$$

where dl is the incremental arc length given by (2.8) and

$$s(\alpha_0) = \int_0^{\lambda_M} \frac{\cos \lambda (1 + 3 \sin^2 \lambda)^{1/2} \, d\lambda}{[1 - \sin^2 \alpha_0 (1 + 3 \sin^2 \lambda)^{1/2}/(\cos^6 \lambda)]^{1/2}}. \qquad (2.24b)$$

A good approximation for $s(\alpha_0)$ is given by (Hamlin et al., 1961)

$$s(\alpha_0) \sim 1.30 - 0.56 \sin \alpha_0. \qquad (2.25)$$

A better one from Wentworth (1960) is

$$s(\alpha_0) \sim 1.38 - 0.32(\sin \alpha_0 + \sin^{1/2} \alpha_0). \qquad (2.26)$$

We see that the bounce period is quite independent of equatorial pitch angle, varying less than a factor of two from $0 \leq \alpha_0 \leq \pi/2$. This is because the spiral path length of the particle's trajectory, dl^*, is related to the field line arc length dl, by

$$dl^* = \frac{dl}{\cos \alpha} = r_0 s(\alpha_0).$$

This shows that it is the spiral path length that is nearly independent of pitch angle and also independent of the particle gyroradius.

The bounce frequency corresponding to the bounce period is

$$\Omega_B = \frac{1}{\tau_B} = \frac{v}{4r_0} \frac{1}{s(\alpha_0)} \qquad (s^{-1}). \qquad (2.27)$$

If field variations are small during a particle bounce period, i.e., if

$$\frac{\tau_B}{B} \cdot \frac{\partial B}{\partial t} \ll 1, \qquad (2.28)$$

a second adiabatic invariant can be obtained from the action integral associated with the particle's bounce back and forth on the field lines (Alfvén and Fälthammar, 1963; Northrup and Teller, 1960). It is given by

$$J = \oint p_\parallel \, dl = 2 \int_{l_{M1}}^{l_{M2}} \gamma_r m v_\parallel \, dl \qquad \text{2nd Adiabatic Invariant} \qquad (2.29)$$

where the integral is taken along the field line between two conjugate mirror points at distances l_{M1} and l_{M2}. From (2.21a) and (2.22b) the integral invariant, I, is defined

$$I = \frac{J}{2\gamma_r m v} = \int_{l_{M1}}^{l_{M2}} \left(1 - \frac{B}{B_M}\right)^{1/2} dl. \tag{2.30}$$

The convenience of using the I integral is that it is conserved in a static magnetic field and depends only on the field configuration. If $\partial B/\partial t \neq 0$ the actual invariant J must be used to describe the particle's motion.

2.5. Drift

It was mentioned earlier that the system in which the charged particle motion is periodic and circular is one moving with the transverse drift velocity of the instantaneous center of gyration. However, under most magnetospheric conditions the drift velocity is much smaller than the particle velocity, and the instantaneous center of gyration can be considered identical to the guiding center. Thus the drift velocities discussed below will be referred to as guiding center drift velocities.

Consider a charged particle in a magnetic field and under the action of an external force, **F**. Its equation of motion is

$$\frac{d\mathbf{p}}{dt} = \mathbf{F} + q\mathbf{v} \times \mathbf{B}. \tag{2.31}$$

Since in the moving system the particles execute circular motion around the field line, the force term $\mathbf{F}'_\perp = \mathbf{F}_\perp$ must be compensated by an induced electric field force, $q\mathbf{E}'$. The fields in the moving system are given by

$$\mathbf{B}' \simeq \mathbf{B} \tag{2.32}$$

$$\mathbf{E}' = \mathbf{E} + \mathbf{V}_D \times \mathbf{B} \tag{2.33}$$

where $\mathbf{V}_D$ is the (transverse) velocity of the moving system (the drift velocity of the guiding center) and $V_D \ll c$. (Primed quantities denote the moving system). For the case of a general external force we have

$$q\mathbf{E}' + \mathbf{F}'_\perp = q\mathbf{V}_D \times \mathbf{B} + \mathbf{F}_\perp = 0. \tag{2.34}$$

Taking the cross product of (2.34) with $\mathbf{B}/qB^2$

$$\mathbf{V}_D = \frac{\mathbf{F}_\perp \times \mathbf{B}}{qB^2} = \frac{\mathbf{F} \times \mathbf{B}}{qB^2}. \tag{2.35}$$

The drift velocity is seen to be perpendicular to both the external force **F** and the magnetic force **B**. Thus there will be no change in the particle's kinetic energy averaged over a gyro-orbit, provided that magnetic field variations are small over the orbit.

We note here that the velocity of the particle in the moving frame is

$$\mathbf{v}' = \mathbf{v} - \mathbf{V}_D \tag{2.36}$$

and that $v'_\perp$ should be used in our calculations. Under the approximations of adiabatic theory we find that for most instances in magnetospheric physics $V_D \ll v$ and we can use $v' = v$. In regions of large electric fields and small magnetic fields (2.36) must be used for particle energies up to several hundred eV. However, in many of these cases the adiabatic approximations are violated and numerical integrations become necessary to obtain the particle's trajectory.

The component of the force parallel to $\mathbf{B}$, $\mathbf{F}_\parallel$ does not contribute to the transverse drift. It determines the guiding center motion along the field line,

$$\frac{d\mathbf{p}}{dt} = \mathbf{F}_\parallel \tag{2.37}$$

For example, if the force term $\mathbf{F}$ is due to magnetic gradients, we have for a magnetic dipole of moment μ_B

$$\mathbf{F} = -\mu_B \nabla \mathbf{B} \tag{2.38}$$

and

$$F_\parallel = -\mu_B \nabla_\parallel B = -\mu_B \frac{\partial B}{\partial \ell}$$

where $\nabla_\parallel$ is the gradient along the field line. Thus

$$\frac{dp_\parallel}{dt} = -\mu_B \frac{\partial B}{\partial \ell}$$

showing that the parallel gradient force drives the particle (of magnetic moment μ_B) away from regions of increasing field strength along a field line, and is, in fact, the mirror force.

We now use (2.35) to obtain expressions for the drift velocity in various external force fields of interest to magnetospheric physics.

2.5.1. ELECTRIC FIELDS

For an electric field we have $\mathbf{F} = q\mathbf{E}$ and consequently

$$\mathbf{V}_{DE} = \frac{\mathbf{E} \times \mathbf{B}}{B^2}. \tag{2.39}$$

Note that this drift is independent of particle charge, mass, and energy and depends only on the electric-magnetic field configuration.

2.5.2. MAGNETIC FIELD GRADIENT

Substitution of (2.38) into (2.35) yields

$$\mathbf{V}_{DM} = -\frac{\mu_B \nabla B \times \mathbf{B}}{qB^2} \tag{2.40}$$

or

$$\mathbf{V}_{DM} = -\frac{mv_\perp^2 \nabla B \times \mathbf{B}}{2qB^3} \tag{2.41}$$

or

$$\mathbf{V}_{DM} = -\frac{K \sin^2 \alpha \, \nabla B \times \mathbf{B}}{qB^3}. \tag{2.42}$$

Unlike electric field drift, the magnetic field gradient drift is dependent on particle energy, charge, and pitch angle. Figure 2.5 presents a convenient way to visualize the particle drift in the fixed reference frame due to a magnetic field gradient. In the region of weaker field (top of Figure 2.5) the particle orbit is larger and in the region of stronger field the orbit is smaller, thereby causing a motion to the right.

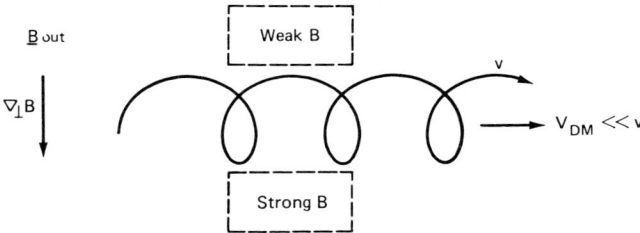

Fig. 2.5. Charged-particle drift in presence of a magnetic field gradient. Particle motion spanning regions of weak field (large gyroradius) and strong field (small gyroradius) produces a drift to the right.

As mentioned earlier, because the drift is perpendicular to **B**, the particle's kinetic energy remains constant. However, this requires that the variation of B over the particle's gyro-orbit must be small; i.e.,

$$\rho \frac{\nabla_\perp B}{B} \ll 1.$$

This together with (2.42) shows that $V_{DM} \ll v$ and $v' \sim v$.

2.5.3. FIELD LINE CURVATURE

The guiding center motion along the field line causes a centrifugal force due to field line curvature given by

$$\mathbf{F} = \frac{mv_\parallel^2}{qR_c} \hat{n} \tag{2.43}$$

where R_c is the radius of curvature of the field line and $\hat{n}$ is a unit vector along the outward normal. This yields a drift velocity

$$\mathbf{V}_{DC} = \frac{mv_\parallel^2}{qR_c} \frac{\hat{n} \times \mathbf{B}}{B^2}. \tag{2.44}$$

For a field with negligible currents, $\nabla \times \mathbf{B} = 0$, the field gradient and curvature are related by (Wentworth, 1960; Roederer, 1970)

$$\frac{1}{\mathbf{R}_c} = \frac{\nabla_\perp B}{B}$$

and

$$\hat{n} = -\frac{R_c \nabla_\perp B}{B}.$$

We can write (2.44) as

$$\mathbf{V}_{DC} = -\frac{mv_\parallel^2 \nabla_\perp B \times \mathbf{B}}{qB^3} \tag{2.45}$$

or

$$\mathbf{V}_{DC} = -\frac{2K \cos^2 \alpha \nabla_\perp B \times \mathbf{B}}{qB^3}. \tag{2.46}$$

Equations (2.45) and (2.46) can be combined with (2.41) and (2.42) respectively to give the following useful expressions for the total drift produced by field gradients and field line curvature, $\mathbf{V}_{DB}$:

$$\mathbf{V}_{DB} = \frac{m}{2qB^3} (v_\perp^2 + 2v_\parallel^2) \mathbf{B} \times \nabla B \tag{2.47}$$

$$\mathbf{V}_{DB} = \frac{K}{qB^3} (1 + \cos^2 \alpha) \mathbf{B} \times \nabla B. \tag{2.48}$$

2.5.4. INERTIAL FORCES

These are forces, such as time-dependent electric fields, that cause the moving frame of reference to be accelerated. If $\mathbf{V}_D$ is the non-inertial force drift velocity, then in the moving frame a force appears due to the acceleration $\dot{\mathbf{V}}_D$,

$$\mathbf{F} = -m\dot{\mathbf{V}}_D \tag{2.49}$$

and

$$\mathbf{V}_{DI} = -\frac{m\dot{\mathbf{V}}_D \times \mathbf{B}}{qB^2}. \tag{2.50}$$

Consider for example a time-dependent perpendicular electric field, $\mathbf{E}_\perp$. From (2.39), $\dot{\mathbf{V}}_D = \dot{\mathbf{E}}_\perp \times \mathbf{B}/B^2$ and

$$\mathbf{V}_{DI} = \frac{m\dot{\mathbf{E}}_\perp}{qB^2}. \tag{2.51}$$

This is called the polarization drift. In general the inertial drift velocity is much smaller than the drift velocities discussed earlier and is usually ignored in most magnetospheric applications.

2.5.5. TOTAL DRIFT

Combining (2.39) and (2.48), we obtain the drift velocity of main interest in magnetospheric physics:

$$\mathbf{V}_D = \frac{\mathbf{B}}{qB^2} \times \left[-q\mathbf{E} + \frac{K}{B}(1 + \cos^2 \alpha)\nabla B \right] \quad (\text{m s}^{-1}). \tag{2.52}$$

The addition of (2.50) includes inertial forces. For other effects such as gravitation or collisional processes, the corresponding drift velocities are obtained from (2.34).

From (2.52) we can obtain a bounce average angular drift frequency and drift period about the magnetic field axis. Consider first the dipole magnetic field drift velocity, $\mathbf{V}_{DB}$, (2.48). To obtain the angular drift frequency divide the angular drift, $\Delta\Psi$, that occurs in one bounce period by the bounce period, τ_B (2.23):

$$\Delta\Psi = 4 \int_0^{\lambda_M} \frac{dl}{d\lambda} \frac{V_{DB}}{r \cos \lambda} \frac{d\lambda}{v_\|} \quad (\text{rad}). \tag{2.53}$$

Using (2.6), (2.8), (2.24a), (2.48), $v_\| = v \cos \alpha$, and the fact that in a dipole field

$$\left| \frac{\mathbf{B} \times \nabla B}{B^3} \right| = \frac{3 \cos^5 \lambda}{r_0 B_0} \frac{(1 + \sin^2 \lambda)}{(1 + 3 \sin^2 \lambda)^2}$$

we get for the angular drift velocity, $2\pi\Omega_{DB}$,

$$\langle 2\pi\Omega_{DB} \rangle = \frac{\Delta\Psi}{\tau_B} = \frac{6K}{qB_0 r_0^2} \frac{Q(\alpha_0)}{s(\alpha_0)} \quad (\text{rad s}^{-1}) \tag{2.54}$$

where $\langle \rangle$ denotes a bounce average and

$$Q(\alpha_0) \equiv \int_0^{\lambda_M} \frac{\cos^3 \lambda (1 + \sin^2 \lambda)}{(1 + 3 \sin^2 \lambda)^{3/2}} \cdot$$

$$\cdot \frac{\left[1 - \frac{(\sin^2 \alpha_0)}{2} \frac{(1 + \sin^2 \lambda)^{1/2}}{\cos^6 \lambda} \right]}{\left[1 - \sin^2 \alpha_0 \frac{(1 + \sin^2 \lambda)^{1/2}}{\cos^6 \lambda} \right]^{1/2}} d\lambda. \tag{2.55}$$

The expression $Q(\alpha_0)/s(\alpha_0)$ has been evaluated by Hamlin et al. (1961) and is approximated by

$$\frac{Q(\alpha_0)}{s(\alpha_0)} \simeq 0.35 + 0.15 \sin \alpha_0. \tag{2.56}$$

We then have

$$\langle 2\pi\Omega_{DB} \rangle \simeq \frac{6K}{qB_0 r_0^2} (0.35 + 0.15 \sin \alpha_0). \tag{2.57}$$

For the bounce average drift period and velocity we have

$$\langle \tau_{DB} \rangle = \frac{1}{\Omega_{DB}} = \frac{\pi q B_0 r_0^2}{3K} \frac{s(\alpha_0)}{Q(\alpha_0)} \quad \text{(s)} \tag{2.58}$$

and

$$\langle V_{DB} \rangle = \frac{2\pi r_0}{\langle \tau_{DB} \rangle} = \frac{6K}{qB_0 r_0} \frac{Q(\alpha_0)}{s(\alpha_0)} \quad (\text{m s}^{-1}). \tag{2.59}$$

It is seen that while Ω_{DB} and τ_{DB} depend only slightly on equatorial pitch angle (as was the case for the bounce-related quantities Ω_B (2.27) and τ_B (2.24a)) they do depend on particle charge, energy, and field geometry.

A similar process can be used to find the angular velocity and drift period due to an ambient electric field. V_{DE} is used in place of V_{DB} in (2.53)

$$\Delta\Psi = 4 \int_0^{\lambda_M} \frac{dl}{d\lambda} \frac{V_{DE}}{r \cos \lambda} \frac{d\lambda}{v_\parallel}. \tag{2.60}$$

If we assume a magnetic dipole field with equipotential field lines, then from conservation of magnetic flux we have

$$\frac{E_\perp}{E_{\perp 0}} = \frac{A}{A_0} = (B_0/B)^{1/2}$$

where A is flux tube area and the subscript '0' refers to equatorial ($\lambda = 0$) values. This gives for the electric field drift velocity

$$V_{DE} = \frac{E_\perp}{B} = \frac{E_{\perp 0}}{B_0} \frac{\cos^9 \lambda}{(1 + \sin^2 \lambda)^{1/2}}$$

and an angular velocity

$$2\pi\Omega_{DE} = \frac{\Delta\Psi}{\tau_B} = \frac{E_{\perp 0}}{r_0 B_0} \frac{Q'(\alpha_0)}{s(\alpha_0)}. \tag{2.61}$$

where

$$Q'(\alpha_0) = \int_0^{\lambda_M} \frac{\cos^7 \lambda (1 + 3 \sin^2 \lambda)^{1/2}}{(1 + \sin^2 \lambda)^{3/4} [1 - \sin^2 \alpha_0 (1 + 3 \sin^2 \lambda)^{1/2}/(\cos^6 \lambda)]} d\lambda$$

and $s(\alpha_0)$ is given by (2.24b). The drift period is given by

$$\tau_{DE} = \frac{1}{\Omega_{DE}} = \frac{2\pi r_0 B_0}{E_{\perp 0}} \frac{s(\alpha_0)}{Q'(\alpha_0)}. \tag{2.62}$$

Expressions (2.61) and (2.62) are of limited use because of the lack of easily used and visualized algorithms for magnetospheric electric fields. Note that an exact expression for electric field drift velocity will depend on both a magnetospheric model and an electric field model. Because of this it usually is most convenient to estimate the electric field drift from (2.39) at some convenient place in the magnetic field (e.g., the equator) and to transform along a field line (using $BA = B_0 A_0$) to the point of interest in one's favorite magnetic field model.

If field changes are small during a particle drift period, a third adiabatic invariant, the flux invariant, can be defined. Considering magnetic variations only, we have that if

$$\frac{\tau_{DB} \dot{B}}{B} \ll 1$$

then

$$\Phi_M \equiv \int B \, ds \qquad 3^{\text{rd}} \text{ Adiabatic Invariant} \tag{2.63}$$

is conserved. Φ_M is the magnetic flux linked in the particle's drift. In a dipole field the drift surface linking Φ_M is the figure of azimuthal revolution of a line of force. Because of the large spatial and temporal scales (relative to μ and J) associated with Φ_M, it is the most easily isolated of the adiabatic invariants.

2.6. Particle Distribution Functions

2.6.1. Phase Space Density and Differential Flux

In the absence of sources and losses, Liouville's Theorem states that the density of particles in phase space f_p is constant along the particle trajectory. That is,

$$f_p = \frac{dN}{dx \, dy \, dz \, dp_x \, dp_y \, dp_z} = \text{const.} \tag{2.64}$$

If we orient the coordinate system with the z axis along the nominal velocity vector $\mathbf{v}$, we have $dx \, dy = dA$ (the area element normal to $\mathbf{v}$), $dz = v \, dt$, and $dp_x \, dp_y \, dp_z = p^2 \, dp \sin\theta \, d\theta \, d\psi = p^2 \, dp \, d\Omega$, where $d\Omega$ = element of solid angle. Using $v \, dp = dK$ we can express (2.64) as

$$\frac{dN}{p^2 \, dA \, dt \, d\Omega \, dK} = \text{const.} \tag{2.65}$$

However $dN/dA \, dt \, d\Omega \, dK$ is just the differential flux, j, (particles/m^2-s-ster-Joule) measured by a detector. We now have

$$f_p \equiv \frac{j}{p^2} = \text{const} \tag{2.66}$$

CHARGED-PARTICLE MOTION IN MAGNETIC AND ELECTRIC FIELDS 21

which is a convenient way of expressing the phase space density of charged particles in the magnetosphere. (See Roederer, 1970, for a discussion of other particle distribution function expressions). The natural units of (2.66) are s^3/m^6 kg^3. An often used form of (2.66) in magnetospheric physics is the velocity space distribution function $m^3 f_p = f$. Common units for j and K are $(cm^2\text{-s-ster-keV})^{-1}$ and keV respectively. With these units, we have for example, for protons

$$f = m^3 f_p = \frac{m^2 j}{2K} = 5.45(10)^{-31} \frac{j}{K} \quad \left(\frac{s^3}{cm^6}\right). \tag{2.67}$$

In a static magnetic field (no electric fields), p is constant and we have from (2.66) that j = constant along the particle trajectory.

As an example of the validity of Equation (2.66) consider a particle distribution traveling along a field line with no losses (Figure 2.6). Under the conditions of adiabatic

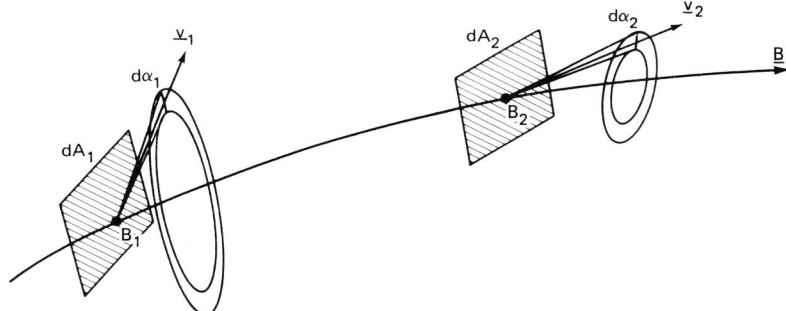

Fig. 2.6. Particle motion from point 1 (dA_1, $\mathbf{B}_1$, $d\alpha_1$, $\mathbf{v}_1$) to point 2 (dA_2, $\mathbf{B}_2$, $d\alpha_2$, $\mathbf{v}_2$) along a magnetic field line.

theory, all particle guiding centers (effectively all particles) passing through a unit area at point 1, dA_1 in Figure 2.6, will also pass through a unit area at point 2, dA_2. Assuming azimuthal symmetry around the field line we have

$$dN = j_1(\alpha_1) \, dA_1 \, \cos\alpha_1 \, 2\pi \, d(\cos\alpha_1) \, dK \, dt$$

$$= j_2(\alpha_2) \, dA_2 \, \cos\alpha_2 \, 2\pi \, d(\cos\alpha_2) \, dK \, dt. \tag{2.68}$$

Conservation of magnetic flux gives

$$B_1 \, dA_1 = B_2 \, dA_2 \tag{2.69}$$

and conservation of the 1st adiabatic invariant (2.15) gives

$$\frac{p_1^2 \sin^2\alpha_1}{B_1} = \frac{p_2^2 \sin^2\alpha_2}{B_2} \tag{2.70}$$

from which we get

$$\frac{p_1^2 \cos \alpha_1 \, d(\cos \alpha_1)}{B_1} = \frac{p_2^2 \cos \alpha_2 \, d(\cos \alpha_2)}{B_2}. \tag{2.71}$$

Combining (2.69) and (2.71) and (2.68) yields

$$\frac{j_1(\alpha_1)}{p_1^2} = \frac{j_2(\alpha_2)}{p_2^2} \tag{2.72}$$

which is identical to (2.66).

2.6.2. Omnidirectional Flux

The omnidirectional flux at any point along a field line (the guiding center trajectory) is given by

$$J_\lambda = 4\pi \int_0^{\pi/2} j(\alpha_\lambda) \sin \alpha_\lambda \, d\alpha_\lambda. \tag{2.73}$$

Using the constancy of j in a static magnetic field, we have

$$j_\lambda(\alpha_\lambda) = j_0(\alpha_0) \tag{2.74}$$

where α_0 and α_λ are related by (2.19), $\sin^2 \alpha_\lambda = B_\lambda/B_0 \sin^2 \alpha_0$. Differentiating and solving for $\sin \alpha_\lambda \, d\alpha_\lambda$ gives

$$\sin \alpha_\lambda \, d\alpha_\lambda = \frac{B_\lambda (1 - \sin^2 \alpha_0)^{1/2}}{[1 - (B_\lambda/B_0) \sin^2 \alpha_0]^{1/2}} \sin \alpha_0 \, d\alpha_0 \tag{2.75}$$

and

$$J_\lambda = \frac{4\pi B_\lambda}{B_0} \int_0^{\alpha^*} j_0(\alpha_0) \frac{(1 - \sin^2 \alpha_0)^{1/2}}{[1 - (B_\lambda/B_0) \sin^2 \alpha_0]^{1/2}} \sin \alpha_0 \, d\alpha_0 \tag{2.76}$$

where $\alpha^* = \arccos(B_0/B_\lambda)$.

Equation (2.76) shows that the omnidirectional intensity at any point along the field line can be obtained from a measurement of the pitch-angle distribution at the equator, $j_0(\alpha_0)$. However, measuring either the omnidirectional or the directional intensity at some point, λ, on the field line does not allow a determination of the equatorial distribution since no information is obtained concerning the number of particles mirroring at lower latitudes than the point of observation. If J_λ is known at *all* points along the field line, then it is possible but still difficult to reconstruct $j_0(\alpha_0)$ (Ray, 1960; Lenchek *et al.*, 1961).

2.7. Summary

We summarize in Table 2.1 a number of expressions useful for estimating charged particle motion in the magnetosphere. The following conventions apply to all expressions:

All energies, K, to be expressed in keV
All equatorial radial distances, R_0, are in units of earth radii
Electric field, E, is in V m^{-1}
subscript p = proton; subscript e = electron
$m_p = 1.67(10)^{-27}$ kg $= 9.38(10)^5$ keV
$m_e = 9.11(10)^{-31}$ kg $= 511$ keV
N_i = mass in units or proton mass
Z = charge in units of electronic charge $1.6(10)^{-19}$ coulombs
ϵ = sign of charge, -1 for electrons
m = particle rest mass

$$\beta = \frac{v}{c} = \left[1 - \left(\frac{mc^2}{K+mc^2}\right)^2\right]^{1/2}$$

$$\gamma_r = (1-\beta^2)^{-1/2} = \frac{K+mc^2}{mc^2}$$

$$B = \frac{0.311}{R_0^3} \text{ G}$$

$$s(\alpha_0) \approx 1.30 - 0.56 \sin \alpha_0$$

$$\frac{Q(\alpha_0)}{s(\alpha_0)} \approx 0.35 + 0.15 \sin \alpha_0.$$

We have included the relativistic correction term even though it can be set to 1 in many cases. For reference we show in Figure 2.7 the variation of γ_r and β with energy K for protons and electrons. Also shown in Figure 2.7 is the variation of magnetic rigidity, $B\rho$, as a function of energy, K.

Table 2.2 shows, for reference, representative gyro, bounce, and drift times of electrons and protons in the earth's dipole field.

TABLE 2.1
Useful expressions for particle motion in the magnetosphere (K in keV, B in G, E in V m^{-1}, R_0 in earth radii)

Item	Units	General expression	At magnetic dipole equator and $\alpha = \pi/2$	Comments
Gyroradius	km	$\rho_p = \dfrac{0.046(K)^{1/2}}{B}\gamma_r \sin\alpha$	$0.148(K)^{1/2} R_0^3 \gamma_r$	For heavy ions multiply ρ_p by $N_i^{1/2}/Z$
		$\rho_e = \dfrac{1.07(10)^{-3}(K)^{1/2}}{B}\gamma_r \sin\alpha$	$3.44(10)^{-3}(K)^{1/2} R_0^3 \gamma_r$	
Gyroperiod	s	$\tau_c^p = \dfrac{6.56(10)^{-4}}{B}\gamma_r$	$2.12(10)^{-3} R_0^3 \gamma_r$	For heavy ions multiply τ_c^p by N_i/Z
		$\tau_c^e = \dfrac{3.75(10)^{-7}}{B}\gamma_r$	$1.15(10)^{-6} R_0^3 \gamma_r$	
Gyrofrequency	(s)$^{-1}$	$\Omega_p = \dfrac{1524\, B}{\gamma_r}$	$\dfrac{474}{R_0^3 \gamma_r}$	For heavy ions multiply Ω_p by Z/N_i
		$\Omega_e = 2.8(10)^6 \dfrac{B}{\gamma_r}$	$\dfrac{8.7(10)^5}{R_0^3 \gamma_r}$	
Bounce period	s	$\tau_B^p = 58.3\, \dfrac{R_0 s(\alpha_0)}{K^{1/2}}$	$43.1\, \dfrac{R_0}{K^{1/2}}$	For heavy ions multiply τ_B^p by $N_i^{1/2}$
		$\tau_B^e = 1.36\, \dfrac{R_0 s(\alpha_0)}{K^{1/2}}$	$\dfrac{R_0}{K^{1/2}}$	Relativistically $\tau_B^e = 8.5(10)^{-2}\,[R_0 s(\alpha_0)]/\beta$

Quantity	Units	Formula	Numerical	Notes
Bounce frequency	$(s)^{-1}$	$\Omega_B^p = 0.017 \dfrac{\hat{v}}{R_0 s(\alpha_0)}$	$0.023 \dfrac{\hat{v}}{R_0}$	For heavy ions multiply Ω_B^p by $N_i^{-1/2}$
		$\Omega_B^e = 0.74 \dfrac{K^{1/2}}{R_0 s(\alpha_0)}$	$\dfrac{K^{1/2}}{R_0}$	Relativistically $\Omega_B^e = 11.8\,\beta/[R_0 s(\alpha_0)]$
Drift velocity	m s^{-1}	$V_{DE} = 10^4 \dfrac{E_\perp}{B}$	$3.2(10)^4 R_0^3 E_\perp$	Electric field drift, all particles drift eastward
		$\langle V_{DB}\rangle = 9.4\,\dfrac{K\gamma_r}{B_0 R_0}\dfrac{Q(\alpha_0)}{s(\alpha_0)}$	$15.2\,K R_0^2 \gamma_r$	Bounce average magnetic drift; electrons drift eastward, protons drift westward. For heavy ions multiply $\langle V_{DB}\rangle$ by Z^{-1}
Drift period	hr	$\tau_{DE} = 1.11\,\dfrac{R_0 B}{E_\perp}$	$\dfrac{0.345}{E_\perp R_0^2}$	
		$\langle \tau_{DB}\rangle = \dfrac{1184\, B_0 R_0^2}{\gamma_r K}\dfrac{s(\alpha_0)}{Q(\alpha_0)}$	$\dfrac{736}{K R_0 \gamma_r}$	For heavy ions multiply $\langle \tau_{DB}\rangle$ by Z
Drift frequency	(hr)$^{-1}$	$\Omega_{DE} = 0.9\,\dfrac{E_\perp}{R_0 B}$	$2.9\,E_\perp R_0^2$	
		$\langle \Omega_{DB}\rangle = 8.4(10)^{-4}\,\dfrac{R_0 B \gamma_r}{E_\perp}$	$1.36(10)^{-3} K R_0 \gamma_r$	For heavy ions multiply $\langle \Omega_{DB}\rangle$ by Z^{-1}

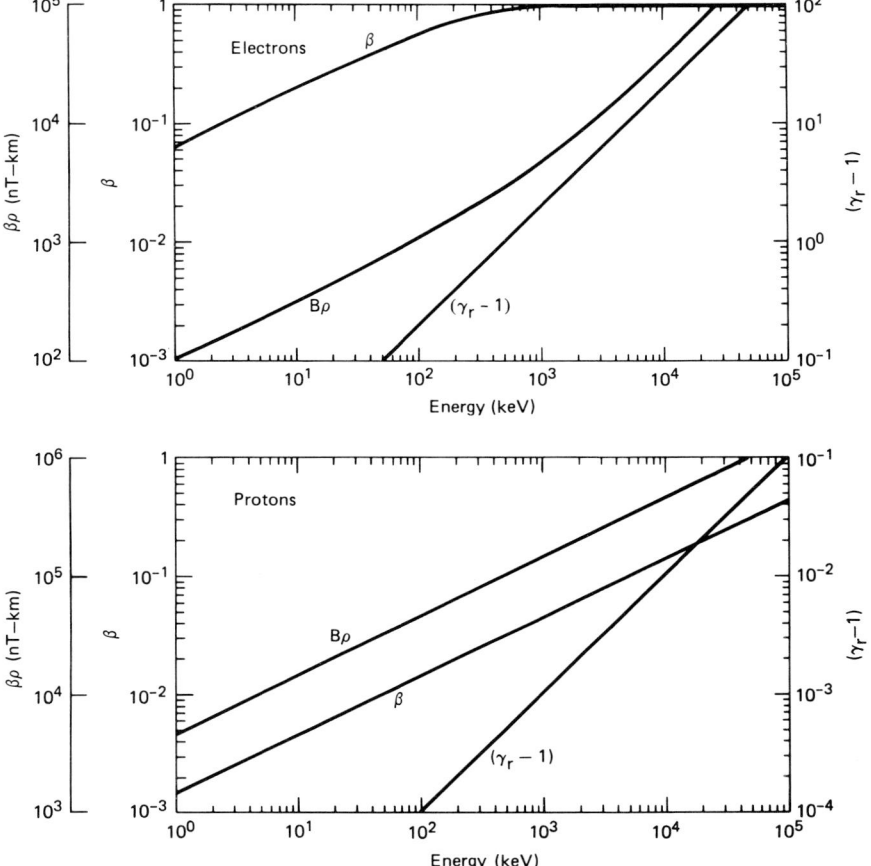

Fig. 2.7. Values of $\beta = v/c$, magnetic rigidity $B\rho$(nT − km), and $\gamma_r - 1 = (1 - \beta^2)^{-1} - 1$ vs energy for electrons and protons.

TABLE 2.2
Characteristic times associated with particles trapped in the earth's magnetic field for $R_0 = 4$ and $\alpha_0 = \pi/2$

		τ_C (s) $\propto R_0^3$	τ_B (s) $\propto R_0$	τ_{DB} (hr) $\propto \dfrac{1}{R_0}$
Electrons	1 keV	$7.4(10)^{-5}$	4.0	184
	10	$7.5(10)^{-5}$	1.3	18.4
	100	$8.8(10)^{-5}$	0.46	1.5
	1000	$2.2(10)^{-4}$	0.27	0.061
Protons	1 keV	0.14	172	184
	10	0.14	54.5	18.4
	100	0.14	17.2	1.8
	1000	0.14	5.45	0.18

References

Alfvén, H., *Cosmical Electrodynamics*, 1st ed. Oxford Univ. Press, London and New York (1950).
Alfvén, H. and C. G. Fälthammar, *Cosmical Electrodynamics*, 2nd ed., Oxford Univ. Press, London and New York (1963).
Hamlin, D. A., R. Karplus, R. C. Vik, and K. M. Watson, 'Mirror and azimuthal drift frequencies for geomagnetically trapped particles', *J. Geophys. Res.* **66**, 1 (1961).
Lenchek, A. M., S. F. Singer, and R. C. Wentworth, 'Geomagnetically trapped electrons from cosmic ray albedo neutrons', *J. Geophys. Res.* **66**, 4027 (1961).
Northrup, T. G., *The Adiabatic Motion of Charged Particles*, Wiley (Interscience), New York, 109 pp. (1963).
Northrup, T. G. and E. Teller, 'Stability of the adiabatic motion of charged particles in the earth's field', *Phys. Rev.* **117**, 215 (1960).
Ray, E. C., 'On the theory of protons trapped in the earth's magnetic field', *J. Geophys. Res.* **65**, 1125 (1960).
Roederer, J. G., *Dynamics of Geomagnetically Trapped Radiation*, Springer-Verlag, New York, Heidelberg, Berlin, 166 pp. (1970).
Spitzer, L., *Physics of Fully Ionized Gases*, Wiley (Interscience), New York, 105 pp. (1956).
Störmer, C., 'Sur les trajectoires des corpuscules electrisiés dans l'éspace sous l'action du magnetism terrestre', *Arch. Sci. Phys.* **24**, No. 5 (1907).
Störmer, C., 'On the trajectories of electrical particles in the field of a magnetic dipole with applications to the theory of cosmic radiation', *Avh. Norske Videnskap. Oslo, Mat. Nat. Kl* **11** (1933).
Störmer, C., *The Polar Aurora*, Oxford Univ. Press, London and New York, 403 pp. (1955).
Wentworth, R. C., 'Lifetimes of geomagnetically trapped particles determined by Coulomb scattering', Ph.D. Thesis, University of Maryland, 112 pp. (1960).

CHAPTER 3

TRAPPING REGION AND CURRENTS DUE TO TRAPPED PARTICLES

3.1. Verification of Geomagnetic Trapping Coordinate Systems

Although early descriptions of charged particle motion in the geomagnetic field were applied mainly to cosmic ray trajectories, geomagnetic storms, and auroral phenomena (Störmer, 1907, 1933, 1955; Chapman and Ferraro, 1931, 1932; Alfvén, 1939, 1940, 1950; Singer, 1956, 1957a, b), the concept of charged particle trapping, the overall trapping region geometry in relation to a dipole magnetic field, and trapped particle motion within such a field were well understood quantitatively (see Chapter 2) before the discovery of the earth's trapped radiation (Van Allen et al., 1958). Verification of the interpretation that the newly observed charged particles above the atmosphere were trapped geomagnetically came from a number of experiments, all of which demonstrated the direct control exerted by the earth's magnetic field on the observed particle intensities. An early example of this is shown in Figure 3.1 (from Yoshida et al., 1960) where the ordering of the data by the earth's magnetic field is clearly demonstrated.

Direct experimental verification of the trapping ability of the geomagnetic field was obtained from the Argus series of high altitude nuclear detonations (Christofolis, 1959;

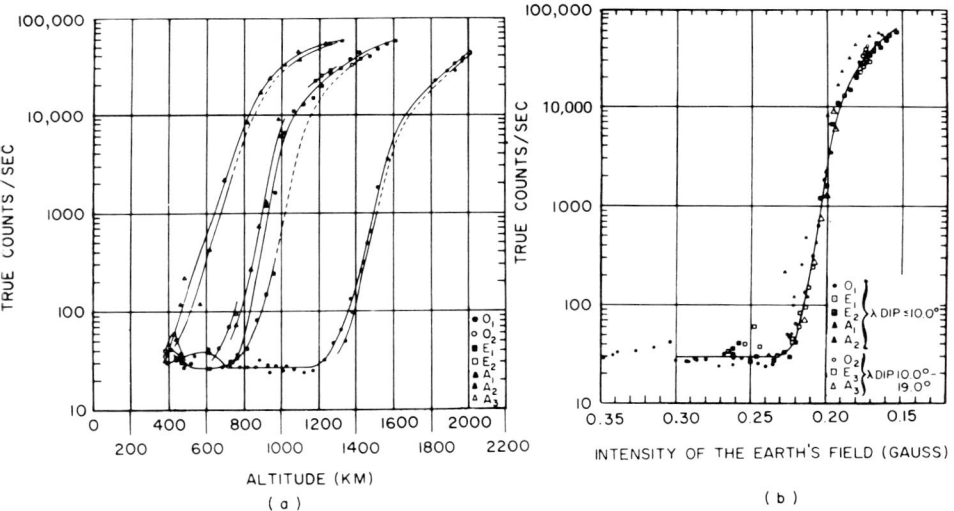

Fig. 3.1. Example of an early observation clearly demonstrating the control of the earth's magnetic field over the observed charged particle population. (Left) Count rate versus altitude curves for several different longitudes near the magnetic dip equator; O_1 and O_2 cover the span 93.4° E to 113.9° E; E_1, E_2, and E_3 cover 12.7° E to 3° W; and A_1, A_2, and A_3 cover 60.9° W to 84.6° W. (Right) Same data plotted versus the intensity of the earth's field (from Yoshida et al., 1960). (© by American Geophysical Union)

Van Allen et al., 1959). These explosions produced electron intensity enhancements that were spatially confined at all longitudes to a particular magnetic shell. These peaked intensity distributions were observed for time periods much greater than the particle drift times and thus directly displayed the high trapping efficiency of the geomagnetic field.

Related to these Argus results was the development of the B, L coordinate system (McIlwain, 1961) based on the adiabatic invariants which describe the motion of a charged particle trapped in the earth's magnetic field. The value of the magnetic field B and the integral invariant I (Equation (2.30)) are obtained for any observational point in space using the earth's magnetic field as represented by a polynomial expansion of the surface field. The equatorial crossing distance L of a dipole field line passing through the point in space having these values of B and I is then obtained. Along most field lines in the earth's field at low altitudes, L varies by $\leqslant 1\%$. This transformation of each observational point in the earth's field to an idealized dipole field yields a coordinate system not only based on the adiabatic invariants but also allowing a convenient spatial conception of the trapping regions. Each point is characterized by its position along a field line, B, and by the equatorial crossing distance of its field line, L. As the particles populating a field line drift in azimuth, they will remain on a drift surface which is simply the azimuthal figure of revolution of the field line characterized by a given L-value, the L-shell.

The formal procedure used to obtain B, L-values for any point in the geomagnetic field is as follows (McIlwain, 1961). From Equations (2.8) and (2.30) we can express the integral invariant, I, as

$$I = r_0 f_1(\lambda). \tag{3.1}$$

We have also

$$B = \frac{M}{r_0^3} \frac{(1 + 3 \sin^2 \lambda)^{1/2}}{\cos^6 \lambda} \equiv \frac{M}{r_0^3} f_2(\lambda). \tag{3.2}$$

From (3.1) and (3.2) we get

$$\frac{I^3 B}{M} = \frac{r_0^3 B}{M} f_1^3(\lambda) = \frac{r_0^3 B}{M} f_3\left(\frac{r_0^3 B}{M}\right) \equiv f_4\left(\frac{r_0^3 B}{M}\right) \tag{3.3}$$

or

$$\frac{r_0^3 B}{M} = f_5\left(\frac{I^3 B}{M}\right). \tag{3.4}$$

Equation (3.4) tells us that in a static dipole field there exist functions of I and B that depend only on r_0 and thus are exactly constant along a line of force. The parameter L is now defined as

$$\frac{L^3 B}{M} = f_5\left(\frac{I^3 B}{M}\right) \tag{3.5}$$

where I and B are evaluated in the earth's real (or model) field and the function f_5 is evaluated for a dipole field.

The success of this coordinate system in ordering trapped particle data from all spatial locations within the geomagnetic field has further substantiated the concept of stable geomagnetic trapping. An example of the ordering power of the B, L coordinate system and an indication of its great usefulness is shown in Figure 3.2 (from McIlwain, 1961). Here are shown both the position in L-value of the charged particles resulting from the Argus high altitude nuclear bursts as a function of longitude and the first of the now well-known B, L representations of trapped particle intensities.

The B, L system orders charged particle data very well in regions of little field distortion due to magnetospheric boundary, tail, and ring current systems. While it remains a valid coordinate system based on adiabatic invariants, it loses its conceptual ordering ability in regions where the field departs significantly from a dipolar form because L no longer remains constant along a magnetic field line. To date there is no generally accepted coordinate system to map charged particles in the distorted geomagnetic field.

Another set of coordinates used in mapping particles in the geomagnetic field is R, λ defined by

$$R = L \cos^2 \lambda \tag{3.6}$$

$$B = \frac{M}{R^3} (1 + 3 \sin^2 \lambda)^{1/2} = \frac{M}{R^3} \left(4 - \frac{3R}{L}\right)^{1/2}. \tag{3.7}$$

The relationship between the R, λ and B, L system is shown in Figure 3.3.

An additional coordinate system occasionally found useful in magnetospheric physics is the α, β system where α and β are the Euler potentials describing the magnetic field (Northrup, 1963). We have for the vector potential

$$\mathbf{A} = \alpha \nabla \beta \tag{3.8}$$

and magnetic field

$$\mathbf{B} = \nabla \alpha \times \nabla \beta. \tag{3.9}$$

For a dipole field $\beta = \Psi$, the magnetic longitude, and $\alpha = M/r_0$. Thus α, β represent equatorial crossing distance and the local time of a magnetic field line. Stern (1967) has presented values of α and β for the distorted magnetic field of Mead (1964).

3.2. Trapping Regions

The B, L coordinate system fails to order charged particle data when magnetosphere distortions due to magnetopause, tail, and ring current effects become significant. This occurs at $L \geqslant 5-6R_e$ during geomagnetically quiet periods. The initial observation of the effect of such magnetospheric distortions on charged particles was that of a large diurnal variation in low-altitude $\geqslant 40$ keV electron intensities (O'Brien, 1963), whereby a given electron intensity was found at a lower altitude at midnight than at noon. However, the use of invariant theory within realistic geomagnetic field models has been able to order much of the high latitude data and indicates that the concept of stable trapping is also supported in the distorted geomagnetic field (Williams and Mead, 1965).

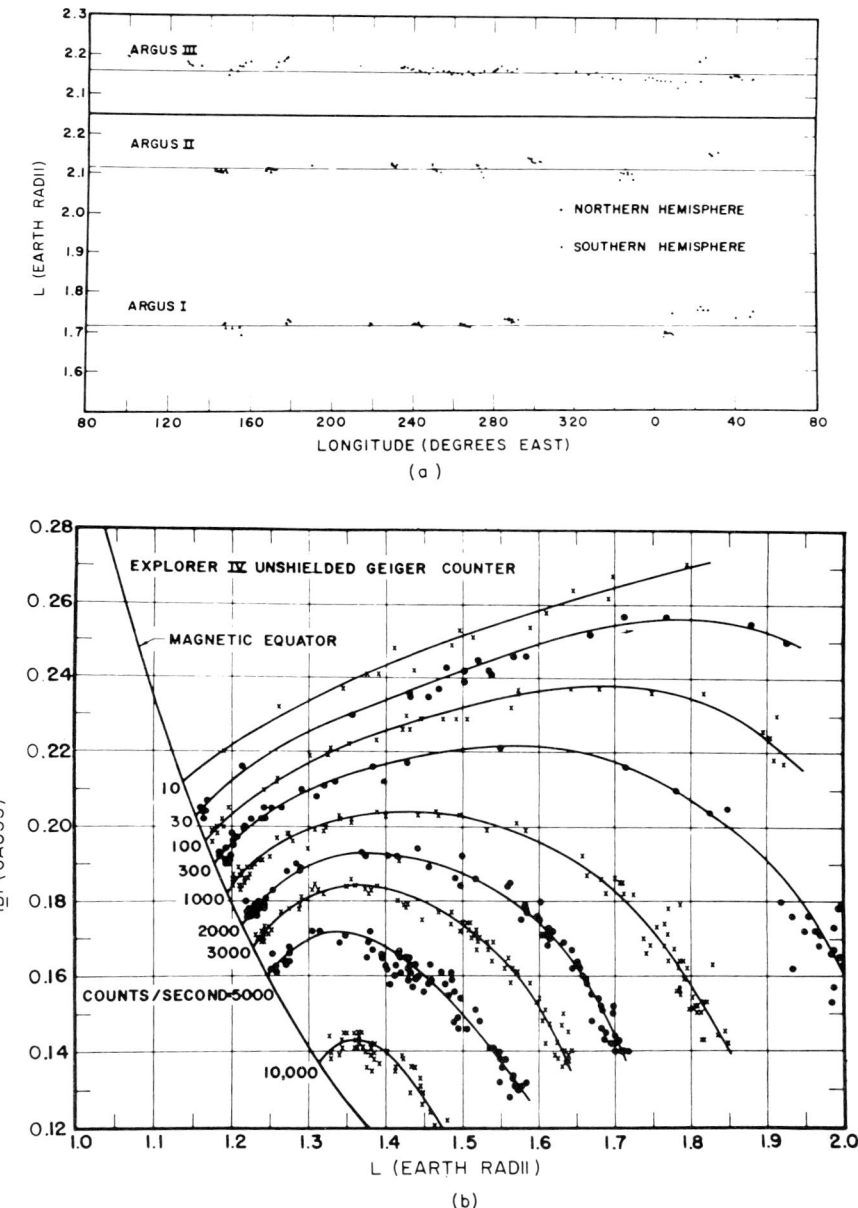

Fig. 3.2. (a) Observed L-values of three Argus shells plotted versus longitude. (b) Contours of constant intensity plotted in B, L coordinate system (from McIlwain, 1961). These initial examples of the B, L system clearly display the natural ordering ability of this coordinate system for charged-particle observations in the earth's magnetic field. Since the B, L system is based on the concepts of particle trapping, its success demonstrates the high trapping efficiency of the geomagnetic field. (© by American Geophysical Union)

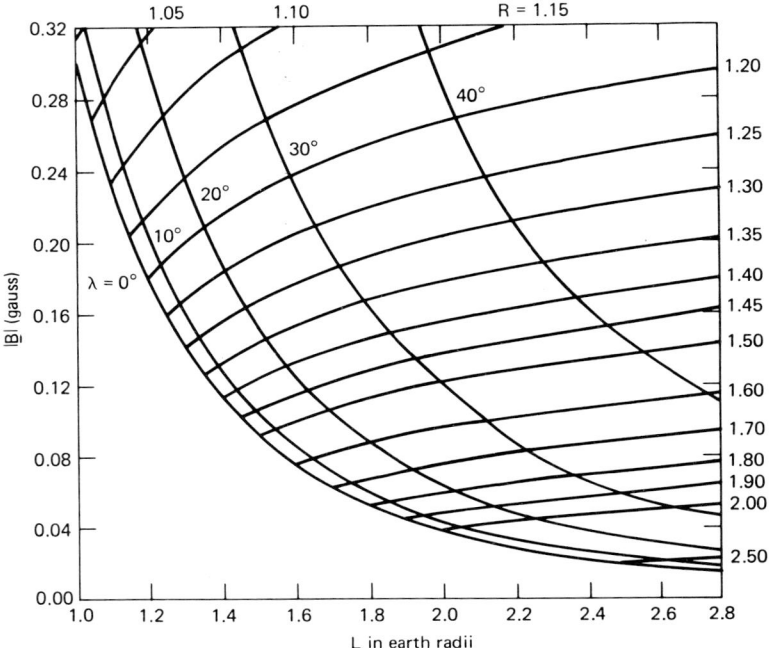

Fig. 3.3. The mapping of the polar coordinates R and λ onto the B, L plane according to the transformation

$$B = \frac{M}{R^3}\left(4 - \frac{3R}{L}\right)^{1/2}, \qquad R = L\cos^2\lambda$$

(from McIlwain, 1961). (© by American Geophysical Union)

Figure 3.4 shows the ordering of energetic electron data obtained at 1100 km in a distorted field. The amount of invariant latitude shift $\Delta\Lambda$, obtained from the latitude profiles shown, is plotted versus the noontime latitude, Λ_D, of observation. The predictions of the latitude shift due to adiabatic motion within a model distorted geomagnetic field are also shown in Figure 3.4 and indicate that these energetic electron spatial distributions are consistent with stable trapping in such a field. The model field used by Williams and Mead (1965), which attempted to include realistically both magnetopause and tail currents, is shown in Figure 3.5.

Calculations using model distorted magnetospheric configurations (Williams and Mead, 1965; Taylor and Hones, 1965; Mead, 1966; Roederer, 1967; Pfitzer et al., 1969; Olson, 1979) have shown the importance of such distortions in the determination of the spatial distributions of the trapped particle population. Not only can many features of these spatial distributions be explained using the simple assumption of adiabatic motion in a realistic field model, but certain time variations in these distributions and in the magnetic field itself also can be explained (Williams and Ness, 1966; Roederer, 1969).

For example, during geomagnetic storms the high latitude edge of the low altitude energetic trapped electron intensity profile displays a characteristic collapse towards low

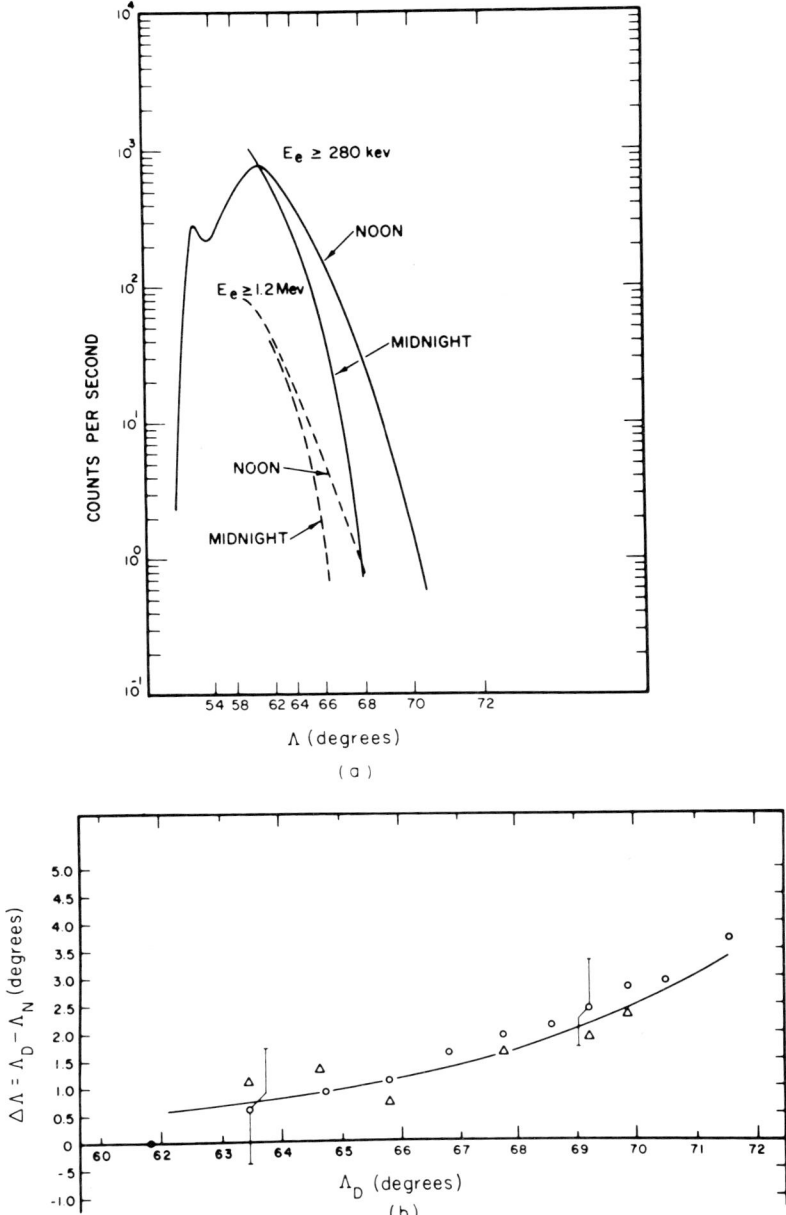

Fig. 3.4. Example of the maintenance of charged particle trapping even in the highly distorted geomagnetic field. (a) Invariant latitude ($\Lambda = \arccos L^{-1/2}$) profiles at 1100 km on the noon-midnight meridian for outer zone energetic electrons. (b) Diurnal variation, $\Delta\Lambda = \Lambda_D - \Lambda_N$ (Λ_D, Λ_N = dayside and nightside latitude of observation, respectively) from data in (a) plotted versus dayside latitude, Λ_D. Solid curve is expected variation based on adiabatic motion of electrons trapped in the distorted geomagnetic field; 0, trapped electrons $E_e \geqslant$ 280 keV-matched pass data; $\triangle$, trapped electrons $E_e \geqslant$ 280 keV – daily averages (from Williams and Mead, 1965). (© by American Geophysical Union)

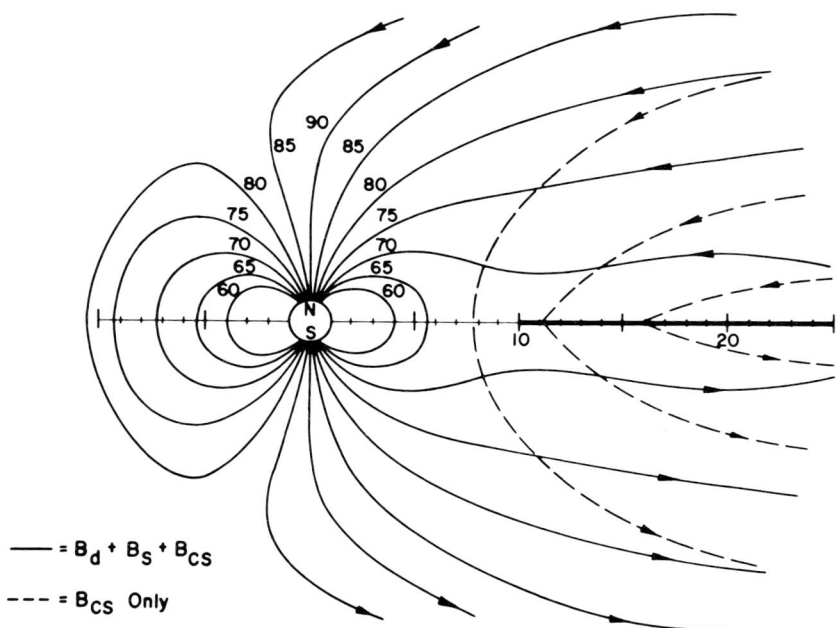

Fig. 3.5. Geomagnetic field model used to explain trapped electron distributions shown in Figure 3.4. Solid lines show the distorted geomagnetic field as made up of components due to the earth's dipole B_d, surface currents at the magnetopause B_S, and a current sheet in the nightside hemisphere B_{CS}. The dashed lines show the field due to the current sheet B_{CS} alone (from Williams and Mead, 1965). (© by American Geophysical Union)

latitudes (Williams, 1966). This is shown in Figure 3.6a where four passes through the outer radiation zone ≥ 280 keV trapped electron region are plotted at various times during a geomagnetic storm. Figure 3.6b shows the close correlation of this boundary collapse with magnetic activity for a several-day period. The boundary collapse during the storm and its subsequent recovery are seen clearly. A possible explanation for such a collapse is the further stretching of field lines on the nightside hemisphere to such an extent that they no longer can support an energetic trapped particle distribution. If these field lines were previously populated, this effect would produce the boundary collapse (disappearance of electrons) observed at low altitudes. In addition, an increase in field strength should be observed simultaneously in the geomagnetic tail.

Figure 3.7 shows one result of a study utilizing simultaneous data on trapped electron intensities at low altitudes and the magnetic field strength in the geomagnetic tail during magnetic storms (Ness and Williams, 1966; Williams and Ness, 1966). Note the increase in tail field intensity during the magnetic disturbance, consistent with additional field lines from lower latitudes extending into the tail. Using the last closed nightside field line in the model of Williams and Mead (1965) as the high latitude boundary for trapped electrons, the observed tail field variations in Figure 3.7 were used to predict the high latitude boundary during the course of the storm. These predictions are also shown and compared with the measured boundary in Figure 3.7. Note that although the absolute

TRAPPING REGION AND CURRENTS DUE TO TRAPPED PARTICLES 35

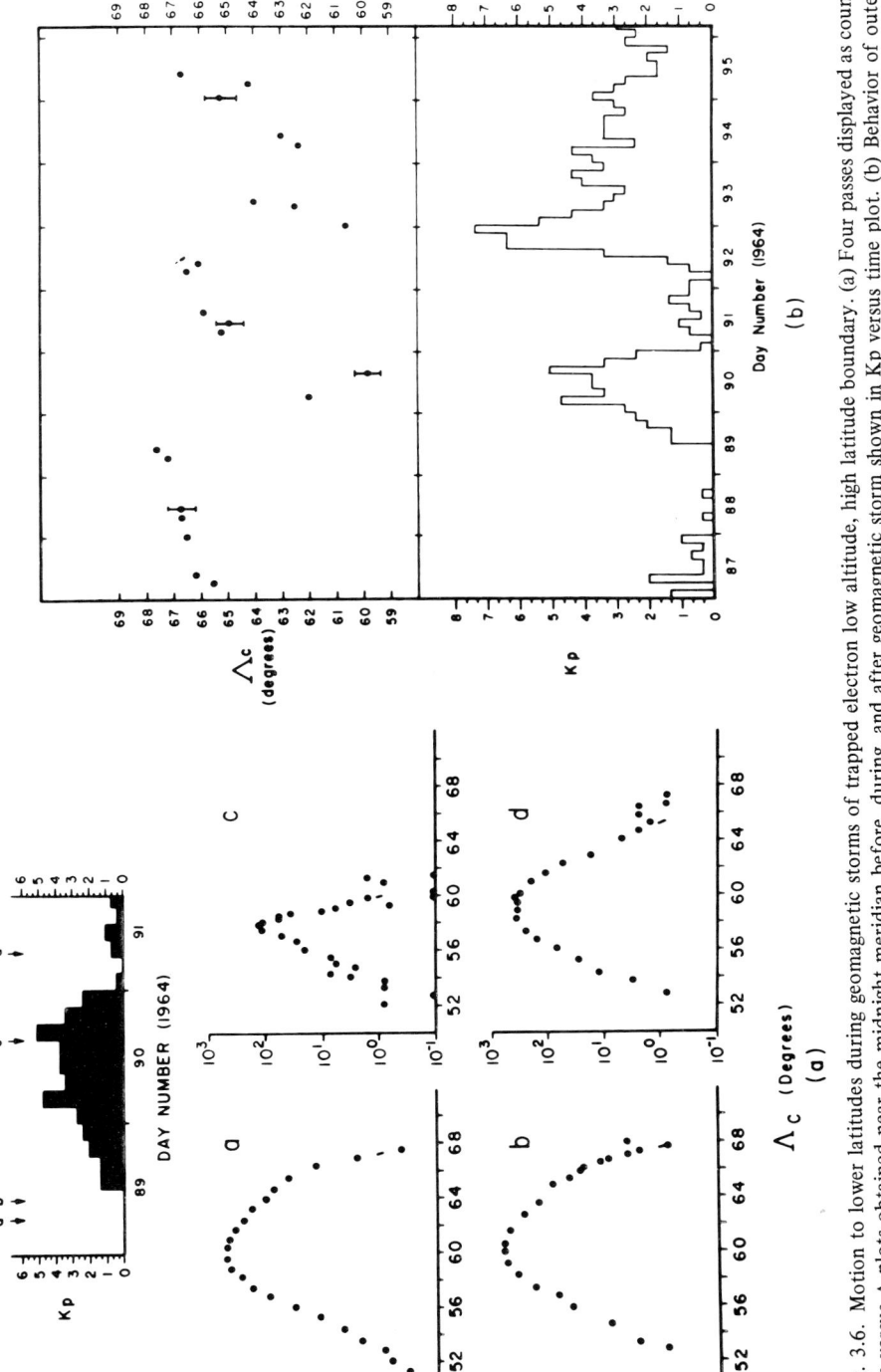

Fig. 3.6. Motion to lower latitudes during geomagnetic storms of trapped electron low altitude, high latitude boundary. (a) Four passes displayed as count rate versus Λ plots obtained near the midnight meridian before, during, and after geomagnetic storm shown in Kp versus time plot. (b) Behavior of outer boundary (defined at one count per second) during period March 29 – April 2, 1964, and its correlation with geomagnetic activity as measured by Kp (from Williams, 1966).

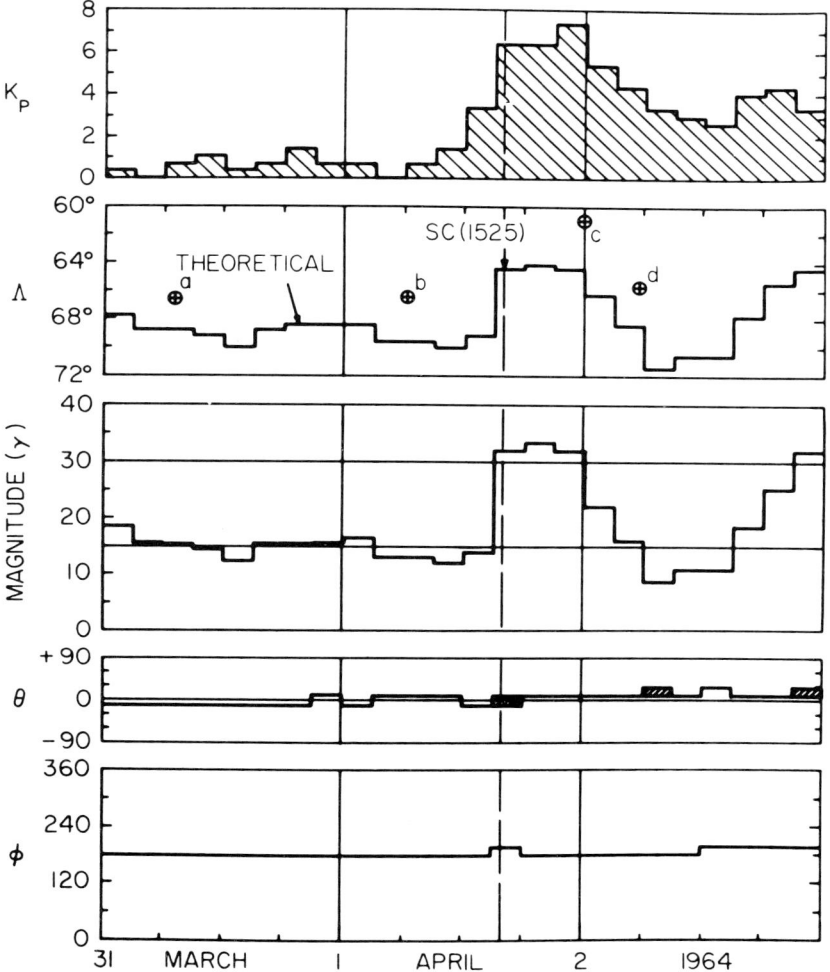

Fig. 3.7. Plot showing Kp variation, tail field magnitude and direction, and the predicted and observed outer trapping boundary throughout the disturbance of April 1, 1964 (from Ness and Williams, 1966). (© by American Geophysical Union)

values of the predicted and measured high latitude boundary differ by a small amount, the time variation of this boundary is produced using measured variations in the distorted field and assuming stable trapping with conservation of the adiabatic invariants.

From calculations using realistic geomagnetic field models have come the concepts of L-shell splitting and pseudo-trapping (Mead, 1966; Roederer, 1967). L-shell splitting refers to the imposition of a strong pitch-angle dependence on invariant drift shells caused by the removal of longitudinal symmetry in the field due to boundary and tail currents. Pseudo-trapping follows immediately from L-shell splitting and refers to particles conserving μ and J, whose longitudinal drift carries them out of the magnetosphere.

In Figure 3.8 the results of shell-splitting calculations of Roederer (1967) are shown as applied to the model field presented by Williams and Mead (1965). The shell-splitting effect of magnetospheric distortions is shown both for particles beginning on given field lines on the noon meridian and drifting to the midnight meridian and for particles initially on field lines on the midnight meridian drifting to the noon meridian. Particles mirroring near the equator will follow contours of constant B-value in their azimuthal drift. Particles with small α_0, which mirror at low altitudes, not only maintain constancy of their mirror point B-value B_M but also through conservation of I maintain a nearly constant field line length between conjugate mirror points.

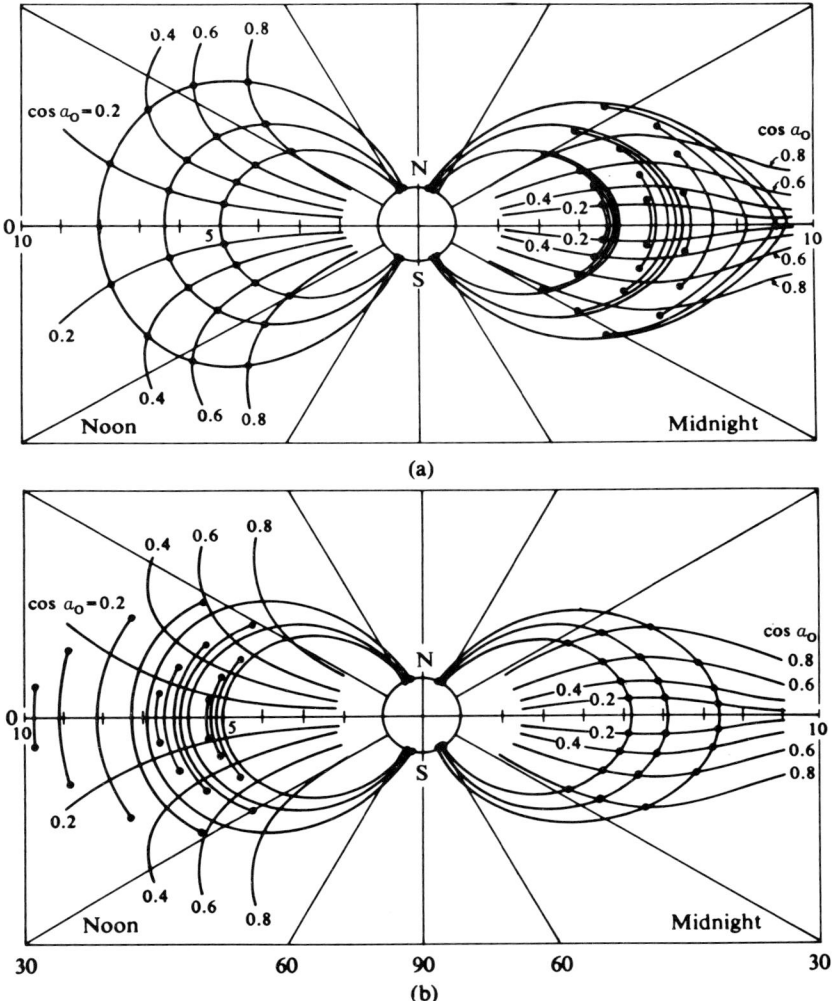

Fig. 3.8. Computed shell splitting (a) for particles beginning on common field lines on the noon meridian and (b) for particles beginning on common field lines in the midnight meridian. Dots represent particles' mirror points, and curves giving position of mirror points for constant equatorial pitch angle α_0 are shown (from Roederer, 1967). (© by American Geophysical Union)

From Figure 3.8 it is seen that particles mirroring near the equator at altitudes $\geq 8R_e$ on the midnight meridian will drift to near-equatorial altitudes of $\geq 10R_e$ on the noon meridian. Since this is outside the magnetopause boundary for this model field, these particles are pseudo-trapped in the sense that their drift will carry them into the magnetopause, and they will be lost from the trapping regions. These particles are unable to complete the azimuthal drift motion characteristic of stably trapped particles. Likewise particles beginning at high altitudes on the noon meridian with small α_0 will drift to nightside field lines which extend into the geomagnetic tail and also will be lost from the trapping regions. This figure shows clearly the strong pitch-angle dependence of drift shells caused by removal of the azimuthal symmetry present in a dipole field.

These effects are demonstrated in Figure 3.9 where the stably trapped and pseudo-trapped regions are shown in a plot of equatorial pitch angle α_0 versus altitude R in geocentric earth radii. Noon and midnight boundaries are assumed to be at $10R_e$. The

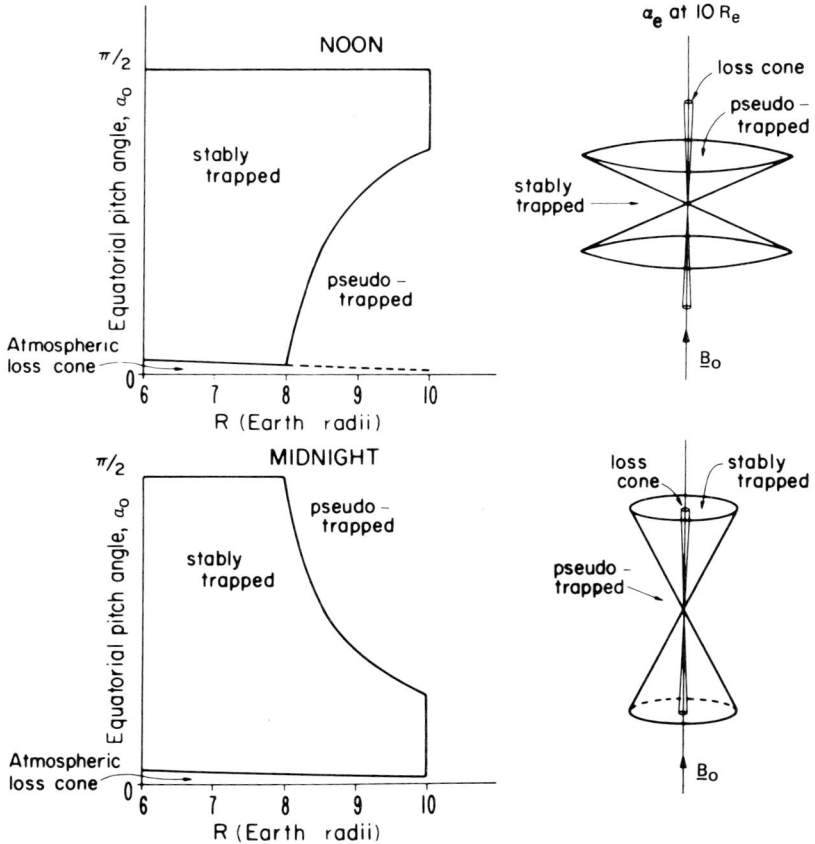

Fig. 3.9. Diagrammatic view of equatorial pitch angle versus altitude showing stably trapped and pseudo-trapped realms. Atmospheric loss cone is also shown. Noon boundary is assumed at $10R_e$.
These various pitch-angle regions are also shown schematically at the equator, $\mathbf{B}_0$, at $10R_e$.

TRAPPING REGION AND CURRENTS DUE TO TRAPPED PARTICLES   39

atmosphere loss cone is shown, and for $10R_e$ altitude the various portions of the equatorial pitch-angle distribution are diagramed on the noon and midnight meridian.

Figure 3.10 presents experimental observations of the earth's magnetic field referred to the equatorial plane (Fairfield, 1968). Contours of constant magnetic field intensity and contours designating the latitude and local time of the point of intersection of a field line with the earth's surface are shown. Many details of trapped particle motion may be obtained from Figure 3.10. Equatorially mirroring particles follow contours of constant B and thus form a stable trapping region within the 65-nT (nT = 10^{-5} G) contour. Those

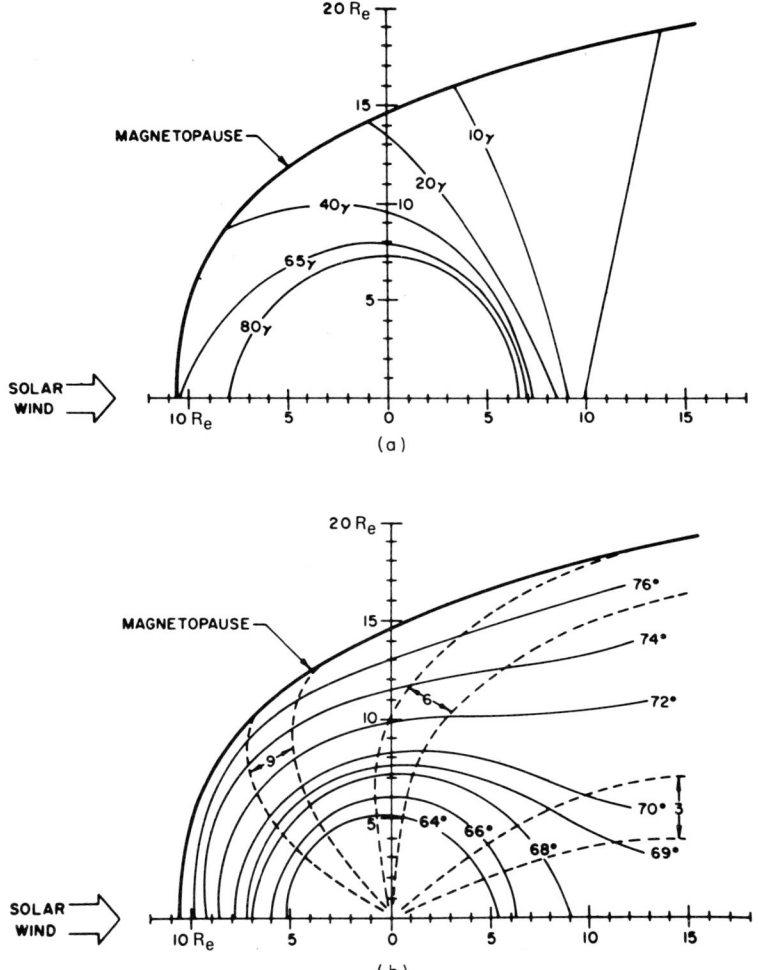

Fig. 3.10. Average magnetic field characteristics projected to the equatorial plane (from Fairfield, 1968): (a) contours of constant magnetic field intensity; (b) contours in the equatorial plane of latitude and local time of the point of intersection of a field line with the earth's surface. Solid and dashed lines are, respectively, the α and β Euler potentials discussed by Northrup (1963) and Stern (1967). (© by American Geophysical Union)

outside this contour ($\geqslant 7R_e$ at midnight) will drift into the magnetopause and be lost, thus defining a pseudo-trapping region for equatorially mirroring particles. That particles impacting the magnetopause are lost from the magnetosphere has been demonstrated using measured distribution functions obtained just earthward of the magnetopause and in the magnetosheath just outside the magnetopause (Speiser *et al.*, 1981; Speiser and Williams, 1982). These authors are able to reproduce the magnetosheath particle distributions by transporting the magnetospheric particle distribution through a model magnetopause boundary by calculating individual particle trajectories using Liouville's Theorem. These results show that particle drift into the magnetopause is a loss process for magnetospheric particles and results in the pseudo-trapping regions discussed above.

The pseudo-trapping region is bounded by the 65-nT contour and a line defining the limit of sustained bounce motion on a given field line (presumably due to significant extension into the tail region). This outer boundary of pseudo-trapping can vary over a large spatial extent and is arbitrarily drawn, for reference, as the solid line beginning at $\sim 10R_e$ in the tail in Figure 3.10. This line also forms the outer boundary of the pseudo-trapping region for low altitude mirroring particles.

From Figure 3.10 it can be seen that particles mirroring at low altitudes and drifting to field lines on the nightside hemisphere which intersect the earth at latitudes $\geqslant 68°-69°$ will be lost into the tail regions. Such particles would start at noon at latitudes $\geqslant 71°-73°$ (Williams and Mead, 1965; Roederer, 1967) or, from Figure 3.10, equatorial crossing altitudes $\geqslant 8-9R_e$. The inner boundary of the pseudo-trapping region for low altitude mirroring particles can thus be visualized as roughly a circular contour running from $\sim 8R_e$ at noon to $\sim 10R_e$ at midnight (Figure 3.10).

The high latitude nightside trapping boundary is seen to be energy dependent (e.g., Figure 3.4) and probably controlled by field line curvature. Therefore, such boundaries are not strict measures of 'last closed' field lines but rather a measure of the effects of field line extension into the tail for a particular particle momentum. These effects can yield a large altitude variation of the trapping boundary at the equator and a rather small ($\sim 1°$) latitude variation at low altitudes for electron energies from ~ 40 keV to a few MeV.

Note that the pseudo-trapping regions (and thus the stable trapping regions) have a marked pitch-angle dependence, and the two extremes discussed here ($\alpha_0 = \pi/2$ and $\alpha_0 \ll \pi/2$, where α_0 = equatorial pitch angle) overlap considerably. As a result, an omnidirectional detector will respond to both trapped and pseudo-trapped particles at high equatorial altitudes.

In this section we have considered only magnetic field effects in describing the expected spatial distribution of charged particles. The energy-dependent effects of electric fields and pitch-angle diffusion significantly modify the first-order approximation presented and are discussed in Chapters 4 and 5.

3.3. Trapped Particle Currents

3.3.1. PARTICLE DISTRIBUTIONS AND CURRENTS

The establishment and behavior of electrical currents within various magnetized plasma environments is a problem of fundamental interest. An example of this type of problem is the development of currents due to arbitrary distributions of trapped particles in the geomagnetic field which are responsible for observed worldwide decreases in the earth's surface magnetic field known as geomagnetic storms. Since the early pioneering studies of Chapman and Ferraro (1931, 1932, 1933), theoretical and experimental work has established the location of and driving force for this current system, known as the ring current (Alfvén, 1955; Parker, 1957; Singer, 1957a, b; Dessler and Parker, 1959; Akasofu and Chapman, 1961; Apel et al., 1962; Hoffman and Bracken, 1965, 1967; Frank, 1967a, b; Smith and Hoffman, 1973; Lyons and Williams, 1976; Williams, 1980).

Although the effects of particle gyration about a field line and particle drift driven by magnetic field geometries occur together, it is instructive to consider them separately to understand the establishment of currents from a magnetically trapped particle distribution. Let us start with a single charged particle trapped in a dipole field at the magnetic equator and having pitch angle of 90° as shown in Figure 3.11.

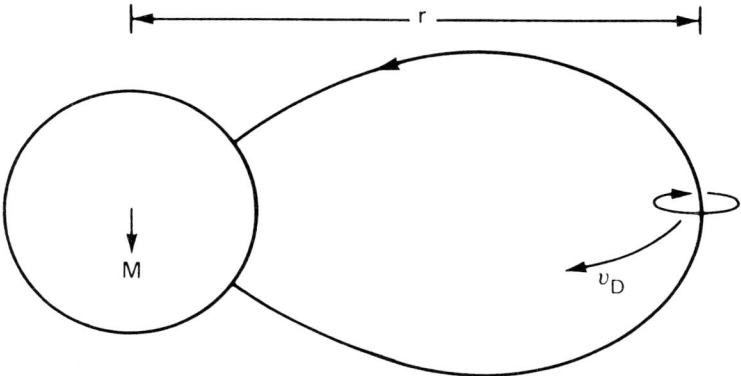

Fig. 3.11. Earth's dipole field. Field line direction and proton gyration and drift are indicated.

First consider the magnetic field effect at the origin of the dipole due to the particle drift velocity, $\mathbf{V}_{DB}$, caused by the main dipole field. From (2.48) and $\mathbf{B} = (M/r^3)\hat{\lambda}$ at the magnetic equator, we have for the drift velocity

$$\mathbf{V}_{DB} = -\frac{3K}{qM} r^2 \hat{\Psi} \tag{3.10}$$

where $\hat{\Psi}$ is a unit vector in the Ψ direction. An application of the Biot-Savart law yields for the field, B_D, at the origin due to the particle's drift

$$B_D = \frac{\mu_0 i}{4\pi} \int_0^{2\pi r} \frac{ds}{r^2} = \frac{\mu_0 i}{2r} \tag{3.11}$$

but

$$i = \frac{qV_{DB}}{2\pi r} \tag{3.12}$$

44   CHAPTER 3

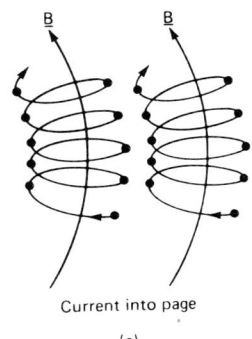

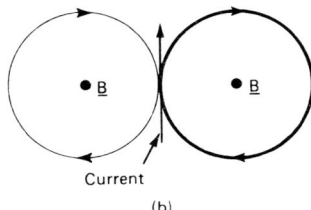

Fig. 3.12. Source of current due to gyration effects on curved field lines. (a) Eastward current between the two curved field lines is due to the crowding of gyro-orbits on the concave side of the lines; (b) source of current due to particle pressure gradient. Particle intensity is indicated by darkness of gyration circles. It is seen that a positive gradient (intensity increasing away from earth) leads to an eastward current.

agreement is quite good for the third-order iteration. The discrepancies seen may be due to measurement inaccuracies and the neglect of electric field effects. Note that while $\mathbf{E} \times \mathbf{B}$ drift produces no net current, the electric field may change the geometry of the current paths in the magnetic field.

3.3.2. THE RING CURRENT

Since it was shown earlier that the magnetic field perturbation at the earth's surface is proportional to the total kinetic energy of the particle population, it is appropriate to consider the ring current in an energy density perspective. Figure 3.14 shows the differential energy density of the ring current measured during the magnetic storm of December 17, 1971 (Smith and Hoffman, 1973). The solid data points (orbit 102) were obtained in the storm recovery phase and at the altitude ($\sim 3.6 R_e$) of maximum ring current intensity. A pre-storm plot of 'quiet' time energy density (orbit 97) is shown for comparison. Very similar results have been reported for a magnetic storm in November 1977 (Williams, 1981a, b).

It is seen that the main injection of ring current particles occurs at energies $\lesssim 200$ keV and particles at energies $\gtrsim 200$ keV show a decrease in intensity. This behavior has

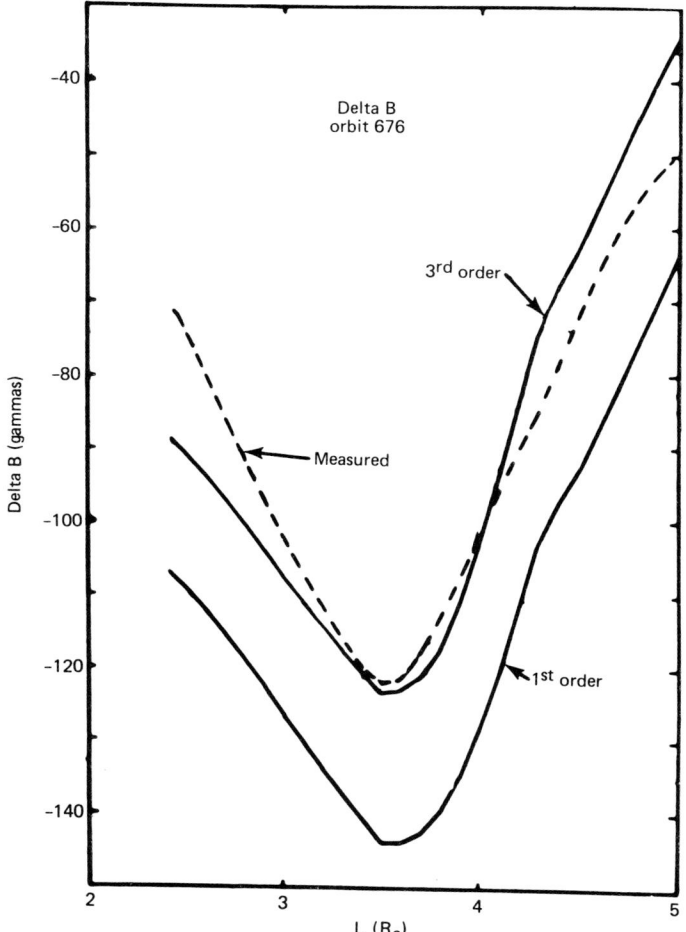

Fig. 3.13. Measured field perturbation, ΔB, and 1st and 3rd order calculated ΔB curves based on Explorer 45 particle measurements (protons assumed) in recovery phase of June 17–18, 1972, magnetic storm (from Berko et al., 1975). (© by American Geophysical Union)

been reported by Lyons and Williams (1976) and Williams (1981a), who show that the decrease of high energy particles is due to an adiabatic response of the energetic particles to the new magnetic field configuration caused by the low energy particle injection (the ring current).

This can be understood qualitatively by assuming that the magnetic field decrease during a geomagnetic storm is slow enough that the first two adiabatic invariants (2.18 and 2.30) are conserved. From (2.18) we have for equatorially mirroring nonrelativistic particles

$$\mu_B = \frac{K}{B} = \frac{K'}{B'} = \text{const} \tag{3.24}$$

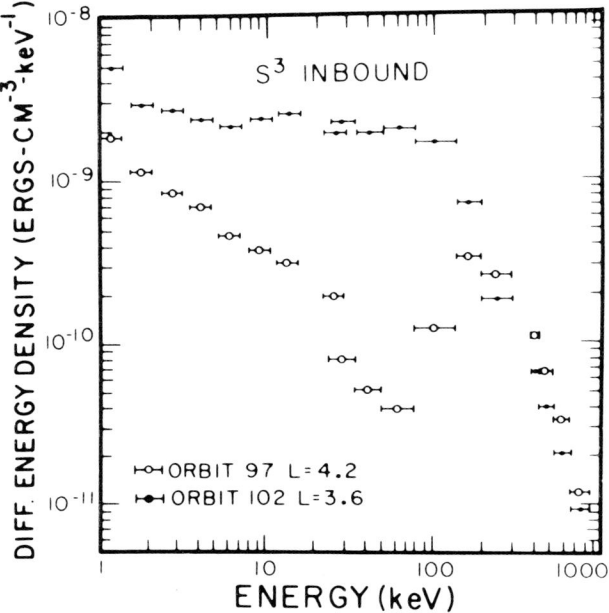

Fig. 3.14. Ion energy density spectra (protons assumed) measured by Explorer 45 during pre-storm quiet (orbit 97) and recovery phase (orbit 102) for December 17, 1971, magnetic storm (from Smith and Hoffman, 1973). (© by American Geophysical Union)

where unprimed and primed values refer to before and during the storm, respectively. Thus particle fluxes measured at an energy K' during the storm appeared at an energy $(B/B')K'$ before the storm. During storms, $B' < B$ and an energy spectrum has fluxes decreasing with increasing energy. Thus due to Liouville's Theorem we expect a flux decrease at a given particle energy during the storm, assuming neither enhanced losses or injection exists at the energy being considered. It is clear that slow variations in the magnetic field can produce particle flux variations which simply are adiabatic in nature. Relativistically, adiabatic variations (insignificant losses and injections on the time scales being considered) preserve phase space density. Thus,

$$f_p = \frac{j(K)}{p^2} = \frac{j'(K')}{(p')^2} = \text{const}, \qquad (3.25)$$

and plots of f vs μ will remain constant in time for adiabatic variations.

An early observation of adiabatic variations during magnetic storms is shown in Figure 3.15 (McIlwain, 1966). Variations in 40–110 MeV proton intensities are shown for several L-shells throughout the magnetic storm of April 17, 1975. The solid lines through the proton intensities are the variations expected from adiabatic effects, as calculated from observed D_{st} values, a ground measure of near-equatorial magnetospheric field variations.

Figure 3.16 shows adiabatic and non-adiabatic magnetic storm variations in particle fluxes at ring current energies and above for the storm of November 25–26, 1977

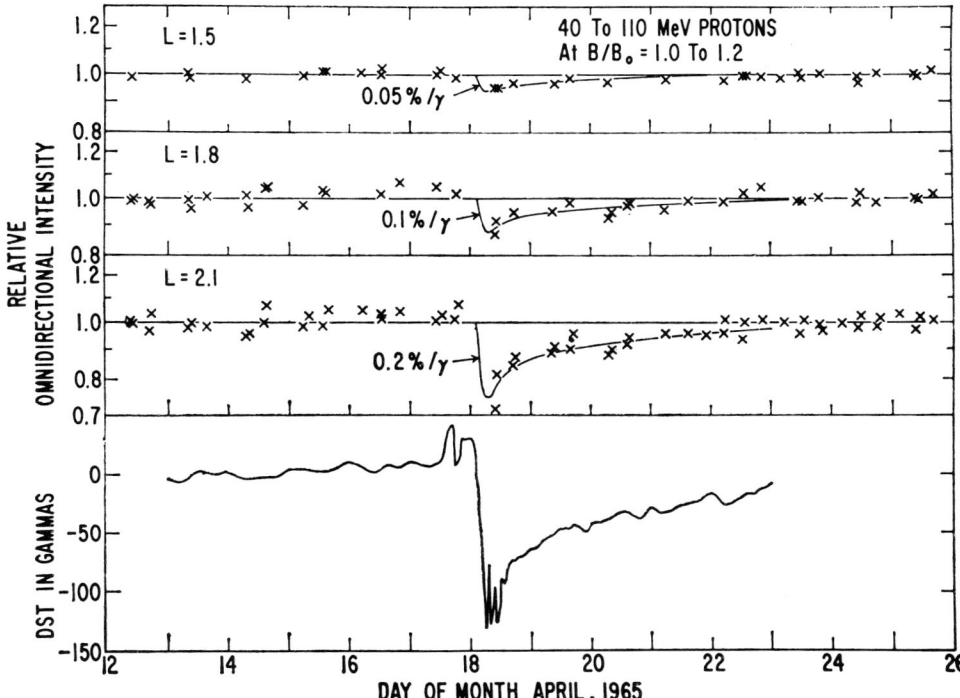

Fig. 3.15. Time history of 40–110 MeV proton intensities during the April 18, 1965, magnetic storm, showing adiabatic readjustments of the trapped proton distribution. The solid lines through the data are the adiabatic intensity variations expected from the observed D_{ST} values (from McIlwain, 1966).
(© by American Geophysical Union)

(Williams, 1981a, b). The left-hand panel, j vs E, shows a large flux increase at energies below ~200 keV and a large decrease at energies above ~200 keV throughout the ring current region, in agreement with Figure 3.14 and earlier observations. The right-hand panel of Figure 3.16 shows measured f- vs -μ plots which display large phase space density increases at low μ values (the ring current injection) and little change in f at high μ values. This shows clearly the adiabatic nature of the high energy decrease shown in the j- vs -E plots. This decrease is due to an adiabatic readjustment of the high energy particle population to the new magnetic field configuration caused by the low energy particle injection.

Integration of the curves of Figure 3.14 yields the results shown in Figure 3.17 (Williams, 1980, 1981b). Here the accumulated percentage of ring current energy density is plotted against energy. Quiet time and recovery phase are shown. The recovery phase solid curve assumes the ring current particles are all protons; the dashed curve assumes all 0^+ ions below 50 keV and all protons above 50 keV. Assuming protons, the median energy in the integral energy density curve is ~85 keV. Energies for which composition measurements have been made in the magnetosphere are indicated.

Figure 3.17 shows that while we know the energy range of the ring current particles, we

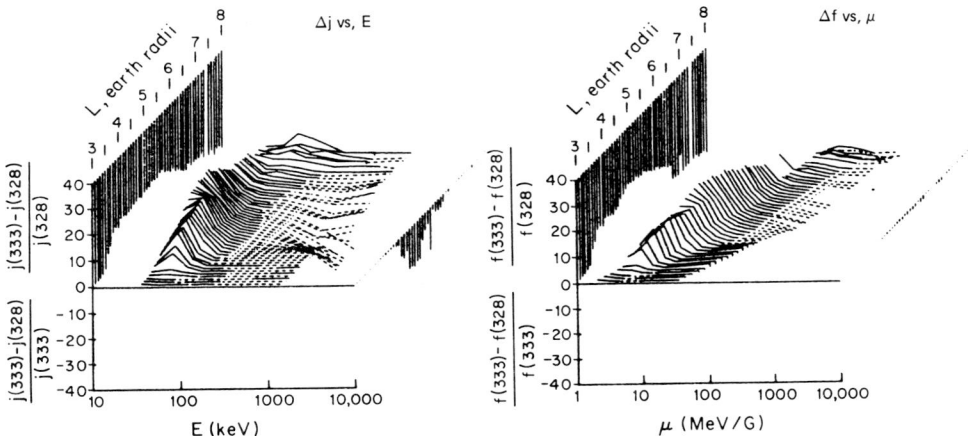

Fig. 3.16. Perspective plot of the factor of increase/decrease of flux, Δj, vs E and phase space density, Δf, vs μ from $L = 3$ to $8R_e$ for the November 25–26, 1977, geomagnetic storm. Δj and Δf are obtained from the pre-storm pass of November 25, 1977, and first post-storm pass of November 27, 1977. The constant phase space densities ($\Delta f \sim 0$) at high μ values show that the large flux decreases at high energies are an adiabatic readjustment of the high energy particle population to the magnetic field distortions produced by the low energy (ring current) particle enhancement (from Williams, 1981b).

do not know the composition of the bulk (80%–90%) of these particles. This effectively prevents us from knowing the source of the ring current particles. The low energy tail ($\leqslant 20$ keV) of the ring current distribution is thought to have an ionospheric origin, and the high energy tail ($\geqslant 600$ keV) is thought to have a solar wind origin. How, or if, these two sources mix to produce the bulk ring current is not known (Williams, 1980, 1981b). In the above context we consider the plasmasphere to be essentially an ionospheric source and the plasma sheet to be an unknown admixture of the ionosphere and solar wind sources.

3.3.3. RING CURRENT GENERATION

Since the composition and thus the source of the ring current particles is unknown, it is not possible to discuss quantitatively generation processes. However we present in this section a brief listing and qualitative description of various processes that have been discussed for many years.

3.3.3.1. *Earthward Convection of the Plasma Sheet*

Here an enhanced dawn-dusk electric field quickly convects (**E** × **B**) the plasma sheet particles towards the earth on the nightside hemisphere. The convecting particles **E** × **B** drift across equipotentials and experience betatron and Fermi acceleration as they are transported into low altitudes (~ 2.5–$4.0R_e$) by the enhanced field to form the ring current. (See, for example, discussions by Frank, 1967b; Axford, 1968; Kivelson and Southwood, 1975; Kivelson et al., 1980.)

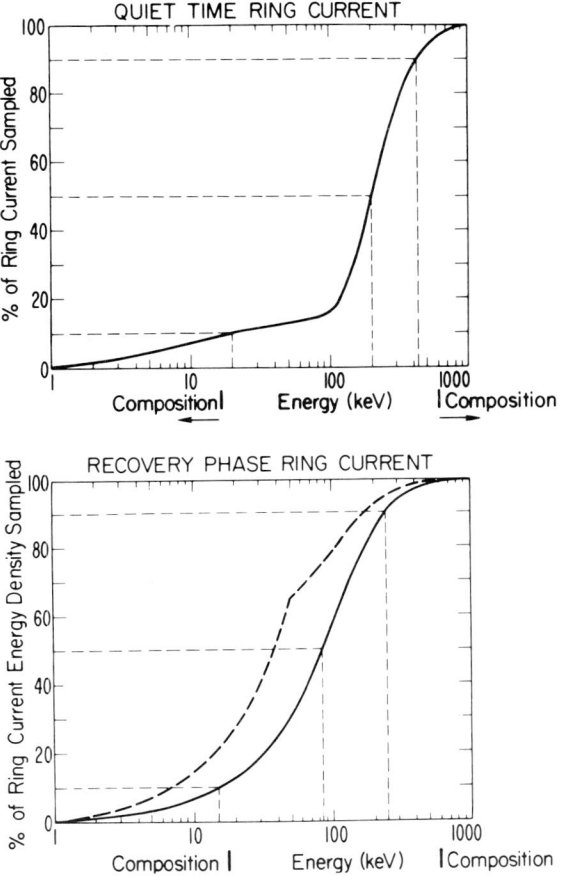

Fig. 3.17. Accumulated percentage of ring current energy density vs energy. Quiet-time and recovery phase curves are from data shown in Figure 3.14. Solid curves assume all ions are protons. The dashed curve in recovery phase shows case for all 0^+ ions below 50 keV and all protons above 50 keV. Energy ranges of available direct composition observations are indicated (from Williams, 1980, 1981b).

3.3.3.2. *Acceleration of Ionospheric Ions by Electric Fields Parallel to Auroral Field Lines*

Since the observation of energetic ($\leqslant 10$ keV) ion beams streaming out of the ionosphere in auroral regions (Shelley *et al.*, 1976) it has been attractive to consider these particles a source for the ring current. Additional processes are required to form an equatorial sin $n\alpha$ ($n \geqslant 0$) pitch-angle distribution, which would then be displaced to low altitude by an enhanced dawn-dusk electric field and accelerated to ring current energies by betatron and Fermi processes.

3.3.3.3. *Source Boundaries*

To explain the complex energy-time signature displayed by magnetospheric plasma at geostationary orbit, McIlwain (1974) invoked the concept that enhanced electric fields cause the appearance of freshly energized particles at locations outside a sharp, well-defined boundary in the nightside hemisphere. After the transient, enhanced electric field, the particles move in the ambient, quiescent magnetic-electric field configuration. Using a realistic magnetic field and tailoring an electric field model inferred from particle data, McIlwain (1974) was able to explain many of the particle energy-time signatures. His hypothesis has been successfully tested by Mauk and McIlwain (1974) and Konradi *et al.* (1975). However, Kivelson (1976) has argued that the injection boundary actually is a sequence of quasi-static energy-dependent surfaces defined by the motion of plasma sheet particles in the combined magnetic and electric (convection plus corotation) field existing at the same time. Observations in the dusk-midnight sector appear to be consistent with this concept (Kivelson and Southwood, 1975; Kivelson *et al.*, 1980). It is apparent that observations in the dusk-midnight sector may not be able to distinguish between the above hypotheses and that a study of geostationary particle signatures at all local times is required to test the models definitively.

3.3.3.4. *Inward Motion of the Outer Radiation Zone*

The same enhanced electric field responsible for rapid convection of the plasma sheet toward the earth will also cause an inward motion of previously trapped particles in the earth's outer radiation zone. Lyons and Williams (1980) quantitatively tested this concept using data from two main phase magnetic storms in 1971 and 1972. They found that the main phase ring current could be accounted for by the inward motion of the pre-storm outer zone trapped particles. Ions and electrons simultaneously were consistent with this hypothesis. Reasonable electric fields giving inward motions consistent with rough plasmapause position measurements were required.

It is clear that these mechanisms bear a resemblance to each other and may represent different aspects of the overall particle transport and acceleration process responsible for the creation of the ring current. For example, mechanism (2) may be closely related to (3), although McIlwain has associated (3) with the plasmapause boundary. Mechanism (4) clearly seems related to all the others with the recognition that (1), (2), and (3) may be providing a fresh supply of trapped particles for the outer zone, to be transported into low altitudes at the next major electric field enhancement, thereby forming the ring current. Much more quantitative work is required to sort through the present qualitative possibilities for ring current generation.

3.3.4. RING CURRENT DECAY

Singer (1957a) was the first to suggest that the main phase ring current was due to the motion of particles trapped in the earth's magnetic field. He suggested the decay of the ring current was due to scattering of the particles either by field changes or atomic collisions. His estimate of an ~ 1 day lifetime was essentially a strong diffusion scattering limit (see Chapter 5 for a discussion of strong pitch-angle diffusion effects). Stuart (1959), Dessler and Parker (1959) and Wentworth *et al.* (1959a, b) were the first to note the importance of charge exchange in removing ring current particles. Charge exchange has since been used extensively in magnetospheric studies, and we show in Figure 3.18 the results of a charge exchange calculation to explain ring current decay after the February 24, 1972, magnetic storm (Smith *et al.*, 1981).

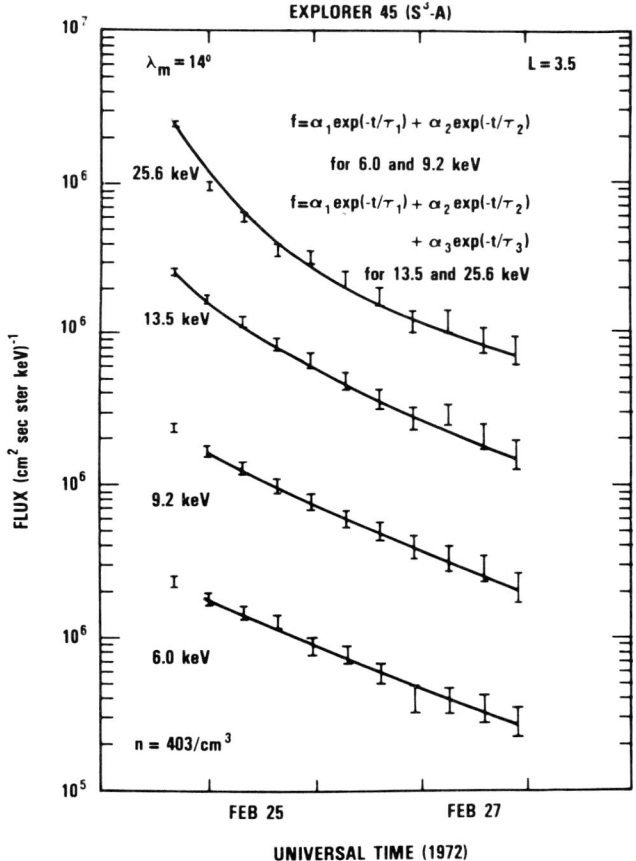

Fig. 3.18. Ion fluxes measured by Explorer 45 at indicated energies during recovery phase of February, 24, 1972, magnetic storm. Best fit curves using multi-ion fit to data are shown as solid lines. Best fit charge exchange lifetimes, τ_1, τ_2, τ_3, are essentially those of H^+, O^+, and He^+. The fits require an admixture of H^+ and O^+ at 6.0 and 9.2 keV and an admixture of H^+, O^+, and He^+ at 13.5 and 25.6 keV (from Smith *et al.*, 1981).

Since we do not know the ring current composition, it is not possible to predict the charge exchange decay time in advance because of the species and energy dependence of the charge exchange cross sections. Smith *et al.* (1981) performed a best fit charge exchange decay analysis of the observed decay by assuming a multi-ion particle population. Parameters extracted from the fits were the relative abundance and charge exchange lifetime of each ion species. The lifetimes obtained from the best fit analysis were essentially the charge exchange lifetimes of H^+, He^+, and O^+. Figure 3.18 shows one set of results for energies from 6 to ~26 keV. Thus proper accounting for a multi-ion ring current shows charge exchange to be a major cause of the overall decay in recovery phase.

However charge exchange is not the sole dissipation mechanism operating on the ring current. Cornwall et al. (1970) predicted intense ion cyclotron wave turbulence generated by the interaction of the ring current particles with the cold plasmaspheric plasma. The resulting pitch-angle diffusion of the ring current particles would represent a major loss process. Williams and Lyons (1974a, b) showed that the pitch-angle evolution of ring current particles in the plasmapause region was consistent with the generation of ion cyclotron waves, although wave growth and resulting particle loss were weak, not strong as predicted. (Ion cyclotron turbulence is discussed in detail in Chapter 5).

Although ring current pitch-angle diffusion is weak instead of strong, sufficient energy is lost from the ring current to cause observable atmospheric effects. Integrating the energy lost from the ring current due to pitch-angle distribution changes observed in the plasmapause region, Williams et al. (1976) obtained the energy deposition per cm^2 into the ionosphere as a function of L-value. Their results along with a comparison of simultaneously observed atmospheric structures (sub-auroral red (SAR) arcs) are shown in Figure 3.19. The energy deposition rate to sustain the two arcs is shown along with upper and lower limit estimates of the ionospheric deposition rate from the ring current. The energy lost from the ring current is seen to be sufficient to sustain the observed SAR arcs,

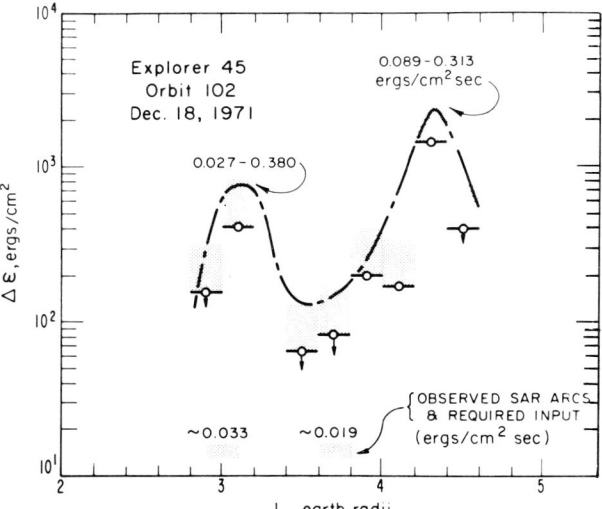

Fig. 3.19. Plot of energy lost from ring current to the ionosphere, $\Delta\epsilon$, because of pitch-angle variations observed in the plasmapause region during the recovery phase of the December 17–18, 1971 magnetic storm. Observed SAR arc locations and required energy input are shown. Ring current loss rate estimates are shown for the peak values. The spatial correspondence is good when field expansion effects are considered. The energy values show that energy lost from the ring current in the plasmapause region due to pitch-angle distribution changes is sufficient to sustain the SAR arcs observed (from Williams et al., 1976). (© by American Geophysical Union)

consistent with a prediction of Cornwall *et al.* (1971). Furthermore, the equatorial location of the two peaks in the ring current loss is consistent with the SAR arc positions when the ring current inflated magnetic field is taken into account.

Thus we see that the general ring current decay is primarily the result of charge exchange and secondarily the result of weak pitch-angle scattering occurring in the plasmapause region, as discussed by Williams and Lyons (1974a, b). The pitch-angle scattering is sufficient to cause observable atmospheric effects.

References

Akasofu, S. I. and S. Chapman, 'The ring current, geomagnetic disturbance, and the Van Allen radiation belts', *J. Geophys. Res.* **66**, 1321 (1961).

Alfvén, H., 'Theory of magnetic storms', 1. *Kgl. Sv. Vetenskapsakad. Handl.* [3] **18**, No. 3 (1939).

Alfvén, H., 'Theory of magnetic storms', 2, 3. *Kgl. Sv. Vetenskapsakad. Handl.* [3] **18**, No. 9 (1940).

Alfvén, H., *Cosmical Electrodynamics*, 1st ed., Oxford Univ. Press, London and New York (1950).

Alfvén, H., 'On the electric field theory of magnetic storms and aurorae', *Tellus* **7**, 54 (1955).

Apel, J. R., S. F. Singer, and R. C. Wentworth, 'Effects of trapped particles on the geomagnetic field', *Advances in Geophysics* **9**, 131 (1962).

Axford, W. I., 'Magnetospheric convection', *Rev. of Geophys.* **7**, 421 (1969).

Berko, F. W., L. J. Cahill, Jr., and T. A. Fritz, 'Protons as the prime contributors to storm time ring current', *J. Geophys. Res.* **80**, 3549 (1975).

Carovillano, R. L. and J. J. Maguire, 'Magnetic energy relationships in the magnetosphere', in *Physics of the Magnetosphere*, R. L. Carovillano, J. F. McClay, and H. R. Radoski (eds.), D. Reidel Publ. Co., Dordrecht, Holland, 290 (1968).

Carovillano, R. L. and G. L. Siscoe, 'Energy and momentum theorems in magnetospheric processes', *Rev. of Geophys. and Space Phys.* **11**, 289 (1973).

Chapman, S. and V. C. A. Ferraro, 'A new theory of magnetic storms', *Terr. Magn. Atmos. Elec.* **36**, 77–97, 171–186 (1931).

Chapman, S. and V. C. A. Ferraro, 'A new theory of magnetic storms', *Terr. Magn. Atmos. Elec.* **37**, 147–156, 421–429 (1932).

Chapman, S. and V. C. A. Ferraro, 'A new theory of magnetic storms', *Terr. Magn. Atmos. Elec.* **38**, 79–96 (1933).

Christofilos, N. C., 'The Argus experiment', *J. Geophys. Res.* **64**, 869–876 (1959).

Cornwall, J. M., F. V. Coroniti, and R. M. Thorne, 'Turbulent loss of ring current protons', *J. Geophys. Res.* **75**, 4699 (1970).

Cornwall, J. M., F. V. Coroniti, and R. M. Thorne, 'A unified theory for SAR arc formation at the plasmapause', *J. Geophys. Res.* **76**, 4428 (1971).

Dessler, A. J. and E. N. Parker, 'Hydromagnetic theory of geomagnetic storms', *J. Geophys. Res.* **64**, 2239 (1959).

Fairfield, D. H., 'Average magnetic field configuration of the outer magnetosphere', *J. Geophys. Res.* **73**, 7329 (1968).

Frank, L. A., 'Several observations of low energy protons and electrons in the earth's magnetosphere with OGO 3', *J. Geophys. Res.* **72**, 1905 (1967a).

Frank, L. A., 'On the extraterrestrial ring current during geomagnetic storms', *J. Geophys. Res.* **72**, 3753 (1967b).

Hoffman, R. A. and P. A. Bracken, 'Magnetic effects of the quiet-time proton belt', *J. Geophys. Res.* **70**, 3541 (1965).

Hoffman, R. A. and P. A. Bracken, 'Higher-order ring currents and particle energy storage in the magnetosphere', *J. Geophys. Res.* **72**, 6039 (1967).

Kivelson, M. G., 'Magnetospheric electric fields and their variation with geomagnetic activity', *Rev. Geophys. Space Phys.* **14**, 189 (1976).

Kivelson, M. G. and D. J. Southwood, 'Approximations for the study of drift boundaries in the magnetosphere', *J. Geophys. Res.* **80**, 3528 (1975).

Kivelson, M. G., S. M. Kaye, and D. J. Southwood, 'The physics of plasma injection events', in *Dynamics of the Magnetosphere*, S.-I. Akasofu (ed.), D. Reidel Publ. Co., Dordrecht, Holland, 385 (1980).

Konradi, A., C. L. Semar, and T. A. Fritz, 'Substorm-injected protons and electrons and the injection boundary model', *J. Geophys. Res.* **80**, 543 (1975).

Lyons, L. R. and D. J. Williams, 'Storm associated variations of equatorially mirroring ring current protons, 1–800 keV, at constant first adiabatic invariant', *J. Geophys. Res.* **81**, 216 (1976).

Lyons, L. R. and D. J. Williams, 'A source for the geomagnetic storm main phase ring current', *J. Geophys. Res.* **85**, 523 (1980).

Mauk, B. H. and C. E. McIlwain, 'Correlation of Kp with the substorm-injected plasma boundary', *J. Geophys. Res.* **79**, 3193 (1974).

McIlwain, C. E., 'Coordinates for mapping the distribution of magnetically trapped particles', *J. Geophys. Res.* **66**, 3681 (1961).

McIlwain, C. E., 'Ring current effects on trapped particles', *J. Geophys. Res.* **71**, 3623 (1966).

McIlwain, C. E., 'Substorm injection boundaries', in *Magnetospheric Physics*, B. M. McCormac (ed.), D. Reidel Publ. Co., Dordrecht, Holland, 143 (1974).

Mead, G. D., 'Deformation of the geomagnetic field by the solar wind', *J. Geophys. Res.* **69**, 1181 (1964).

Mead, G. D., 'The motion of trapped particles in a distorted field', in *Radiation Trapped in the Earth's Magnetic Field*, B. M. McCormac (ed.), D. Reidel Publ. Co., Dordrecht, Holland, 481 (1966).

Ness, N. F. and Williams, D. J., 'Correlated magnetic tail and radiation belt observations', *J. Geophys. Res.* **71**, 322 (1966).

Northrup, T. G., *The Adiabatic Motion of Charged Particles*, Wiley (Interscience), New York, 109 pp. (1963).

O'Brien, B. J., 'A large diurnal variation of the geomagnetically trapped radiation', *J. Geophys. Res.* **68**, 989 (1963).

Olbert, S., G. L. Siscoe, and V. M. Vasyliunas, 'A simple derivation of the Dessler-Parker-Sckopke relation', *J. Geophys. Res.* **73**, 1115 (1968).

Olson, W. P. (ed.), *Quantitative Modelling of Magnetospheric Processes*, American Geophysical Union, Geophysical Monograph 21 (1979).

Parker, E. N., 'Newtonian development of the dynamical properties of ionized gases at low density', *Phys. Rev.* **107**, 924 (1957).

Pfitzer, K. A., Lezniak, T. W., and Winckler, J. R., 'Experimental verification of drift-shell splitting in the distorted magnetosphere', *J. Geophys. Res.* **74**, 4687 (1969).

Roederer, J. G., 'On the adiabatic motion of energetic particles in a model magnetosphere', *J. Geophys. Res.* **72**, 981–992 (1967). See also *Dynamics of Geomagnetically Trapped Radiation*, Springer, Heidelberg (1970).

Roederer, J. G., 'Quantitative models of the magnetosphere', in *Magnetospheric Physics*, D. J. Williams and G. D. Mead (eds.), 77–96 (1969), Am. Geophys. Union, Washington, D.C. [also in *Rev. Geophys.* **7**, Nos. 1, 2.]

Sckopke, N., 'A general relation between the energy of trapped particles and the disturbance field near the earth', *J. Geophys. Res.* **71**, 3125 (1966).

Shelley, E. G., R. D. Sharp, and R. G. Johnson, 'Satellite observations of an ionospheric acceleration mechanism', *Geophys. Res. Letters* **3**, 654 (1976).

Singer, S. F., 'Trapped orbits in the earth's dipole field'. (Abstr.), *Bull. Amer. Phys. Soc.* [2] **1**, 229 (1956).

Singer, S. F., 'A new model of magnetic storms and aurorae', *Trans. Am. Geophys. Union.* **38**, 175 (1957a).

Singer, S. F., 'Project Far Side', *Missiles Rockets* **2**, Part 2, 120 (1957b).

Smith, P. H. and R. A. Hoffman, 'Ring current particle distribution during the magnetic storms of Dec. 16–18, 1971', *J. Geophys. Res.* **78**, 4731 (1973).

Smith, P. H., N. K. Bewtra, and R. A. Hoffman, 'Inference of the ring current ion composition by means of charge exchange decay', *J. Geophys. Res.* **86**, 3470 (1981).

Speiser, T. W., D. J. Williams, and H. A. Garcia, 'Magnetospherically trapped ions as a source of magnetosheath energetic ions', *J. Geophys. Res.* **86**, 723 (1981).

Speiser, T. W. and D. J. Williams, 'Magnetopause modelling: Flux transfer events and magnetosheath quasi-trapped distributions', *J. Geophys. Res.* **87**, 2177 (1982).

Stern, D., 'Geomagnatic Euler potentials', *J. Geophys. Res.* **72**, 3995 (1967).

Störmer, C., 'Sur les trajectoires des corpuscules electrisiés dans l'éspace sous l'action du magnetism terrestre', *Arch. Sci. Phys.* **24**, No. 5, 113, 221, 317 (1907).

Störmer, C., 'On the trajectories of electrical particles in the field of a magnetic dipole with applications to the theory of cosmic radiation', *Avh. Norske Videnskap. Akad, Oslo, Mat., Nat. Kl.* **11** (1933).

Störmer, C., *The Polar Aurora*, Oxford Univ. Press, London and New York, 403 pp. (1955).

Stuart, G. W., 'Satellite measured radiation', *Phys. Rev. Letters* **2**, 417 (1959).

Taylor, H. E. and Hones, E. W., Jr., 'Adiabatic motion of auroral particles in a model of the electric and magnetic fields surrounding the earth', *J. Geophys. Res.* **70**, 3605 (1965).

Van Allen, J. A., McIlwain, C. E., and Ludwig, G. H., 'Satellite observations of electrons artificially injected into the geomagnetic field', *J. Geophys. Res.* **64**, 874 (1959).

Van Allen, J. A., Ludwig, G. H., Ray, E. C., and McIlwain, C. E., 'Observation of high intensity radiation by satellites', *Jet Propul.* **28**, 588 (1958).

Wentworth, R. C., W. M. MacDonald, and S. F. Singer, 'Lifetimes of trapped auroral radiation belt particles', *Bull. Am. Phys. Soc. Ser 2*, **4**, 7 (1959a).

Wentworth, R. C., W. M. MacDonald, and S. F. Singer, 'Lifetimes of trapped radiation belt particles determined by Coulomb scattering', *Phys. of Fluids* **2**, 499 (1959b).

Williams, D. J., 'Outer zone electrons', in *Radiation Trapped in the Earth's Magnetic Field*, B. M. McCormac (ed.), D. Reidel Publ. Co., Dordrecht, Holland, 263 (1966).

Williams, D. J., 'Magnetospheric Proton Dynamics', *EOS* **55**, 1021 (1974).

Williams, D. J., 'Ring current composition and sources', in *Dynamics of the Magnetosphere*, S.-I. Akasofu (ed.), D. Reidel Publ. Co., Dordrecht, Holland, 407 (1980).

Williams, D. J., 'Phase space variations of equatorially mirroring ring current ions', *J. Geophys. Res.* **86**, 189 (1981a).

Williams, D. J., 'Ring current composition and sources: An update', *Planet. Space Sci.* **29**, 1195 (1981b).

Williams, D. J. and L. R. Lyons, 'The proton ring current and its interaction with the plasmapause: Storm recovery phase', *J. Geophys. Res.* **79**, 4195 (1974a).

Williams, D. J. and L. R. Lyons, 'Further aspects of the proton ring current interaction with the plasmapause: Main and recovery phases', *J. Geophys. Res.* **79**, 4791 (1974b).

Williams, D. J. and Mead, G. D., 'Nightside magnetosphere configuration as obtained from trapped electrons at 1100 km', *J. Geophys. Res.* **70**, 3017 (1965).

Williams, D. J. and Ness, N. F., 'Simultaneous trapped electron and magnetic tail field observations', *J. Geophys. Res.* **71**, 5117 (1966).

Williams, D. J., G. Hernandez, and L. R. Lyons, 'Simultaneous observations of the proton ring current and stable auroral red arcs', *J. Geophys. Res.* **81**, 608 (1976).

Yoshida, S., Ludwig, G. H., and Van Allen, J. A., 'Distribution of trapped radiation in the geomagnetic field', *J. Geophys. Res.* **65**, 807 (1960).

CHAPTER 4

ELECTRIC FIELDS

4.1. Introduction

The magnetospheric plasma is essentially collisionless. In such a plasma, electric fields aligned along the magnetic field direction (referred to as parallel electric fields, $E_\parallel$) freely accelerate particles. Electrons and ions are accelerated in opposite directions, giving rise to a current along the magnetic field lines. When the scale of magnetic field gradients is much larger than a particle gyroradius, so that the guiding center approximation is valid as is generally the case throughout the magnetosphere, electric fields normal to the magnetic field (perpendicular electric fields, $E_\perp$) cause particles to drift normal to the magnetic field with a velocity given by $\mathbf{V}_{DE} = (\mathbf{E} \times \mathbf{B})/B^2$ (see Chapter 2). All particles drift with the same velocity so that perpendicular electric fields do not cause currents under the guiding center approximation. Thus electric potential differences along magnetic field lines give rise to currents in the direction required for decreasing the potential differences, while perpendicular potential variations generally do not give rise to such currents.

Thus processes that give rise to non-zero charge densities within the magnetosphere can give rise to large-scale, perpendicular electric fields and resulting plasma flows throughout the magnetosphere, as postulated by Gold (1959) and called convection. Significant parallel electric fields, on the other hand, can exist only along magnetic field lines where charge densities are continually generated at some point along the field line. Thus large-scale parallel electric fields should not exist throughout the magnetosphere, but should exist only along field lines where such a generating process exists.

4.2. The Convection Electric Field

4.2.1. PROPOSED LARGE-SCALE CONVECTION ELECTRIC FIELD

Dungey (1961) noted that the existence of an interplanetary magnetic field implies that the magnetic field lines emanating from the earth's polar regions may be directly connected to the interplanetary magnetic field. For a southward-directed interplanetary magnetic field, he inferred the magnetic field topology schematically illustrated in Figure 4.1 for the noon-midnight meridian plane. This topology can be obtained by simply adding the interplanetary magnetic field to the earth's dipole field, though Figure 4.1 includes a geomagnetic tail.

There are four classes of field lines in Figure 4.1: The closed field lines (1) connected to the earth in both hemispheres, the interplanetary field lines (2) not connected to the earth, and the open field lines with one end connected to the northern (3) or southern (4) polar cap and the other end directly connected to the interplanetary field. In the frame of reference of the earth, the solar wind flows across the open polar cap field lines

ELECTRIC FIELDS

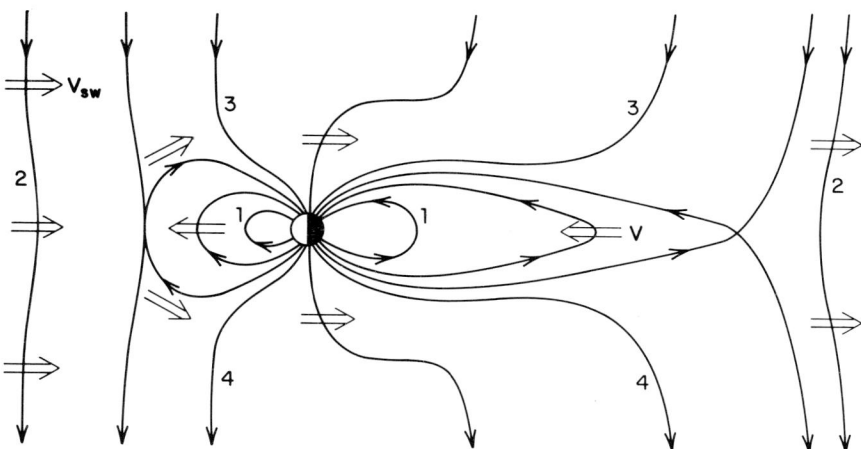

Fig. 4.1. Magnetic field topology for the noon-midnight meridian plane obtained by Dungey (1961) by adding the interplanetary magnetic field to the earth's dipole field. This figure is for a southward directed interplanetary field and incudes a geomagnetic tail. The four classes of field lines discussed in the text are labeled, and the plasma flow direction is indicated by the open arrows.

giving an electric field $\mathbf{E} = -\mathbf{V}_{sw} \times \mathbf{B}$ across the portion of the polar cap field lines within the interplanetary medium, where $\mathbf{V}_{sw}$ is the solar wind velocity. This electric field is in the dawn-to-dusk direction. Assuming large potential differences do not exist along the open polar cap field lines, the electric field will map along the open field lines through the polar magnetosphere to the polar caps, giving rise to an anti-sunward flow of plasma over both polar caps.

The basic topology of the magnetic field configuration obtained from adding the earth's dipole field to the interplanetary field remains the same if the interplanetary magnetic field is oriented in an arbitrary direction, except for the case of a purely northward interplanetary field for which no open field lines exist (Dungey, 1963; Stern, 1973). Figure 4.2 (adapted from Stern, 1973) schematically shows the sum of a dipole field and a uniform external field oriented normal to the dipole axis compared with the case of a uniform southward external field. It can be seen from Figure 4.2 that the four classes of field lines exist independent of the orientation of the external field. Of particular importance are (1) the roughly circular regions over both polar caps where open field lines emanate and (2) the approximately dawn-to-dusk electric field over the polar caps. Figure 4.2 shows the external magnetic field as normal to the solar wind velocity so that the figure represents the dawn-dusk meridian plane. The topology will also remain unchanged for other angles of the external field with respect to the solar wind; however, when the field is parallel or antiparallel to the solar wind, $\mathbf{E} = -\mathbf{V}_{sw} \times \mathbf{B} = 0$ across the interplanetary field.

The topology illustrated in Figure 4.2 is valid not only in the dawn-dusk meridian plane, but also for field lines emanating from the earth along any circular path that penetrates the open field line regions in both polar caps. Under the assumption that the solar wind flow does not penetrate onto closed field lines, the interplanetary electric field

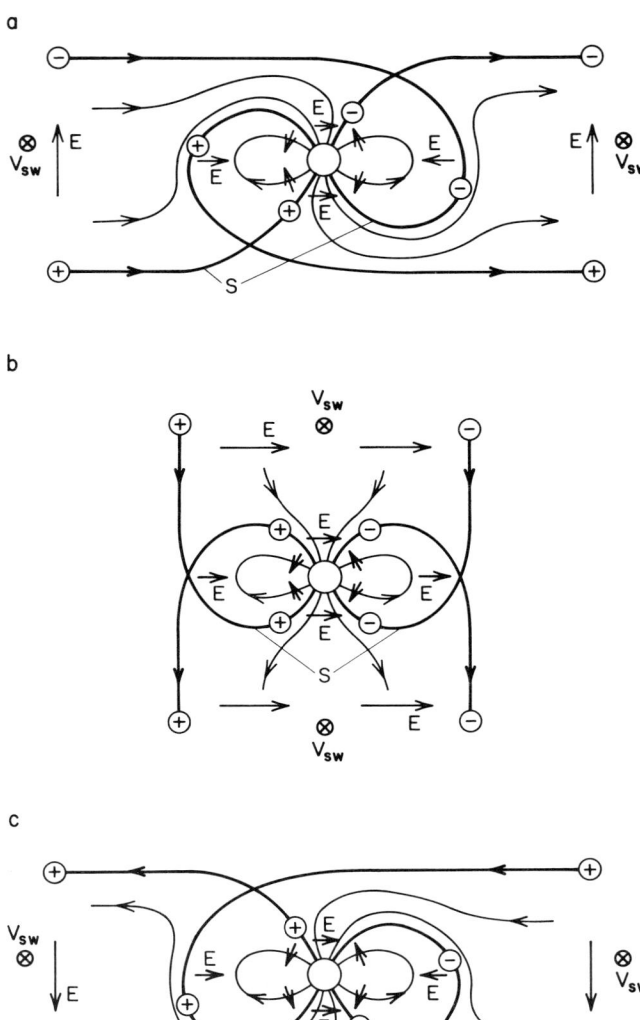

Fig. 4.2. Schematic illustration of the sum of a dipole magnetic field and a uniform external field oriented normal to the dipole axis, compared with the case of a uniform southward external field. Taking each panel to represent the dawn-dusk meridian plane, the figure also illustrates the solar wind vector V_{sw}, the electric field orientation with each magnetic field region, and the charge along the boundaries S separating the field regions (adapted from Stern, 1973). (© by American Geophysical Union)

exists across the open polar cap field lines but not the closed field lines. Under these conditions, the surface S separating the closed and open field lines must be positively charged on the dawn side of the earth and negatively charged on the dusk side as indicated in the figure. In the equatorial plane, these charges give a dawn-to-dusk electric field across the region of closed magnetic field lines, which implies there must be anti-sunward plasma flow throughout the closed field line region. In addition, the potential difference between a field line on the dawn side of S and one on the dusk side of S must be the same across the closed field line region as across the open polar cap field lines.

The overall magnetospheric convective flow V_{DE} in the equatorial plane resulting from the superposition of the interplanetary and the earth's magnetic field as described above is illustrated in Figure 4.3. The surface S, positively charged on the dawn side and negatively charged on the evening side, separates the regions of sunward and anti-sunward plasma convection. If the solar wind does not penetrate onto the closed magnetic field region, the surface S also separates the regions of open and closed geomagnetic field lines.

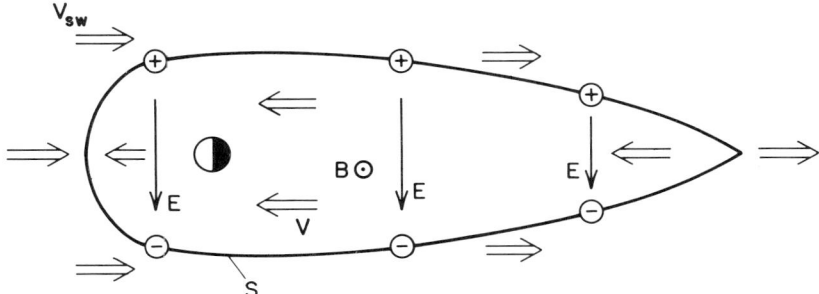

Fig. 4.3. Illustration of convective flow V in the equatorial plane resulting from the superposition of the interplanetary and the earth's magnetic field. The surface S, charged as indicated, separates the regions of sunward and anti-sunward convection.

It has been proposed (Lemaire, 1977; Heikkila, 1979) that flowing solar wind plasma can penetrate onto closed magnetic field lines. If such were the case, then part of the cross-polar-cap potential difference would be generated by the solar wind flow on closed field lines and part by the solar wind flow over the open polar cap field lines. The overall convection pattern of Figure 4.3 would be the same. However, the surface S would be within the closed field line region, and the boundary between open and closed field lines would be within the region of anti-sunward convection.

In addition, essentially the same convective pattern as that shown in Figure 4.3 was proposed by Axford and Hines (1961) to result from the interaction of the solar wind plasma with the magnetosphere plasma. They suggested that a viscous-like interaction between the solar and magnetosphere plasma could drive the plasma convection, and they did not specifically consider the possibility of open geomagnetic field lines. However, they emphasized that the flow pattern should result independent of how solar wind energy was transferred to the magnetosphere plasma.

4.2.2. Mapping of the Convection Electric Field to the Ionosphere and Resulting Ionosphere Currents

The surface S separating the regions of sunward and anti-sunward convection maps along field lines to the ionosphere, forming a roughly circular curve surrounding both polar caps as illustrated in Figure 4.4 for the northern hemisphere. Assuming the magnetic field lines are approximately equipotentials, it can be seen from Figure 4.2 that the convection electric field maps to the ionosphere so as to be discontinuous across both the dawn and dusk sides of S. The resulting signs of the charge on S in the ionosphere are shown in Figure 4.4 and are consistent with those in Figure 4.2. Dungey's (1961) schematic mapping of the equipotential lines of the convection electric field to the ionosphere is also shown in Figure 4.4. The ionospheric electric field will be normal to these equipotential lines, converging at the negatively charged boundary on the dusk side and diverging from the positively charged region on the dawn side.

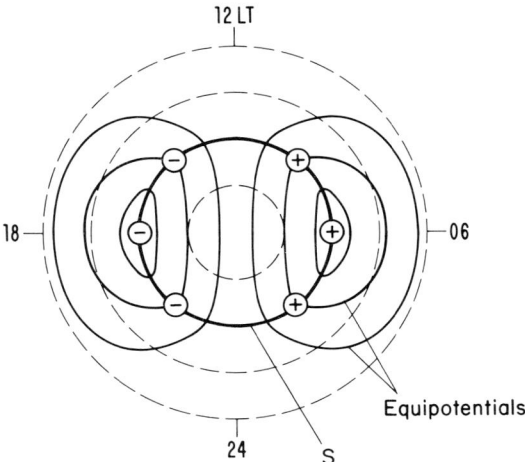

Fig. 4.4. Mapping of the convection electric potentials and the surface S from the magnetosphere to the ionosphere, as seen from above the north pole.

The ionospheric plasma is far more dense than the magnetospheric plasma and is not collisionless. The collisions inhibit the electric field drift of particles in a manner such that electric fields normal to **B** cause differential motion of ions and electrons, giving rise to ionospheric currents. It is often convenient to divide the ionospheric currents into two components, the Pedersen current $j_p = \sigma_p E$ along the direction of the electric field and the Hall current $j_H = \sigma_H E$ in the direction of $\mathbf{B} \times \mathbf{E}$ normal to the electric field. The constants relating j to E, σ_p and σ_H, are known as the Pedersen and Hall conductivities, respectively. They depend on the ionospheric collision frequencies as discussed, for example, by Rishbeth and Gariott (1969).

Ionospheric currents are readily detectable from magnetic field measurements on the ground. Under the assumption that the ionospheric currents vary over a horizontal scale greater than the altitude of the ionospheric currents (~ 100 km), the ground magnetic field response is essentially to the total ionospheric current above the point of the magnetic field measurement. This total ionospheric current is referred to as the height-integrated ionospheric current, which is the integral over height of the ionospheric current above a point on the earth. The height-integrated Hall conductivity is generally comparable to or greater than the height-integrated Pedersen conductivity, so that the height-integrated ionospheric current is generally dominated by the Hall current. Thus the height-integrated current is generally dominated by the ionospheric height region where the Hall current is largest. This occurs in the altitude range ($\sim 90-130$ km; Rishbeth and Garriott, 1969) where the ion gyromotion is significantly inhibited by collision while the electron gyromotion is essentially unaffected, so that the electrons move with their $(\mathbf{E} \times \mathbf{B})/B^2$ velocity while the ion drift in the $\mathbf{E} \times \mathbf{B}$ direction is significantly inhibited.

The dominance of the height-integrated Hall current by the electron $\mathbf{E} \times \mathbf{B}$ drift implies that the ionospheric currents will approximately follow the ionospheric equipotentials but will be in the opposite direction from the convective flow (Dungey, 1961; Levy et al., 1964). Thus the ionospheric currents will be in the sunward direction over the polar caps for the equipotential contours shown in Figure 4.4, and a return current will flow equatorward of the convection boundary S.

Indirect evidence for the existence of the overall convection pattern as mapped to the ionosphere can be obtained from the auroral observation of Davis (1960, 1962). The visible aurora studied by Davis results from electron precipitation into the ionosphere, and these electrons drift in the direction of the convective flow. He found a definite pattern in the average alignment of auroral forms and of the motion of visual irregularities along the auroral forms. This pattern is illustrated in Figure 4.5. The pattern was fixed with respect to magnetic coordinates, local midnight being at the bottom of Figure 4.5. Ground auroral observing stations rotate under the auroral pattern; the figure shows the location of 13 ground stations at 1020 UT.

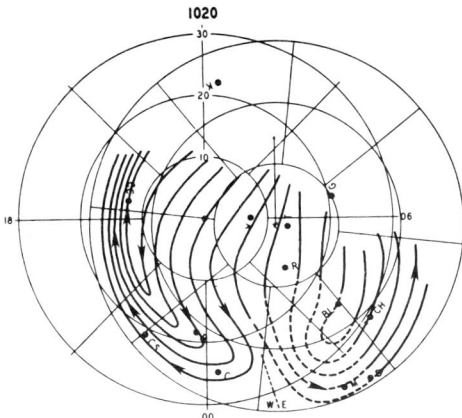

Fig. 4.5. Average alignment of auroral forms and the movement of visual irregularities along the auroral forms. This pattern was found to be fixed with respect to magnetic coordinates, local magnetic midnight being at the bottom of the figure. Locations of 13 ground stations at 1020 UT are shown by dots. Arrowheads indicate the direction of motion of the irregularities (from Davis, 1962). (© by American Geophysical Union)

The heavy solid and dashed lines in Figure 4.5 represent the alignment of the auroral forms, and the arrowheads on the lines indicate the direction of motion of the irregularities. The thin, straight, dashed line near midnight separates the regions of primarily westward and primarily eastward auroral motion in the region of maximum auroral occurrence (the auroral zone). The structure of the pattern observed by Davis is strikingly similar to the convection pattern in Figure 4.4 suggested by Dungey. In particular, Davis's observations show anti-sunward flow over the polar cap which turns around in the vicinity of the auroral zone to become sunward flow at lower latitudes. In addition, Davis noted that observed ionospheric currents were generally parallel to the alignment of the auroral forms and in the opposite direction to the observed motion of the irregularities, as expected.

Dungey noted the similarity of his predicted ionospheric potential pattern to an established ionospheric current system known as the S_D (also referred to as DP and DS) current system (Chapman, 1935; Chapman and Bartels, 1940; Silsbee and Vestine, 1942; Fukushima and Oguti, 1953). The structure of this current system was obtained from analysis of ground magnetic field variations during geomagnetically disturbed conditions. The current system as obtained by Fukushima and Oguti (1953) is shown in Figure 4.6, and the similarity to Dungey's equipotential convection pattern can be seen at latitudes above 60°.

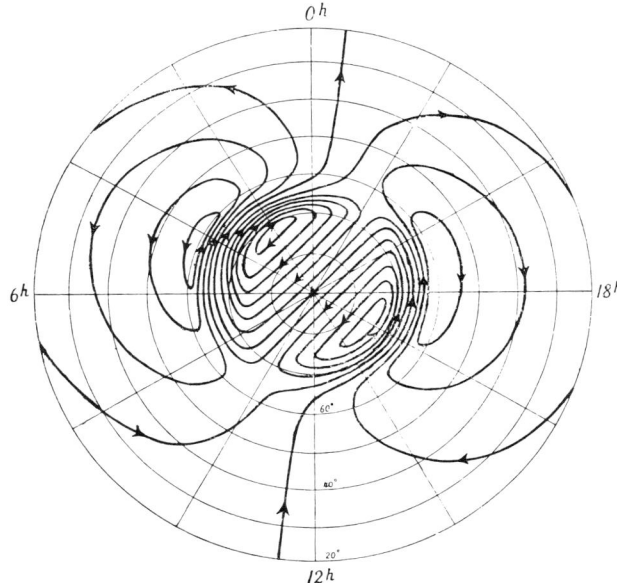

Fig. 4.6. S_D ionospheric current system for the equinoctial season (from Fukushima and Oguti, 1953).

However, it must be remembered that the S_D current system is obtained during periods of geomagnetic disturbances so that it may not represent a relatively permanent magnetospheric convection pattern. Subsequent studies of the S_D current system found it to be directly associated with the auroral electrojet, a narrow area of great current intensity directly associated with the auroral zones (Akasofu, 1968; Nishida and Kokubun, 1971). The association with aurora and geomagnetic activity led to the question of whether the S_D current (or at least part of it) is directly associated with the auroral processes rather than a direct result of the overall convection electric field.

Nagata and Kokubun (1962) recognized a current distribution over the polar caps that is not associated with the auroral electrojet or disturbed geomagnetic conditions (see also Nishida and Kokubun, 1971). They analyzed ground magnetic field data from geomagnetically quiet periods and obtained the ionospheric current system shown in Figure 4.7. This current system was referred to as S_q^p (q for quiet time, p for polar) to distinguish it from the well-known S_q current system which dominates the ionosphere current at lower latitudes during quiet times and is due to the dynamo driven by ionospheric tides. Figure 4.7 shows the S_q^p current system to be more intense in summer than in winter, as expected since the ionospheric density, and thus the ionospheric conductivity, is higher in summer than in winter. The S_q^p current system is probably not completely distinct from the S_D current system. Specifically, the S_D system may contain the S_q^p system, but the S_q^p system does not include the part of S_D associated with the auroral electrojet.

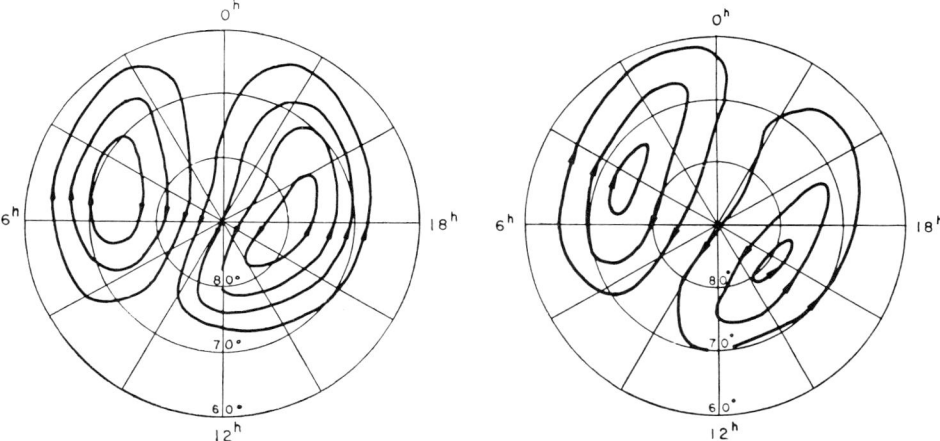

Fig. 4.7. S_q^p ionospheric current system for summer conditions (left) and winter conditions (right). The current between adjacent streamlines is 2×10^4 A in the left figure and 1×10^4 A in the right figure (from Nagata and Kokubun, 1962).

The current system obtained by Nagata and Kokubun was confined to latitudes $>55°$, since they included no data from lower latitudes in the analysis. This was done to separate the current system due to solar wind interactions with the magnetosphere from the S_q current system due to the ionospheric dynamo. The S_q^p currents are significantly more irregular in time than are the S_q currents. This is expected if S_q^p is due to solar wind interactions, since the solar wind and interplanetary magnetic field are more variable than are atmospheric tides. Nishida et al. (1966) took advantage of these fluctuations to map the S_q^p current system from the poles to the equator. An example of his current system is shown in Figure 4.8.

The S_q^p current system, as shown in Figures 4.7 and 4.8, is just what is expected from mapping the predicted magnetospheric convection electric field to the ionosphere as illustrated in Figure 4.4. Nishida et al.'s extension of the current system from the poles to the equator indicates that magnetospheric convection does indeed involve the entire magnetosphere.

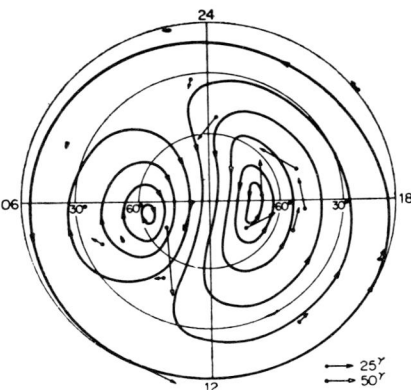

Fig. 4.8. Mapping of S_q^p current system from the poles to the equator for December 2, 1963. The current between adjacent contours is 5×10^4 A (from Nishida, 1968). (© by American Geophysical Union)

These current systems inferred from ground magnetic field observations and the auroral observations gave the first evidence for the existence of the large-scale magnetospheric convection electric field and its mapping to the ionosphere. Now, however, this convection electric field distribution has been verified by direct electric field measurements on low altitude, polar-orbiting satellites (e.g., Cauffman and Gurnett, 1971; Heppner, 1972a). Figure 4.9 (from Heppner, 1972b) gives two typical examples of the electric field observed along low altitude satellite passes approximately along the dusk-dawn meridian. The electric field over the polar cap is observed to be ~ 20 m V m^{-1} and is in the dawn-to-dusk direction as required to give anti-sunward convection. At lower latitudes the electric field reverses, giving the region of sunward convection. In the region of sunward convection, the electric field peaks at higher latitudes and decreases towards lower latitudes. Heppner (1972b) found this general pattern of the observed electric field to be almost always present for satellite trajectories from near dusk to near dawn and going within 10° to 15° of the geomagnetic pole. The basic pattern was found to be independent of the level of magnetic activity, though the magnitude of the electric fields increases with increasing magnetic activity.

Using the low altitude measurements of electric fields, together with the drift rates of Ba$^+$ clouds artificially injected from rockets into the auroral region near midnight, Heppner (1977) constructed average, empirical models of the high latitude electric field in the ionosphere. His model for average geomagnetic conditions (Figure 4.10) provides direct verification of the overall, convection electric field pattern as mapped into the ionosphere.

4.2.3. EVIDENCE FOR AN OPEN, POLAR CAP MAGNETIC FIELD

The preceding discussion establishes the existence of the overall convection electric field throughout the magnetosphere. We now address the question of whether the polar cap magnetic field is directly connected to the interplanetary magnetic field, so that at least

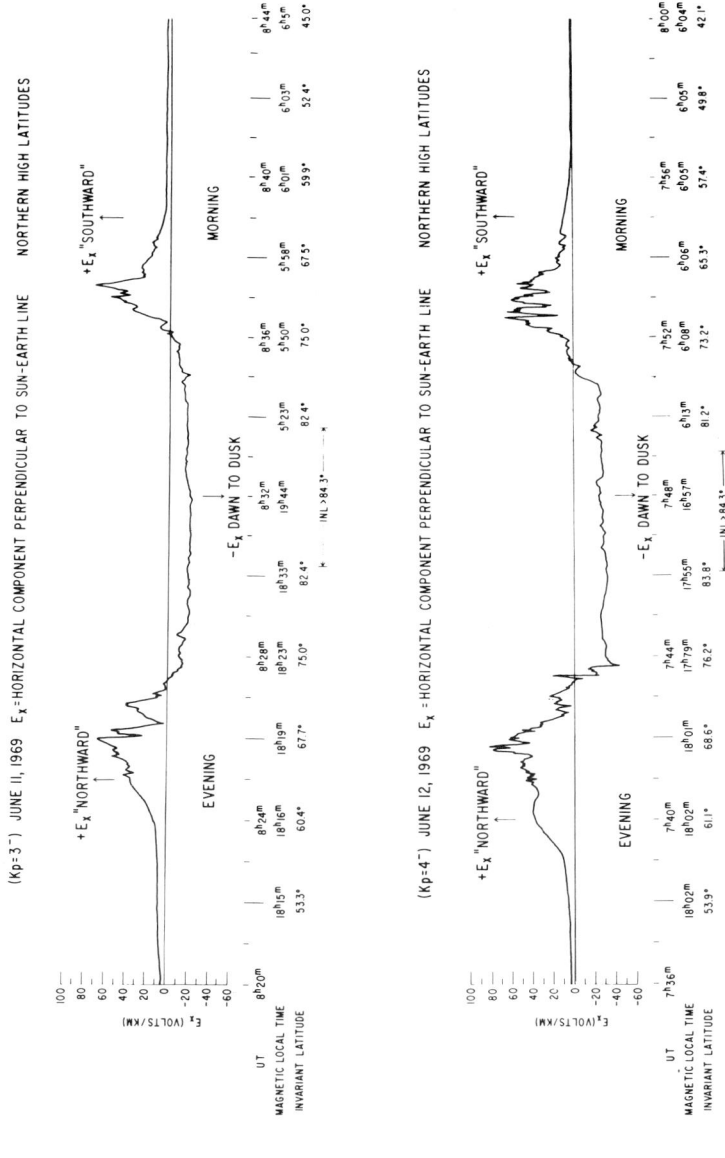

Fig. 4.9. Two typical examples of the electric field observed along low-altitude, polar satellite passes approximately along the dusk-dawn meridian (from Heppner, 1972b).

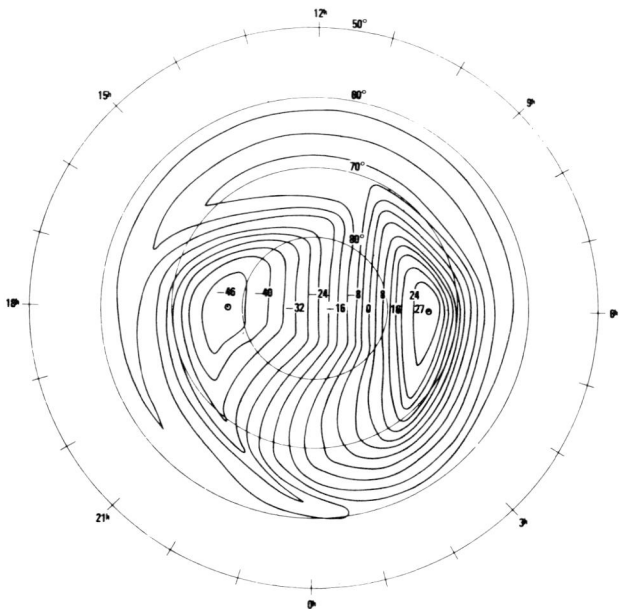

Fig. 4.10. Model of ionospheric electric field distributions for average geomagnetic conditions obtained from low altitude satellite measurements and the drift rates of Ba$^+$ clouds artificially injected from rockets into the auroral region near midnight (from Heppner, 1977). (© by American Geophysical Union)

some of the convection electric field results directly from the mapping of $\mathbf{E} = -\mathbf{V}_{sw} \times B$ into the magnetosphere. Much evidence exists that this is the case, the most definitive coming from observations of energetic particles over the polar caps.

Energetic particles with a velocity along the magnetic field much greater than the electric field drift velocity and the perpendicular magnetic drift velocity will nearly follow magnetic field lines. Such particles with a sufficiently small pitch angle, and with a parallel velocity headed along field lines towards the earth, will precipitate into the ionosphere. Thus if polar magnetic field lines are open, particles in the interplanetary medium with a speed much greater than V_{sw} will precipitate directly into the polar ionosphere, provided the angular distribution of the interplanetary particles with respect to $\mathbf{B}$ includes particles with velocities along $\mathbf{B}$. Despite the fact that the loss cone for particle precipitation into the ionosphere is exceptionally small in interplanetary space (the ratio of the magnetic field in interplanetary space to that in the polar ionosphere is $\approx 1/10\,000$, giving a loss cone angle of $\approx 0.6°$), Liouville's theorem assures that the particles will have the same distribution function when they reach the ionosphere as they have outside the magnetosphere. Thus, assuming the particles' energy is not altered as they propagate from the interplanetary medium to the ionosphere, the differential flux j for particles that reach the ionosphere will be the same in the ionosphere as in the interplanetary medium.

Initial evidence for the direct access of energetic solar particles to the polar caps was obtained from riometer measurements of the ionospheric absorption of cosmic radio noise (Reid and Leinbach, 1959). The measured absorption was shown to be an efficient detector of 5–50 MeV proton precipitation into the ionosphere. A pronounced uniformity of the absorption was observed over the northern polar cap, and Reid and Leinbach suggested this uniformity could be explained if the magnetic field emanating from the polar cap were directly connected to the interplanetary magnetic field.

West and Vampola (1971) compared simultaneous measurements of electrons in the interplanetary medium (from the OGO–V satellite) and over the polar caps (from the OV1–19 satellite) during a solar particle event. A large increase (3 orders of magnitude at ~ 100 keV) of energetic electrons in the interplanetary medium resulted from a flare on the sun, which gave easily observable electron fluxes over the energy range 79–1530 keV, and the measured fluxes at low altitudes (< 2000 km) over the polar caps increased in conjunction with the interplanetary fluxes. A comparison of the electron fluxes as a function of energy from near the time of the peak in the event is shown in Figure 4.11. It can be seen that the agreement between the fluxes in the interplanetary medium and over the polar caps is excellent at all energies, and such agreement was found throughout the particle event.

A uniform flux of ~ 0.1 keV electrons was later found by Winningham and Heikkila (1974) to precipitate over the entire polar cap, and they suggested that this precipitation was present at all times. Fennell et al. (1975) verified this possibility by studying the precipitation of 0.3–8 keV electrons over the polar caps. They found such electrons to be precipitating over the polar caps at all times during geomagnetically quiet periods analyzed for 1972 and 1973, and they compared the polar cap fluxes (from the STP 72–1 satellite) with simultaneously measured fluxes in the interplanetary medium (from the Vela satellites). Comparisons between the polar cap (0.3–2 keV) and interplanetary (0.2–1 keV) fluxes are given in Figure 4.12 for times just prior to and during a change in the interplanetary electron intensity. Again the agreement is excellent, the increase in the interplanetary fluxes being directly reflected in the polar cap observations. Electron fluxes observed in the high latitude magnetotail are also shown in Figure 4.12. These fluxes agree well with the interplanetary and polar cap fluxes, which suggests that open field lines extend into the magnetotail as expected from the overall magnetic field topology shown in Figure 4.1.

Agreement has also been found between proton fluxes in the interplanetary medium and over the polar caps as shown in Figure 4.13 (from Fennell, 1973). This figure shows the intensity of $\geqslant 300$ keV protons over a 25-day period in April 1969 that includes the same solar particle event analyzed by West and Vampola (Figure 4.11).

The above comparisons between the interplanetary and polar cap particle intensities give strong evidence that the polar cap magnetic field is open to the interplanetary magnetic field. However, an asymmetry in the precipitation over the two polar caps often occurs, and this asymmetry has been used to provide even more convincing evidence that the polar cap field lines are open. Reid and Sauer (1967) noticed a definitive asymmetry in the precipitation of 10–30 MeV protons over the polar caps during a solar particle event, the fluxes of protons precipitating into the southern polar cap being greater than those into the northern polar cap. They proposed a remarkably simple explanation of the asymmetry under the assumption that the polar cap field lines were open.

This explanation is illustrated in Figure 4.14. Magnetic field lines emanating from the sun will connect directly to field lines emanating from only one polar cap of the earth,

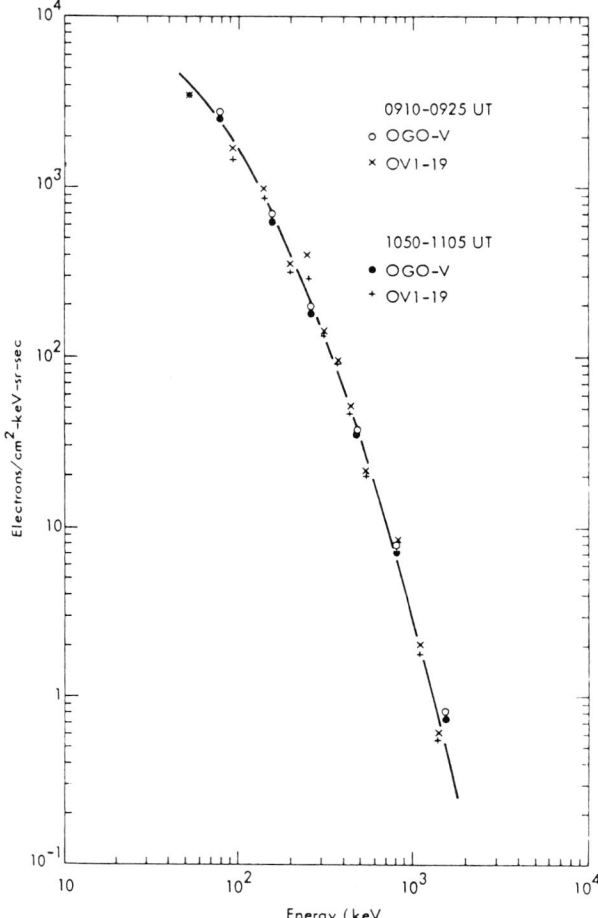

Fig. 4.11. Simultaneous measurements of electrons in the interplanetary medium from the OGO-5 satellite and over the polar caps from the OV1-19 satellite near the peak of a solar particle event on April 13, 1969. For the 0910-0925 UT data OV1-19 was at an altitude > 4500 km over the south polar cap; for the 1050-1105 UT data, it was over the north polar cap at an altitude < 2000 km (from West and Vampola, 1971). (© by American Physical Society)

that polar cap being determined by the direction of the interplanetary magnetic field. If the interplanetary magnetic field points away from the sun, as illustrated in Figure 4.14, then field lines from the northern polar cap will connect to interplanetary field lines coming from the sun while field lines from the southern polar cap will connect to interplanetary field lines going away from the sun. The opposite will occur when the interplanetary magnetic field is pointed toward the sun. Having a southward-pointed interplanetary magnetic field as illustrated in Figure 4.14 is not significant to this argument. If the flux of particles in the interplanetary medium is anisotropic and more particles are going away from the sun along field lines than toward the sun as illustrated in the figure, then more particles will strike the polar cap that has open field lines directly

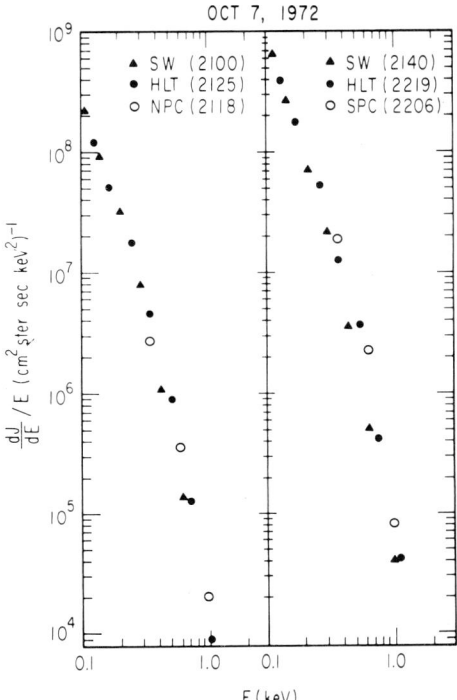

Fig. 4.12. Comparison of polar cap (NPC, SPC), solar wind (SW), and high latitude magnetotail (HLT) electron fluxes for times just prior to and during a change in the solar wind electron intensity. The polar cap observations are from the STP 72-1 satellite, and the solar wind and magnetotail observations are from the Vela satellites (from Fennell et al., 1975; Mizera and Fennell, 1978). (© by American Geophysical Union)

connected to field lines coming from the solar direction than will strike the other polar cap. The particle fluxes over the polar cap whose field lines are coming from the solar direction should equal the interplanetary fluxes of field-aligned particles going away from the sun, while the particle fluxes over the other polar cap should equal the interplanetary fluxes of field-aligned particles going toward the sun.

A definitive observation of this effect was presented by Van Allen et al. (1971). They reported simultaneous measurements of the flux of 0.3–9.2 MeV protons in the interplanetary medium (from Explorer 33) as a function of direction relative to the Sun and the flux of 0.3–10.0 MeV protons incident upon both polar caps (from Injun-5). The measurements (Figure 4.15) were taken during a solar particle event. The interplanetary magnetic field in the plane perpendicular to the ecliptic plane and parallel to the earth-sun line was measured to be oriented as shown in Figure 4.14. Particles coming along the magnetic field from the solar direction should thus come from the direction shown as sector 4 in the figure, and these particles should strike the northern polar cap. Particles coming along the magnetic field from the anti-solar direction should be within sector 2 and should strike the southern polar cap. Figure 4.15 shows good agreement of the

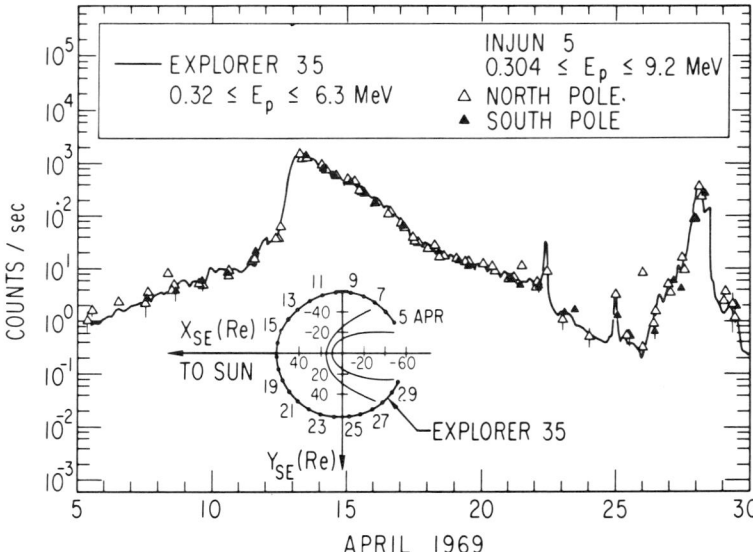

Fig. 4.13. Comparison between proton fluxes in the interplanetary medium and those over the polar caps (from Fennell, 1973). (© by American Geophysical Union)

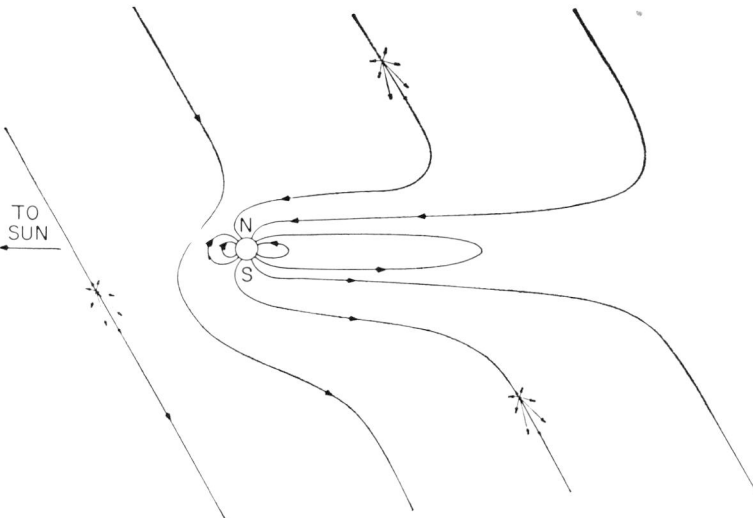

Fig. 4.14. Magnetic field topology in the non-midnight meridian plane. Assuming that the polar cap magnetic field is open, an anisotropic particle flux in the interplanetary medium leads to an asymmetry in the polar cap particle precipitation, as illustrated (from Van Allen et al., 1971). (© by American Geophysical Union)

ELECTRIC FIELDS

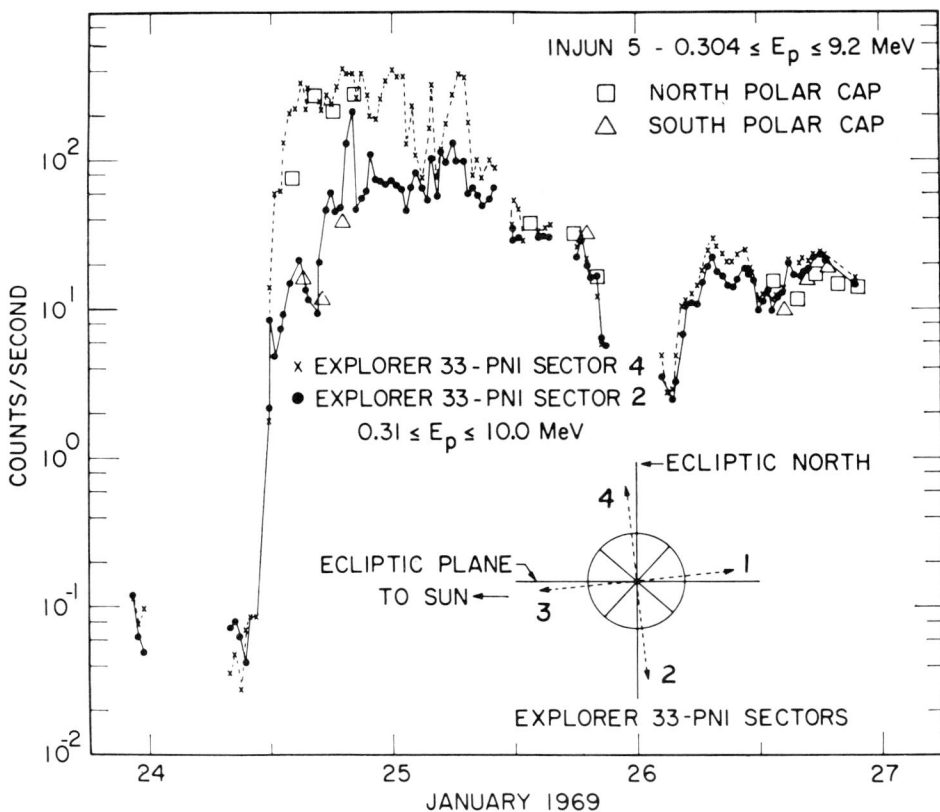

Fig. 4.15. Simultaneous measurements of the flux of 0.3–9.2 MeV protons in the interplanetary medium (from Explorer 33) as a function of direction relative to the sun and the flux of 0.3–10.0 MeV protons incident upon both polar caps (from Injun-5). Particles coming along magnetic field lines from the solar directions are within sector 4, as illustrated, and these particles should strike the northern polar cap. Particles coming along field lines from the anti-solar direction should be within sector 2 and should strike the southern polar cap (from Van Allen et al., 1971). (© by American Geophysical Union)

sector 4 interplanetary fluxes with the northern polar cap fluxes and of the sector 2 interplanetary fluxes with the southern polar cap fluxes during the solar particle event. The interplanetary fluxes showed a marked anisotropy during the early stages of the event (late part of January 24 and early part of January 25) and became isotropic later in the event. The same proton flux anisotropy was seen in the polar cap fluxes, and the transition to isotropy was also reproduced in the polar cap observations.

The precipitating proton fluxes as a function of geomagnetic latitude for a satellite pass over both polar caps are shown in Figure 4.16 (from Van Allen et al., 1971) for two energy ranges, 0.304–9.2 MeV and 3.44–79 MeV. These observations were obtained near the peak of the solar particle event of Figure 4.15. Note the uniformity of the fluxes over each polar cap at latitudes $\gtrsim 80°$, which is what is expected if the magnetic field was

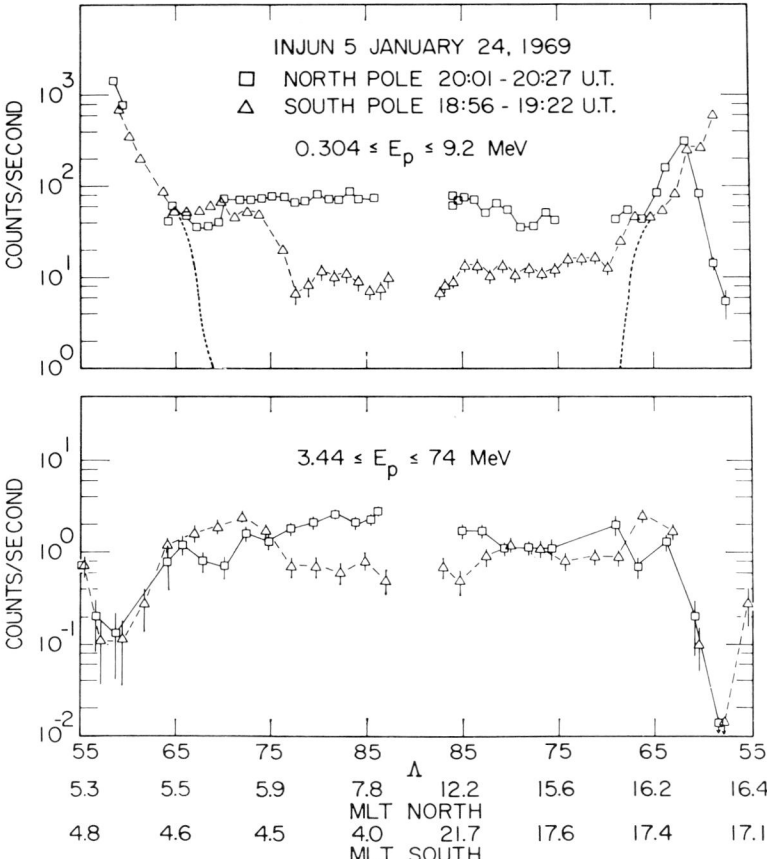

Fig. 4.16. Precipitating proton fluxes as a function of geomagnetic latitude for a satellite pass over both polar caps for two energy ranges, 0.304–9.2 MeV and 3.44–74 MeV. These observations were obtained near the peak in the solar particle event of Figure 4.15 (from Van Allen et al., 1971). (© by American Geophysical Union)

open over this latitude range and the interplanetary fluxes were uniform over the time period of the satellite pass over each polar cap.

Fennell et al. (1975) extended the polar cap asymmetry analysis to 0.3–8.0 keV electrons. They recognized that solar wind electrons are generally anisotropic at these energies, and that the anti-sunward-flowing fluxes are greater than the sunward-flowing fluxes by a factor of 2–5 (Frank and Gurnett, 1972; Feldman et al., 1975). Figure 4.17 shows intensity profiles of 0.62 and 2.6 keV precipitating electrons for two groups of orbits, October 5–6 and October 8–9, 1972, over the polar caps during very quiet geomagnetic conditions. The relatively constant fluxes of electrons at low levels over the polar caps are presumably directly from the solar wind. All the data in Figure 4.17 were taken while the interplanetary magnetic field was directed toward the sun, so that the southern polar cap should have been exposed to anti-sunward-flowing electrons. As can

ELECTRIC FIELDS

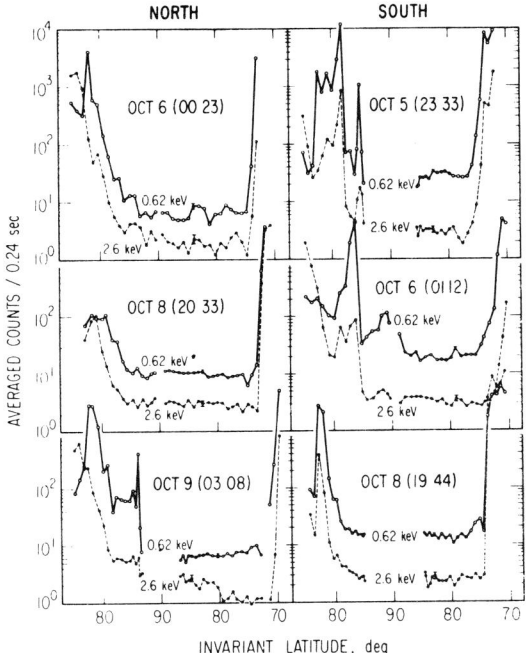

Fig. 4.17. Intensity profiles of 0.62 and 2.6 keV precipitating electrons for two groups of STP 72-7 satellite orbits, October 5-6 and October 8-9, 1972, over the polar caps during very quiet geomagnetic conditions. Local noon is at the left and local midnight is at the right in each panel (from Fennell *et al.*, 1975).

be seen in the figure, the southern polar cap intensities were consistently greater than the northern polar cap intensities as expected if the polar cap field lines were open. In addition, the magnitude of the polar cap asymmetry at 0.62 keV is consistent with a factor of ~ 2 anisotropy in the 0.5 keV interplanetary electrons observed at this time (Fennell *et al.*, 1975). Fennell *et al.* took the ratio of the fluxes over the two polar caps from 51 sets of consecutive passes over the two polar caps and compared the results with the interplanetary magnetic field direction as shown in Figure 4.18. The majority (42 out of 51) of the ratios were found to be in agreement with that expected from the direct access of interplanetary electron to the earth's polar caps. A similar conclusion was reached by Yeager and Frank (1976) using 0.3-0.5 keV electrons.

The above is strong evidence that polar cap magnetic field lines are generally open to the interplanetary medium over a significant portion of the polar cap. Thus a significant portion of the magnetospheric convection electric field must result directly from a mapping of the interplanetary electric field along field lines into the magnetosphere. This does not imply that all of the magnetospheric convection electric field is generated on open field lines. It is possible that some part of the electric field is generated on closed field lines (Axford and Hines, 1961; Lemaire, 1977; Heikkila, 1979); however, it has not yet been determined how much, if any, is so generated.

Given an estimate for the area of the polar cap from which the magnetic field is

Interplanetary Magnetic Sector

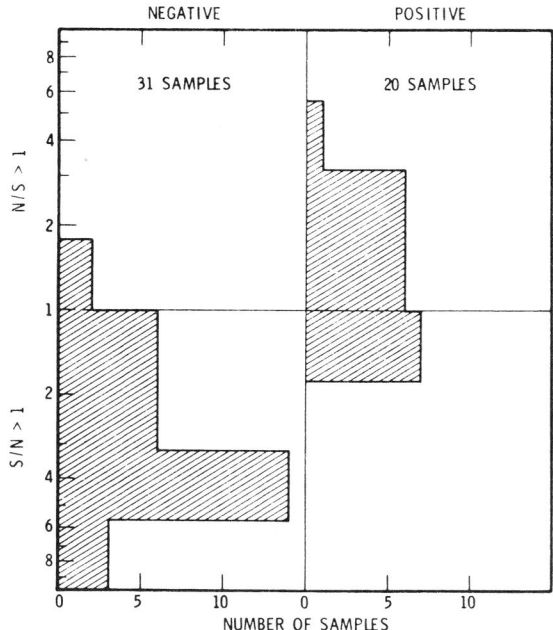

Fig. 4.18. Comparison of the ratio of the 0.62 keV precipitating electron fluxes over the two polar caps from 51 sets of constructive STP 72–1 satellite passes over the polar caps with the interplanetary magnetic field direction. N/S is the north/south polar cap ratio, S/N is the south/north polar cap ratio, and negative sector indicates a solar directed magnetic field (from Fennell et al., 1975; Mizera and Fennell, 1978). (© by American Geophysical Union)

directly connected to the interplanetary field and an estimate of the potential difference across the open polar cap field lines in the dawn-to-dusk direction, it is possible to estimate the length ℓ_I along the solar wind direction and the width d normal to the solar wind direction for the area in the interplanetary medium where the magnetic field is directly connected to each polar cap. Following Dungey (1965) and Stern (1973), we approximate the region of open field lines to be a circle of radius 1300 km (which subtends a latitudinal angle of 24°). The polar magnetic field strength at the earth's surface is $\approx 6.2 \times 10^{-5}$ T (0.62 G), so that the total flux of open magnetic field emanating from each polar cap is $\approx 3.3 \times 10^8$ T m². Since $\nabla \cdot \mathbf{B} = 0$, this flux must be the same across any area away from the earth through which all the open field lines from one polar cap pass but not other field lines. Thus, for an interplanetary field strength of 5 nT, the magnetic flux from each polar cap must penetrate an area normal to the magnetic field in the interplanetary medium of 3.3×10^8 T m²/5×10^{-9} T = 6.6×10^{16} m² = $1620\, R_e^2$. Approximating this area in the interplanetary medium as a rectangle of area $d \cdot \ell_I$ as illustrated in Figure 4.19, the potential difference across the polar cap generated by the solar wind flow across open field lines is $\Phi_{pc} = V_{sw} \cdot B \cdot d$. We take $V_{sw} = 300$ km s^{-1}

ELECTRIC FIELDS 75

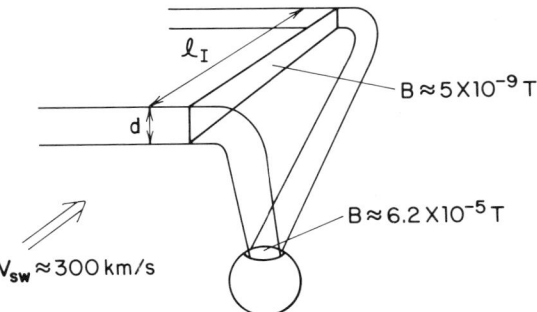

Fig. 4.19. Mapping of the polar cap open magnetic field region into the interplanetary medium.

and assume that the entire polar cap potential difference is generated by the solar wind flow across open polar cap field lines so that Φ_{pc} = 65 keV (a rough average from Figure 4.10). This gives $d = \Phi_{pc}/(V_{sw}B) \approx 7R_e$ and $\ell_I = 1620/7 \approx 230 R_e$.

This value of ℓ_I is a rough estimate for the length of the geomagnetic tail and indicates that the tail extends ~ 4 times beyond the orbit of the moon. The $\approx 7R_e$ estimate of the width of the open field line region is much less than the $\approx 40 R_e$ diameter of the tail. These widths give the estimate that only ~ 18% of the interplanetary magnetic field, intercepted by the magnetosphere in a plane normal to the solar wind flow, is directly connected to the geomagnetic field. Stern (1973) referred to the long narrow regions of open magnetic field along the geomagnetic tail as 'windows'.

4.2.4. COLD PLASMA CONVECTION AND THE PLASMAPAUSE

The convection electric field throughout the magnetosphere is extremely important in controlling the distribution of particles throughout the magnetosphere. Cold plasma particles, such as those from the ionosphere, undergo negligible magnetic gradient and curvature drifts, so that their velocity normal to the magnetic field is given simply by $\mathbf{V}_{DE} = (\mathbf{E} \times \mathbf{B})/B^2$. If the magnetic field lines are equipotentials, $\mathbf{V}_{DE}$ lies along lines of constant electric potential ϕ, where $\mathbf{E} = -\nabla\phi$.

Sunward convection of cold plasma extends throughout most of the equatorial magnetosphere, as illustrated in Figure 4.3. However, mappings of the electric field throughout the equatorial magnetosphere are not yet available from measurements. A useful, first-approximation model of the equatorial convection electric field $\mathbf{E}_c$ is simply to take it to be uniform. Equipotential contours from this exceptionally simple model are shown in the upper left panel of Figure 4.20 for the equatorial plane of the magnetosphere.

The magnetospheric electric field must include a contribution from the earth's rotation in addition to that from the convection electric field. This gives a corotation electric field $\mathbf{E}_R = -\mathbf{v}_R \times \mathbf{B}$ in the non-rotating frame of reference, where $\mathbf{v}_R$ is the corotation velocity. Assuming again that the geomagnetic field lines are equipotentials, this corotation electric field must exist throughout the entire magnetosphere. At the earth's surface $E_R = 2\pi R_e \cos \lambda (3.1 \times 10^{-5} \text{ T}) (1 + 3 \sin^2 \lambda)^{1/2}/(8.64 \times 10^4 \text{ s}) = 1.44 \times 10^{-2} \cos \lambda (1 + 3 \sin^2 \lambda)^{1/2}$ V m^{-1}, and is directed radially inward at the equator. Here λ is geomagnetic latitude, and the geomagnetic field is taken to be dipolar and to have an equatorial field value of 3.1×10^5 T and a dipole axis coincident with the earth's rotation axis. The corotation electric field decreases as L^{-2} with increasing distance from the earth so as to maintain corotation. Equipotential contours for the corotation electric field are shown in the upper right panel of Figure 4.20, spaced 3 kV apart.

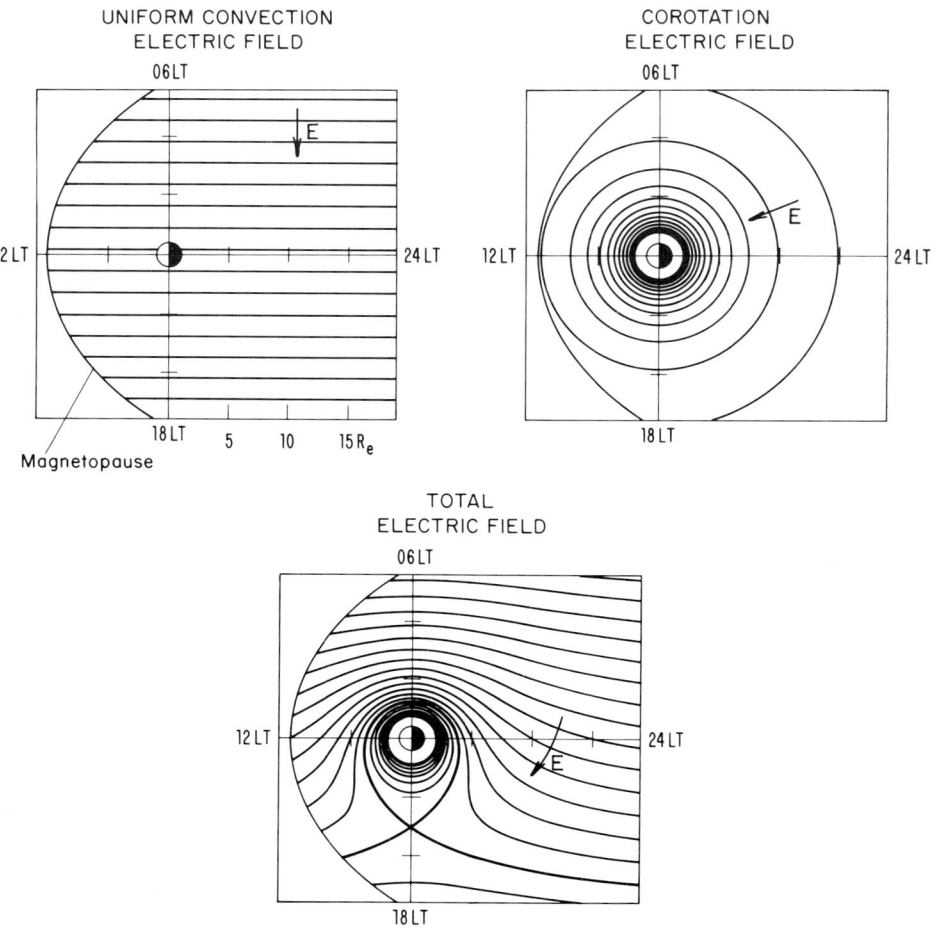

Fig. 4.20. Equipotential contours for the magnetospheric electric field in the equatorial plane. Upper left: first-order approximation for the convection electric field $\mathbf{E}_c$ as uniform. The contours are spaced 3 kV apart for $E_c = 2.5 \times 10^{-4}$ V m^{-1}. Upper right: the corotation electric field, contours spaced 3 kV apart. Lower: sum of convection and corotation electric fields. The heavy contour separates the closed and open convection regions.

Adding the convection and corotation electric potentials gives a total magnetospheric potential in the equatorial plane

$$\phi = -1.44 \times 10^{-2} R_e/L + E_c R_e L \sin \psi \qquad (4.1)$$

where ψ is the longitude angle measured eastward with respect to 2400 LT. To estimate an average value for E_c we again use the average cross-polar-cap potential difference of ~ 65 kV from Figure 4.10 and note that this potential difference must also exit across the tail region of sunward convection. Taking a tail width of $40R_e$ gives $E_c = 2.5 \times 10^{-4}$ V m^{-1}. The potential contours in the upper left panel of Figure 4.20 are spaced 3 kV apart for this value of E_c. The lower panel of Figure 4.20 shows contours of constant ϕ from Equation (4.1) every 3 kV for $E_c = 2.5 \times 10^{-4}$ V m^{-1}. Notice that the heavy contour in this panel separates contours that give a closed convection of plasma around the earth from contours that give a basically sunward convection of particles from the tail through the dayside magnetopause. The region of sunward convection is assumed to extend all the way to the dayside magnetopause in this and subsequent similar figures of particle convection patterns; however, this assumption is not critical to our discussion.

Nishida (1966) proposed that the convection pattern shown in Figure 4.20 could account for the discontinuity in the magnetospheric cold plasma distribution known as the plasmapause. These cold plasma particles have energies near those of ionospheric particles, and the plasmapause is a relatively abrupt, field-aligned transition at $L = 3-6$ from cold plasma densities $\gtrsim 10^8$ m^{-3} (10^2 cm^{-3}) to densities $\lesssim 10^6$ m^{-3} (Carpenter, 1963, 1966). Sample radial distributions of the plasma density across the plasmapause are shown in Figure 4.21 (from Chappell et al., 1970). The densities within the plasmapause are in agreement with those expected from an equilibrium distribution of ionospheric plasma along closed magnetic field lines, while the densities outside the plasmapause are well below those for such an equilibrium (Angerami and Carpenter, 1966).

Nishida suggested that plasma within the open convection region would be convected across the dayside magnetopause (or to the deep magnetotail if anti-sunward convection

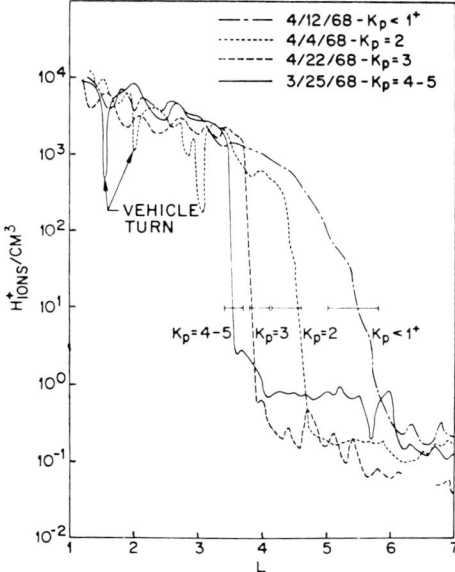

Fig. 4.21. Sample radial distributions of the plasma density across the plasmapause (from Chappell et al., 1970). (© by American Geophysical Union)

extends within the magnetopause) so that such plasma should be lost from the closed field line part of this region within the inner magnetosphere. Plasma within this open convection region should thus be well below the equilibrium value, since the rate of replenishment from the ionosphere was estimated to be much less than the time scale of the plasma convection. On the other hand, ionospheric plasma within the closed convection region should always be prevented from escaping since it will always be on field lines that close within the inner magnetosphere. Thus this plasma can reach its equilibrium value, so that the boundary between the two convection regions should correspond to the plasmapause under steady state conditions.

The boundary between closed and open convection paths from Figure 4.20c is compared in Figure 4.22 with the average position of the plasmapause obtained by Chappell et al. (1971) from more than 150 satellite passes near the equator. Considering the simplicity of the convection electric field model and that E_c has not been adjusted to maximize the agreement, the agreement in Figure 4.22 is excellent. The convection pattern suggested by Nishida (1966) reproduces the average location of the plasmapause to within $1R_e$ and the variation of its distance from the earth with local time, which shows a minimum near 0600 LT and a maximum near 1800 LT.

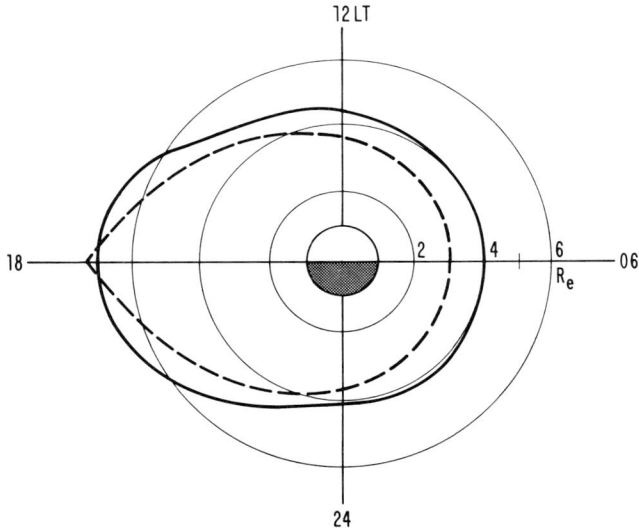

Fig. 4.22. Comparison of the boundary between the closed and open convection regions (dashed line) with the average position of the plasmapause (solid lines) obtained by Chappell et al. (1971) from more than 150 OGO–5 profiles of the plasmapause. Coordinates are L versus local time.

The L-value of the plasmapause is known to vary with geomagnetic activity as shown in Figure 4.21 for 0000–0040 LT. This figure shows the plasmapause at $L = 4-5$ for average geomagnetic conditions ($K_p = 2-3$), at $L < 4$ for moderately disturbed conditions ($K_p = 4-5$), and at $L > 5$ for quiet conditions ($K_p < 2$). Such a variation is expected since the cross polar cap potential drop, and thus E_c, varies with geomagnetic activity.

The variation of the plasmapause location with E_c can easily be obtained from the above analysis of the magnetospheric electric field. As can be seen from the lower panel of Figure 4.20, the electric field is zero at the 1800 LT point along the boundary between the regions of closed and open convection. This occurs where the corotation and convection components of the electric field are equal and opposite, i.e., where $1.44 \times 10^{-2}/L^2 = E_c$. Substituting $L = (1.44 \times 10^{-2}/E_c)^{1/2}$ and $\sin \psi = -1$ into Equation (4.1) gives the potential ϕ_{pp} of the plasmapause:

$$\phi_{pp} = -2R_e(1.44 \times 10^{-2} E_c)^{1/2},$$

and substituting ϕ_{pp} back into (4.1) gives the L-value L_{pp} of the plasmapause as a function of local time:

$$L_{pp} = \left(\frac{1.44 \times 10^{-2}}{E_c}\right)^{1/2} \left[\frac{\sqrt{1 + \sin \psi} - 1}{\sin \psi}\right].$$

This gives $L_{pp} = 0.5(1.44 \times 10^{-2}/E_c)^{1/2}$ at 1200 and 2400 LT ($\sin \psi = 0$), a minimum value $L_{pp} = (\sqrt{2} - 1)(1.44 \times 10^{-2}/E_c)^{1/2}$ at 0600 LT, and the maximum value $(1.44 \times 10^{-2}/E_c)^{1/2}$ at 1800 LT.

Using Heppner's (1977) estimates of the cross-polar-cap potential differences, we estimate that E_c may vary from approximately a factor of 3 below 2.5×10^{-4} V m^{-1} during quiet times to a factor of 2 above during disturbed conditions. Electric fields obtained from the drift of whistler ducts within the plasmasphere (Carpenter, 1970) are in agreement with these estimates. Thus, near midnight, we estimate that L_{pp} should vary from 2.65 during disturbed condition to 6.50 during quiet times and have an average value near 3.75. These predicted plasmapause locations are compared in Figure 4.23 with the locations observed near local midnight by Taylor et al. (1968). The observations are shown versus the maximum value of the K_p index during the preceding 24 hr, since it takes several days for a region to refill from the ionosphere once the location of the plasmapause has been displaced to an L-value within that region (Park, 1970).

It can be seen from Figure 4.23 that the simple convection theory outlined above reproduces the observed variation of the plasmapause location quite well. The predicted mean location is $\sim 0.5-1.0 R_e$ too close to the earth as is also the case near midnight for the comparison shown in Figure 4.22. This suggest that 2.5×10^{-4} V m^{-1} may be a slight overestimate of the convection electric field strength in the inner magnetosphere.

Thus the distribution and motion of thermal plasma, at least within the inner magnetosphere, appears to be governed by the convection and corotation electric fields as suggested by Nishida (1966). The agreement between the predicted and observed plasmapause location supports the concept that the convection electric field exists throughout the magnetosphere, and the agreement also indicates that the assumption of a uniform, dawn-to-dusk convection electric field throughout the magnetospheric equatorial plane is a useful, first-order approximation.

4.2.5. ENERGETIC PLASMA CONVECTION

The motion of energetic plasma in the magnetospheric electric field is particularly important in bringing plasma from the tail and magnetospheric boundaries to the inner regions of the magnetosphere. This inward convection of plasma causes a significant energization

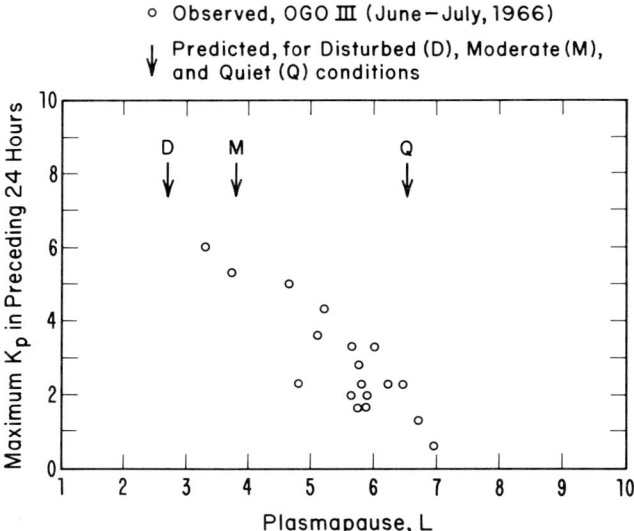

Fig. 4.23. Comparison of the plasmapause locations predicted from the boundary between open and closed convection regions with locations observed from the OGO-3 satellite near local midnight by Taylor et al. (1968). The predicted locations are for convection electric field values of 0.83×10^{-4} V m^{-1} (quiet), 2.5×10^{-4} V m^{-1} (moderate), and 5.0×10^{-4} V m^{-1} (disturbed).

of the plasma as required by the conservation of the first two adiabatic invariants. For a dipole magnetic field, the particle energy varies as L^{-n}, where n varies from 2 to 3 as the equatorial particle pitch angle is increased from 0° to 90° (see Cowley and Ashour-Abdalla, 1975).

However, the trajectories of energetic plasma particles are not the same as those for the low energy plasma, owing to the importance of the gradient and curvature magnetic drifts. For electrons, the magnetic drifts add to the corotation velocity. Thus the trajectories of energetic electrons are similar to those of low energy particles, except the region of closed trajectories will extend to increasing radial distances from the earth with increasing electron energy. The magnetic drifts for energetic ions are opposite in direction to the corotation velocity, which leads to far more complicated particle trajectories.

It is possible to evaluate the trajectories of equatorially mirroring energetic particles in a model for the magnetospheric magnetic and electric fields without having to calculate the actual particle drift velocities (Roederer, 1970). This is accomplished by using the conservation of the total particle energy $W = K + q\phi$ for a particle of charge q and kinetic energy K and conservation of the first adiabatic invariant $\mu = p_\perp^2/(2mB)$. Non-relativistically $\mu = K/B$ for equatorially mirroring particles. Conserving W and substituting μB for K in the expression for W shows that non-relativistic particles of a given magnetic movement will follow trajectories given by

$$\mu B + q\phi = \text{const.} \tag{4.2}$$

These trajectories are especially straightforward to evaluate for a dipole magnetic field and the simple model for the electric potential (Equation (4.1)) used in the plasmapause calculation. For a particle initially with kinetic energy K_i at $L = L_i$ and $\psi = \psi_i$, we obtain for the trajectories

$$K_i \left(\frac{L_i}{L}\right)^3 + q\left(\frac{-1.44 \times 10^{-2} R_e}{L} + E_c R_e L \sin \psi \right)$$
$$= K_i + q\left(\frac{-1.44 \times 10^{-2} R_e}{L_i} + E_c R_e L_i \sin \psi_i \right), \quad (4.3)$$

from which ψ can easily be obtained as a function of L for the trajectories of any particle for which an initial kinetic energy, L-value, and local time are specified.

Since particles conserve μ as they follow the trajectories given by (4.2) and (4.3), it is convenient to display trajectories as a function of μ. Figures 4.24 and 4.25 show

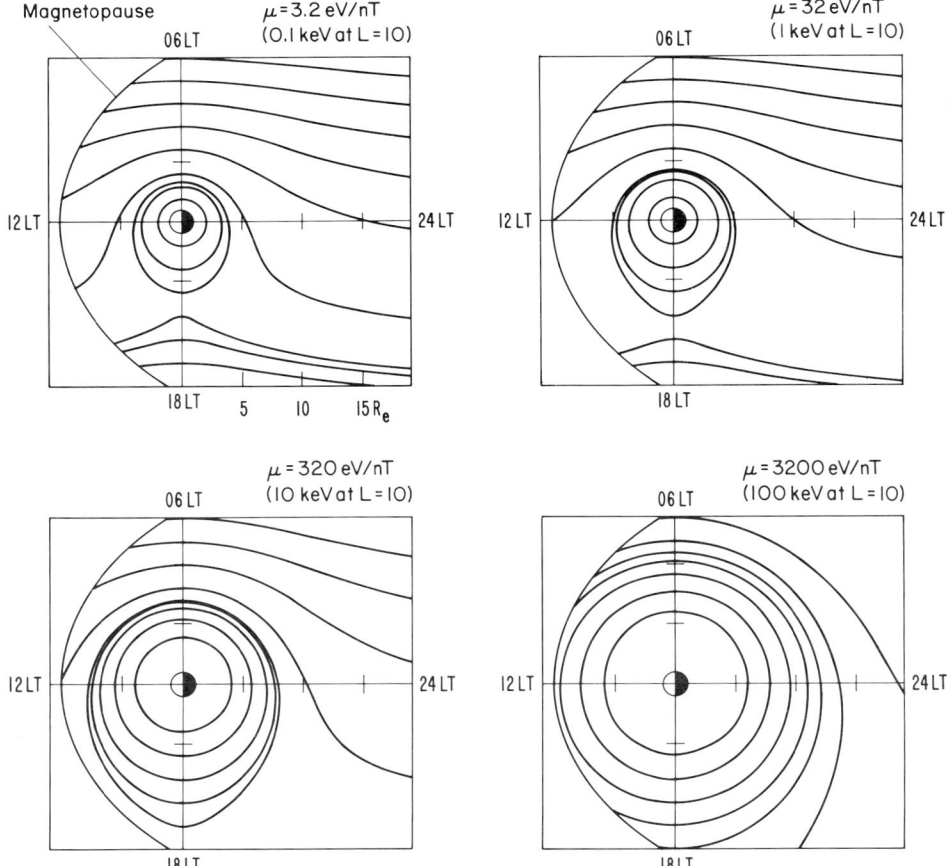

Fig. 4.24. Electron drift trajectories as obtained from Equation (4.3). All trajectories in each panel are for the value of μ indicated. No trajectories are shown at L-values where the electron kinetic energy exceeds $m_e c^2$.

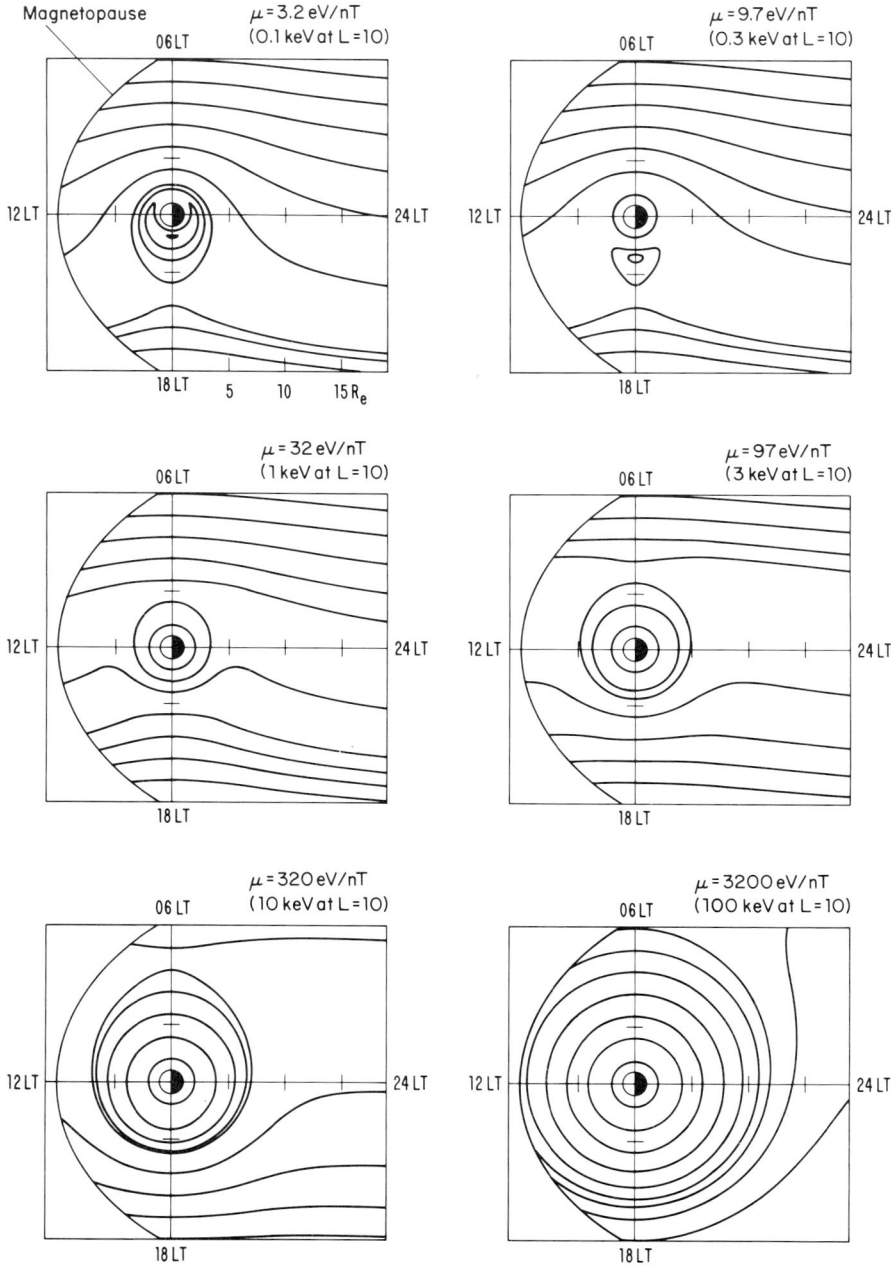

Fig. 4.25. Singly charged ion drift trajectories as obtained from Equation (4.3). All trajectories in each panel are for the value of μ indicated.

representative electron and singly charged ion trajectories for equatorially mirroring particles as obtained from (4.3). Each panel shows trajectories for the value of μ indicated on that panel, and panels are shown for values of μ from 3.2 eV/nT (corresponding to 0.1 keV at $L = 10$) to 3200 eV/nT (corresponding to 100 keV at $L = 10$). No trajectories are shown for electrons at L-values where the kinetic energy exceeds the rest energy $m_e c^2 = 512$ keV. This occurs at $L \leqslant 2.7$ for 320 eV/nT electrons and $L \leqslant 5.8$ for 3200 eV/nT electrons.

For all values of μ, the electron and ion trajectories show a clear separation between the closed trajectories at lower radial distances and the open trajectories at larger radial distances. The electron trajectories for finite μ are similar to those for the cold particles, except the region of closed trajectories monotonically extends to larger radial distance with increasing μ, owing to the increasing magnetic drift, which adds to the corotation velocity. The ion trajectories, on the other hand, show the regions of closed trajectories decreasing in size with increasing μ at the lower values of μ, owing to the increase in the magnetic drift velocity, which subtracts from the corotation velocity. At larger values of μ, where the magnetic drifts dominate corotation, the region of closed trajectories increases with increasing μ as for the electrons. The bulge in the closed ion trajectories occurs at dawn, with the minimum radial distance occurring at dusk. This is opposite to that for the electrons since the magnetic drifts are in opposite directions. In addition to the closed trajectories encircling the earth, the lower μ ions show closed trajectories that do not cross the dawn meridian. Ions follow the corotation direction along the outer portion of these trajectories and the magnetic drift direction along the inner portion.

While the trajectories for fixed values of μ are physically meaningful, satellite particle observations are made of particles at fixed energies. Thus observations at different locations within the magnetosphere are generally at fixed particle energies rather than at fixed μ. Figures 4.26 and 4.27 show electron and singly charged ion trajectories where the particle energy is the same for all the trajectories in each panel at the point indicated by the heavy dots where the trajectories cross the dawn-dusk meridian. The trajectories given by the solid lines are for particles having the energy indicated on each panel at 1800 LT, and the trajectories given by the dashed lines are for particles having the indicated energy at 0600 LT. The particle energy varies along each trajectory in accordance with the conservation of μ; however, the trajectories in Figure 4.26 and 4.27 should be of value in interpreting equatorial particle measurements at fixed energies obtained along the dawn-dusk meridian.

The electron trajectories in Figure 4.26 show a significant overlap on the dawn side between the closed trajectories for the indicated energies at 1800 LT and the open trajectories for that energy of 0600 LT. In addition, there is a very large asymmetry in the radial distances to which closed trajectories extend for a given energy at 1800 LT and the same energy at 0600 LT. For example, 3 keV electrons at 1800 LT follow closed trajectories for all $L \lesssim 11$, but 3 keV electrons at 0600 LT follow closed trajectories only at L-values < 4.

The overlap and dawn-dusk asymmetry of the trajectories becomes significantly greater for the ion trajectories. For example, the 30 keV ion trajectories in Figure 4.27 are closed to beyond $L = 14$ at 0600 LT, while the trajectories are open for all $L \geqslant 2$ at 1800 LT. An additional complication arises for the ions as can be seen in the 10 keV trajectories. At 0600 LT, the trajectories are closed at $L \leqslant 2$ because of the dominance of

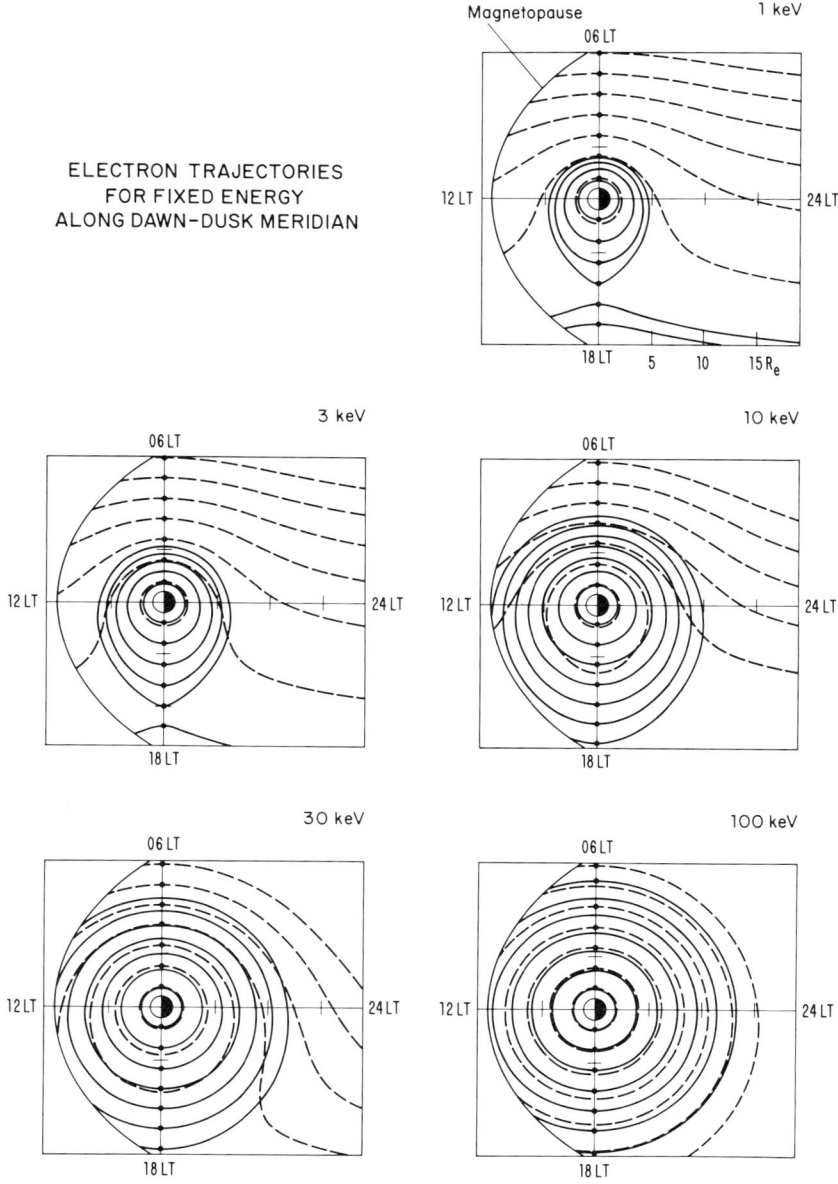

Fig. 4.26. Electron drift trajectories as obtained from Equation (4.3). Within each panel, the particle energy is the same for all trajectories at the points indicated by the heavy dots where the trajectories cross the dawn-dusk meridian. The trajectories given by the solid lines are for particles having the indicated energy at 1800 LT, and the trajectories given by the dashed lines are for particles having that energy at 0600 LT.

ELECTRIC FIELDS 85

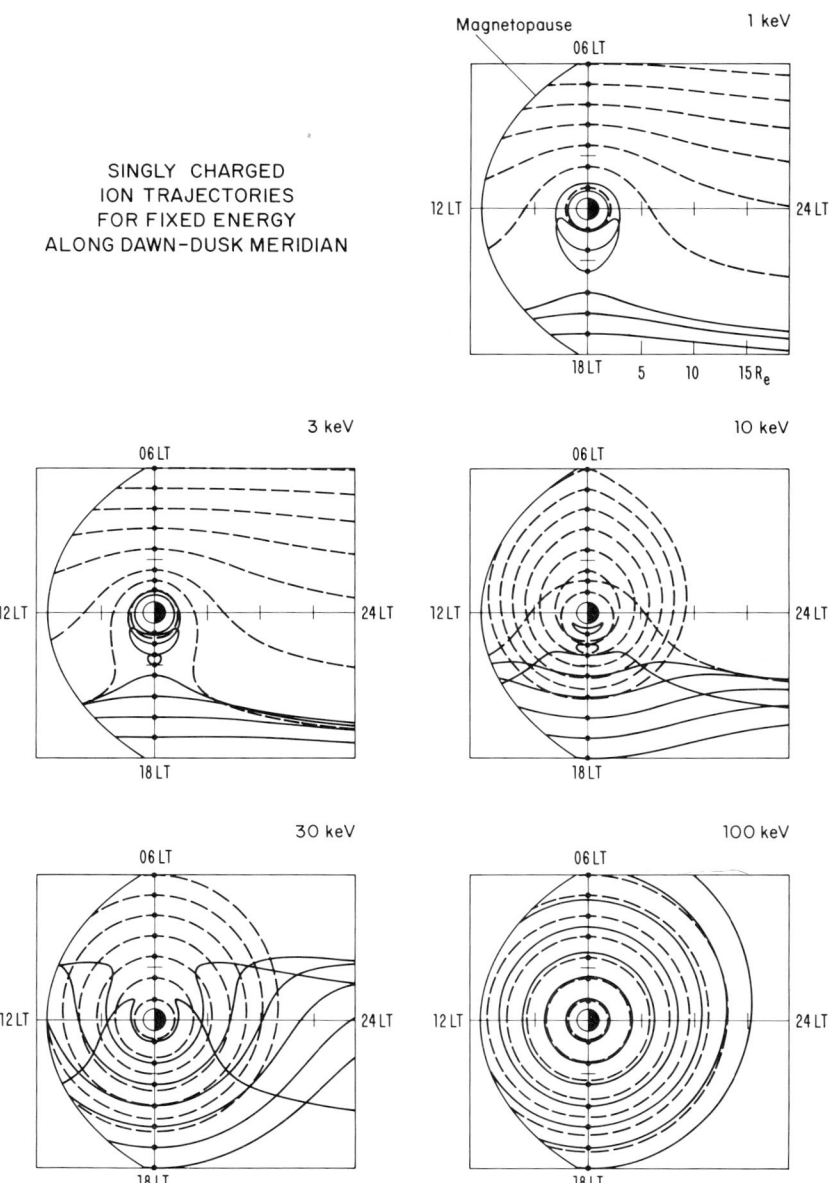

Fig. 4.27. Singly charged ion drift trajectories as obtained from Equation (4.3). Within each panel, the particle energy is the same for all trajectories at the points indicated by the heavy dots where the trajectories cross the dawn-dusk meridian. The trajectories given by the solid lines are for particles having the indicated energy at 1800 LT, and the trajectories given by the dashed lines are for particles having that energy at 0600 LT.

corotation and at $L \geqslant 4$ because of the dominance of the magnetic drifts. However, in the vicinity of $L = 3$, the two drifts cancel, giving a region of open trajectories. (See also Cowley and Ashour-Abdalla (1976a, b)).

Two points should be kept in mind when applying trajectories such as those in Figures 4.24—4.27:

(1) The uniform convection electric field assumed in the calculation is only a simple approximation of the real convection electric field. In particular, modifications of the electric field due to the charge separation of particles drifting with the electric field are completely ignored. An important example of such charge separation is that the convection electric field is partially shielded from the inner magnetosphere ($L \lesssim 5$) during geomagnetically quiet periods (Schield *et al.*, 1969; Swift, 1971; Vasyliunas, 1972; Stern, 1975; Volland, 1975; Wolf *et al.*, 1982).

(2) The trajectories are for a time-independent convection electric field (2.5×10^{-4} V m^{-1} in the figures). In reality, significant time variations do occur so that particles can be transferred from open to closed trajectories and vice versa. Open trajectories will penetrate to decreasing L-values as the convection electric field is increased. Thus the increase in E_c associated with increasing geomagnetic activity can inject particles from outside the normally closed convection region to within that region. In addition, increases in E_c can move particles on closed trajectories closer to the earth and can also move them to the region where trajectories are normally open. Thus a temporary increase in E_c can cause the injection of particles onto closed trajectories from regions not normally accessible to the region of closed trajectories, and particles on closed trajectories can be redistributed within that region or be ejected from the closed trajectory region.

4.3. Current Sheet Energization in the Tail

4.3.1. ESTIMATE OF TOTAL PARTICLE ENERGIZATION RATE

The current in the geomagnetic tail associated with the reversal of the magnetic field from one tail lobe to the other (illustrated in Figure 1.1b) is in the dawn-to-dusk direction. Thus the convection electric field is parallel to the tail current, so that the electric field must energize the particles carrying the tail current.

A straightforward estimate of the rate at which particles are energized within the tail current sheet can be obtained as follows: The magnetic field in the tail lobes is ~ 20 nT (Ness, 1965), so that the magnetic field change across the current sheet is ~ 40 nT. Assuming a uniform current sheet, this magnetic field change implies that the dawn-to-dusk current I_t across the tail, per unit distance along the tail length, is $\sim 40 \times 10^{-9}/(4\pi \times 10^{-7}) = 3.2 \times 10^{-2}$ A m^{-1}. In Section 4.2.4 we estimated a value of 2.5×10^{-4} V m^{-1} for the convection electric field E_c by assuming a tail width of $40 R_e$ and a cross-tail potential difference of 65 kV. Using this value of E_c, we obtain $\mathbf{E}_c \cdot \mathbf{I}_t = 8.0 \times 10^{-6}$ W m^{-2}. Multiplying this value by the width and the length of the current sheet region in the tail will give the total particle energization. A width of $40 R_e$ is reasonable, but the length is more difficult to estimate. A maximum estimate for the length is $200 R_e$, since the current sheet has been observed to begin as close as $10 R_e$ from the earth (Speiser and Ness, 1967) and the open field line regions from the polar caps map to an $\sim 230 R_e$ region in the interplanetary medium (see Section 4.2.3). However, the distance to which a

ELECTRIC FIELDS

current sheet of the above magnitude extends has not been determined, though it extends to at least $30R_e$ from the earth (Speiser and Ness, 1967). We thus, somewhat arbitrarily, choose $20R_e$ as a minimum estimate for the current sheet region. This gives a total energization rate of $\sim 8.0 \times 10^{-6} \, (40R_e) \, (20-200R_e) = 3 \times 10^{11} - 3 \times 10^{12}$ W (see also Heikkila *et al.*, 1979; Stern, 1980).

This energization of current sheet particles is $\sim 2-20\%$ of the total energy flux carried by the solar wind through an area equal to the cross-section of the dayside magnetopause (based on an estimate given by Stern, 1980), and this energization is an important, and perhaps the dominant, means by which solar wind energy is transferred to magnetospheric particles. Thus particle energization within the current sheet is a crucial part of magnetospheric electrodynamics. Assuming the convection electric field is generated directly by the solar wind particles, energy transfer to the current sheet particles comes directly from a slowing down of solar wind particles.

4.3.2. Particle Motion in a Current Sheet

In order to understand the physical process by which current sheet particles are accelerated in the geomagnetic tail, it is necessary to understand the particle trajectories in the vicinity of the current sheet. The usual guiding center theory (Chapter 2) which conserves the first two adiabatic invariants is not valid within the current sheet since the magnetic field changes significantly in distances less than a gyroradius (Speiser, 1965, 1967).

Consider first an idealized current sheet model as illustrated in Figure 4.28. This model

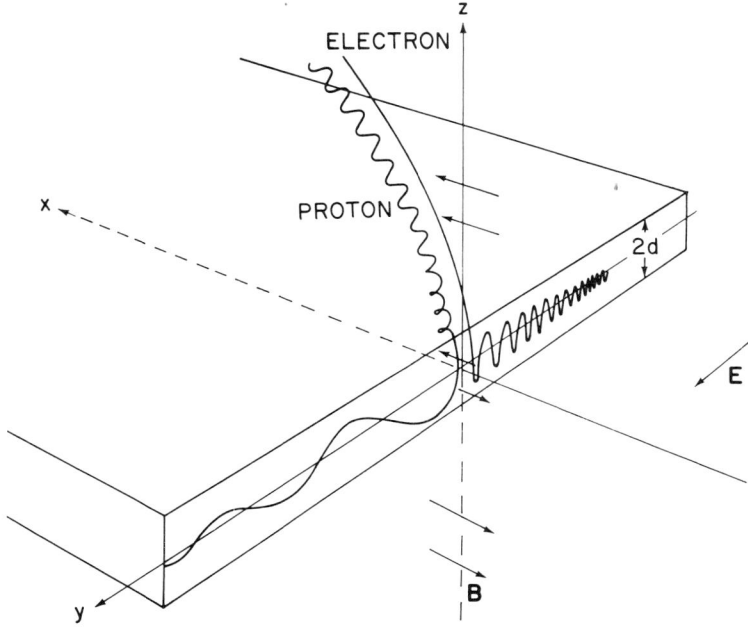

Fig. 4.28. Illustration of particle motion in an idealized current sheet; $B = 0$ along the midplane of the sheet (from Speiser, 1965). (© by American Geophysical Union)

has no magnetic field normal to the current sheet, so that the magnetic field reverses across the $z = 0$ plane and is of zero magnitude along that plane. In the figure, the current sheet is shown to have a width $2d$ centered along the $z = 0$ plane. Particles away from the current sheet with $|z| > d$ will undergo their normal circular motion about the magnetic field. However, particles sufficiently close to $z = 0$ will be unable to execute a complete orbit about the magnetic field without encountering the magnetic field reversal. This will occur within any current sheet where $B = 0$ along the midplane of the current sheet. The circular motion of such particles will be of one sense at $z > 0$ and of the other sense at $z < 0$, so that these particles will undergo oscillatory motion about the $z = 0$ plane as illustrated in the figure. Electrons will move in the negative y-direction (toward dawn) and ions in the positive y-direction (toward dusk) giving a dawn-to-dusk current as required for the magnetic field reversal. The convection electric field in the dawn-to-dusk direction continually accelerates both electron and ions within the current sheet in the direction of their velocity along the y-axis, so all particles undergoing the oscillatory motion about the midplane of the current sheet are energized by the convection electric field (Dungey, 1953; Alfvén, 1968; Alfvén and Fälthammar, 1971). As these particles are energized in this model of a current sheet with a strictly neutral plane, the amplitude of their oscillations decreases as illustrated in Figure 4.28, owing to the increase in the magnitude of the y-component of their velocity, and the particles are unable to leave the current sheet.

Above the current sheet, particles undergoing their normal circular motion will drift toward the current sheet because of the convection electric field until they reach the current sheet and are energized. For the case of a strictly neutral plane along the center of the current sheet, all particles reaching the current sheet will be continually energized until they reach the edges of the tail current sheet. Under such conditions, the energized particles would be lost from the tail and might not contribute to the energetic particle population within the magnetosphere.

In general, however, the current sheet is not strictly neutral along its midplane, but has a small magnetic field normal to the sheet. Speiser (1965, 1967) has calculated particle trajectories in a current sheet with such a normal magnetic field across the sheet, and his results are illustrated in Figure 4.29 for a northward normal magnetic field (the direction required for field lines connected to the earth). He found that particles oscillate about the neutral sheet and are energized by a dawn-to-dusk electric field as in the case of a strictly neutral current sheet midplane. However, a non-zero, northward magnetic field was found to deflect particles toward the earth and to eject them from the current sheet. (A southward, normal magnetic field would have the same effect, except that particles would be deflected away from the earth). Ejection can be either above or below the current sheet. Thus particles can be energized in the current sheet by the cross-tail electric field and then be ejected toward the earth along closed magnetic field lines within the magnetosphere. This suggests that current sheet energization should be considered an important source of energized particles within the magnetosphere.

A simple and very useful model for the geomagnetic tail current sheet is the one-dimensional model illustrated in Figure 4.30. Uniformity in the y and x directions is assumed. B_x reverses within the current sheet of width $2d$ and is constant within the tail lobes ($|z| > d$). B_z and $\mathbf{E} = E_y \hat{y}$ do not vary. Exact particle trajectories within such a current sheet model can be evaluated numerically. However the particle energization can

ELECTRIC FIELDS

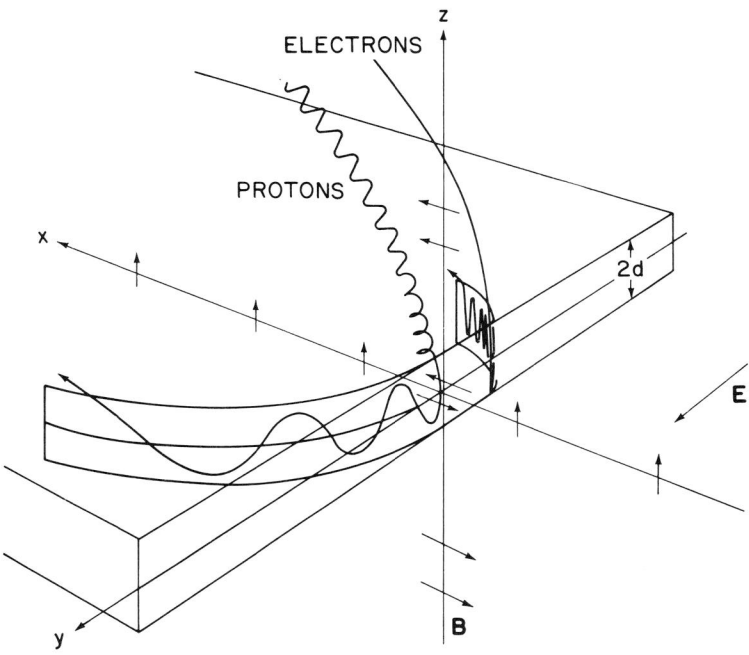

Fig. 4.29. Illustration of particle motion in a current sheet with a small magnetic field normal to the sheet (from Speiser, 1965). (© by American Geophysical Union)

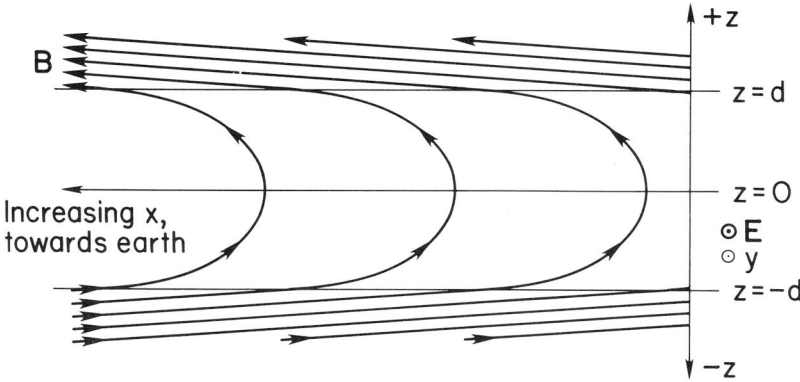

Fig. 4.30. One-dimensional model for current sheet fields. There are no variations in the x and y directions. B_x reverses within the current sheet region of width $2d$ and is constant at $|z| > d$. B_z and $\mathbf{E} = E_y \hat{y}$ do not vary (from Lyons and Speiser, 1982). (© by American Geophysical Union)

be simply estimated, since the electric field can be removed by transforming to a frame of reference moving with velocity E_y/B_z in the x-direction (Speiser, 1965; Cowley, 1980). In this frame, particle energy must be conserved in the absence of collisions and particle interactions with plasma waves.

Let subscripts 'i' and 'f', respectively, refer to particles prior to and after their interaction with the current sheet. Energy conservation in the frame moving with velocity E_y/B_z in the x-direction gives

$$\left(v_{xi} - \frac{E_y}{B_z}\right)^2 + v_{yi}^2 + v_{zi}^2 = \left(v_{xf} - \frac{E_y}{B_z}\right)^2 + v_{yf}^2 + v_{zf}^2$$

which can be rewritten as

$$v_i^2 - 2v_{xi}\frac{E_y}{B_z} = v_f^2 - 2v_{xf}\frac{E_y}{B_z}. \tag{4.4}$$

To be useful, Equation (4.4) requires information about v_{xf} versus v_{xi}, which cannot be obtained from the above analysis. However, Speiser (1965, 1968) has obtained the simple, but useful, approximation that the x-component of a particle's velocity in the transformed frame reverses sign with its magnitude and the other velocity components remain unchanged during the particle's interaction with the current sheet; i.e.,

$$v_{xf} \approx -v_{xi} + 2E_y/B_z. \tag{4.5}$$

In addition, $v_{zf} = -v_{zi}$ for particles injected into the current sheet from above (below) the sheet and ejected above (below) the sheet. Otherwise $v_{zf} = v_{zi}$. The condition for validity of this approximation is basically that the particle motion in the vicinity of the current sheet be dominated by the oscillatory motion about the midplane of the current sheet over the gyration about the current sheet field lines. In other words, the particle must undergo a number of oscillations during its interaction with the current sheet. The validity of this approximation has been examined by Speiser and Lyons (1983).

Using the approximation (4.5), Equation (4.4) becomes

$$v_f^2 \approx v_i^2 + 4\frac{E_y}{B_z}\left(\frac{E_y}{B_z} - v_{xi}\right).$$

This gives an energy increase during a particle's interaction with the current sheet ΔK of

$$\Delta K \approx 2m\frac{E_y}{B_z}\left(\frac{E_y}{B_z} - v_{xi}\right). \tag{4.6}$$

Equation (4.6) gives the result that $\Delta K < 0$ for $v_{xi} > E_y/B_z$. However, such particles cannot reach the current sheet since their component of parallel velocity along the z-axis in the direction away from the current sheet exceeds their electric field drift toward the current sheet. For the one-dimensional current sheet model of Figure 4.30, all particles with a component of velocity along **B**, $v_\parallel$, such that $v_\parallel < (E_y/B_z)\cos\xi$ reach the current sheet and are energized. Here $\xi = \cos^{-1}(B_x/B)$ is the angle the lobe magnetic field makes with the x-axis. Thus $v_\parallel = (E_y/B_z)\cos\xi$ separates the regions in velocity space where particles move toward and away from the current sheet. Averaging over the gyromotion, which gives $v_x = v_\parallel \cos\xi + E_y B_z/B^2$, we find that the condition $v_\parallel = (E_y/B_z)\cos\xi$ is equivalent to $v_x = E_y/B_z$.

Using $\cos \alpha_f = (\mathbf{v}_f \cdot \mathbf{B})/v_f B$ and the approximate relations above between $\mathbf{v}_i$ and $\mathbf{v}_f$, we obtain an estimate for the pitch angle α_f of the ejected particle:

$$\cos \alpha_f = \frac{\left(-v_{xi} + 2\frac{E_y}{B_z}\right) \cos \xi - v_{zi} \sin \xi}{\left[v_i^2 + 4\frac{E_y}{B_z}\left(\frac{E_y}{B_z} - v_{xi}\right)\right]^{1/2}}. \tag{4.7}$$

Here $\cos \alpha$ is assumed to be negative (positive) for particles with $v_\parallel$ away from (toward) the earth. From (4.6) and (4.7), we obtain that for low energy incident particles, where $|v_i| \ll E/B_z$,

$$\Delta K \approx 2m \left(\frac{E_y}{B_z}\right)^2$$
$$\cos \alpha_f \approx \cos \xi, \tag{4.8}$$

and for high energy incident particles, where $|v_{xi}| \gg E_y/B_z$,

$$\Delta K \approx 2m|v_{xi}|\frac{E_y}{B_z}$$
$$\cos \alpha_f \approx -\cos \alpha_i. \tag{4.9}$$

Equation (4.8) shows that all low energy particles of a given mass are accelerated by the same energy increment and for small ξ are ejected with α near 0°. The energy increment is proportional to m and (E_y/B_z), but is independent of charge state. However, caution must be used in applying (4.8), since the particle gyroradius decreases with decreasing energy so that gyromotion about the small normal magnetic field near the midplane of the current sheet becomes increasingly important relative to the oscillatory motion about the current sheet midplane. High energy particles, on the other hand, are accelerated by an energy increment proportional to $K_i^{1/2}$, the energy increment for a given K_i also being proportional to $m^{1/2}$ and E_y/B_z. Pitch angle is approximately conserved for these particles. The gyromotion varies v_{xi} as a particle approaches the current sheet, but the effects of this variation on applying the approximation (4.5) have not yet been explicitly evaluated. Neglecting the effects of this variation of v_{xi}, we obtain that $\Delta K \sim \cos \alpha_i$ for high energy incident particles of a given energy. At all energies, ions will be energized far more than electrons.

4.3.3. EFFECTS OF CURRENT SHEET ENERGIZATION OF IONS IN THE GEOMAGNETIC TAIL

In order to calculate the distribution of particles resulting from the energization within the current sheet, it is necessary to specify the particle distribution incident upon the current sheet. Lyons and Speiser (1982) calculated this distribution using a distribution for the incident particles based on a suggestion by Swift (1977), Cowley (1980), and

Cowley and Southwood (1980). A boundary layer of plasma exists between the magnetotail lobes and the magnetosheath plasma flowing in the anti-sunward direction just within the tail magnetopause (Hones *et al.*, 1972; Akasofu *et al.*, 1973; Rosenbauer *et al.*, 1975). This boundary layer is known as the 'plasma mantle' or at lower latitudes as the 'magnetotail boundary layer'. The convection electric field should cause the plasma within this mantle to drift toward the center of the tail as it flows down the tail away from the earth. This mantle plasma should then be energized within the current sheet and ejected somewhat closer to the earth than its location of incidence upon the current sheet.

The mantle ions will be energized much more than the mantle electrons, and the ejection from the current sheet should form an earthward-streaming distribution of energized ions. The earthward-streaming ions outside the loss cone will mirror and return to the current sheet, but the earthward stream may be well separated from the returning mirrored ions by the electric field drift during the time it takes an ion to travel along field lines to its mirror point and return to the midplane of the geomagnetic tail. Consider, for example, an ion with a parallel energy of 3 keV and a $200R_e$ path length from its location of ejection from the current sheet to its point of return to the midplane of the tail. For a convection electric field of 2.5×10^{-4} V m^{-1} and a lobe magnetic field strength of 20 nT, V_{DE} = 12.5 km s^{-1} and such an ion will drift $\sim 3R_e$ across field lines in the tail lobes as it traverses the $200R_e$ distance. With B_z = 1 nT along the midplane of the current sheet, these mirroring ions will cross the midplane of the current sheet $\sim 60R_e$ closer to the earth than the distance from which they were initially ejected from the current sheet. If B_z were of the order of 1 nT or less over $\gtrsim 60R_e$ of distance along the tail, then significant multiple energization of individual ions within the current sheet would be possible. However, if the normal component of B increases significantly over distances $\lesssim 60R_e$ along the tail in the direction toward the earth, then significant energization would be expected to occur only during the first interaction with the current sheet. Earthward-streaming ions within the loss cone will precipitate into the ionosphere, and as shall be seen, are an important source of auroral ion precipitation.

The electric field drift of the ions implies that earthward-streaming ions should be on the outer boundary of energetic ions trapped on closed magnetic field lines within the tail and that the mirroring of the earthward-streaming ions should be an important source for the energetic ions in the tail. The region of energetic particles in the tail is known as the plasma sheet, and earthward-streaming ion distributions are often observed on the outer boundary of the plasma sheet (Lui *et al.*, 1977; DeCoster and Frank, 1979; Williams, 1981). Williams found that the earthward-streaming ion distributions were consistent with adiabatic particle motion from a source of energized ions $\sim 100R_e$ from the earth, and he specifically observed the mirroring of the initially earthward-streaming ions. The energization of mantle ions and the effects of the energized ions are illustrated in Figure 4.31.

Lyons and Speiser (1982) specified an ion distribution for the particles incident upon the current sheet based on observations of the plasma mantle. Akasofu *et al.* (1973) and Rosenbauer *et al.* (1975) reported mantle ion densities of $\sim 0.2 - 50 \times 10^5$ m^{-3}, temperatures of 1.5×10^6 K, and tailward flow velocities of ~ 100 km s^{-1}. Lundin *et al.* (1981) presented energy spectra for a number of PROGNOZ-7 satellite crossings of the mantle. Energy spectra for the highest density portion of the mantle from four of

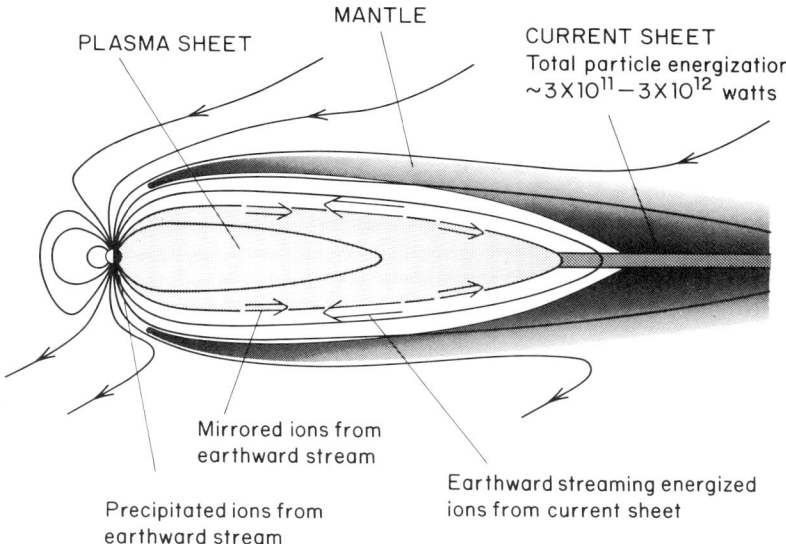

Fig. 4.31. Energization of mantle ions and the effects of the energized ions within the magnetosphere.

the mantle crossings reported by Lundin *et al.* (1981) are shown by the solid circles in Figure 4.32. Lundin *et al.* (1981) reported densities and flow velocities for the mantle in agreement with those of Akasofu *et al.* (1973) and Rosenbauer *et al.* (1975), but Lundin *et al.* (1981) temperatures ($\sim 10^7$ K) are consistently somewhat greater. This temperature difference is probably due to the fact that Lundin *et al.*'s (1981) measurements extended up to 30 keV in energy while the earlier measurements extended up to a much lower energy, and Lundin *et al.*'s (1981) energy spectra show a definite high energy tail having 30 keV fluxes well above that expected for a Maxwellian distribution.

Energy spectra within the tail mantle at energies > 30 keV have not been specifically reported, to the best of our knowledge. However, Williams (1979) reported anti-sunward-streaming ions in the 100–400 keV energy range. The ions were claimed to be in the magnetosheath, but the magnetosheath was not distinguished from the mantle so that the ions may have been part of the mantle. The fluxes reported by Williams (1979) are also shown in Figure 4.32, and they can be seen to match well with an extrapolation of the higher energy fluxes of Lundin *et al.* (1981). This match of the two energy spectra led Lyons and Speiser to speculate that the tailward-streaming ions observed by Williams may be the high energy portion of the mantle plasma distribution, and they used Williams's fluxes as a guide to the 100–1000 keV mantle fluxes. Less information has been reported concerning the angular distribution of the mantle plasma than concerning the energy distribution.

As the initial, unaccelerated distribution function f, a flowing Maxwellian with an $f \sim K^{-\gamma}$ (K is kinetic energy) high energy tail was used as shown in Figure 4.32. Distribution 1 is based on the parameters given by Akasofu *et al.* (1973) and Lundin *et al.* (1981). The streaming velocity is 100 km s^{-1} in the $-x$-direction, and the thermal energy K_{th} is 10^7 K. The density N of 1.87×10^5 m^{-3} was adjusted to give a maximum value of

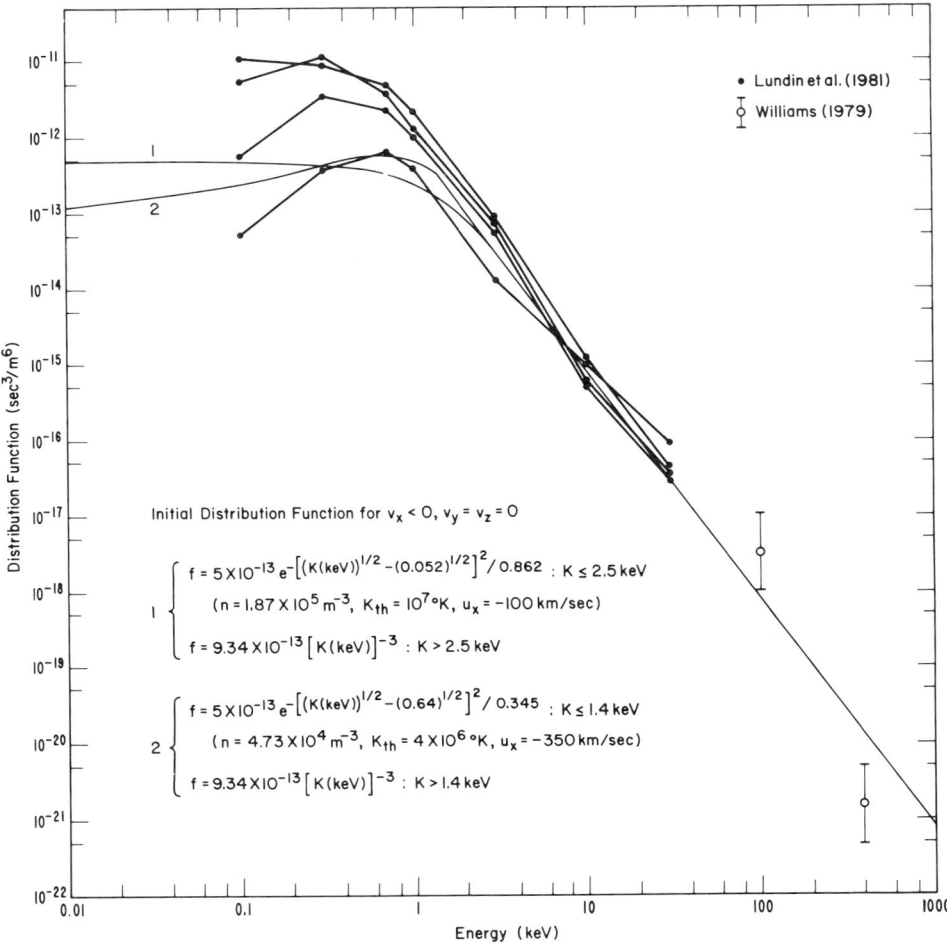

Fig. 4.32. Ion distribution function versus energy for the tail plasma mantle. Observations are from Lundin et al. (1981) and Williams (1979). Models 1 and 2 are Maxwellian distributions with a bulk flow and a high energy tail (from Lyons and Speiser, 1982). (© by American Geophysical Union)

$f = 5 \times 10^{-13}$ s^3 m^{-6} to match the maximum value of f for a particular observation at the other edge of the plasma sheet (Figure 4.33). The high energy tail was taken as $f = 9.34 \times 10^{-13} [K(\text{keV})]^{-3}$ to give an order-of-magnitude estimate for the 2.5–1000 keV mantle fluxes. The ions were assumed to be entirely protons, though Lundin et al.'s (1981) observations show that other ions are present. At energies $\leqslant 2.5$ keV, the angular distribution given by a streaming Maxwellian was used. At energies > 2.5 keV, all protons with a velocity component along **B**, $v_\parallel$, in the tailward direction were taken to have the value of f given by the $K^{-\gamma}$ distribution above. This value was arbitrarily reduced by a factor of 2 for all earthward-directed $v_\parallel$'s. Lyons and Speiser found that the ejection velocity plane is filled mostly by particles with $v_\parallel$ initially tailward so that the assumption concerning the distribution of particles with earthward-directed initial $v_\parallel$'s is not critical.

It is evident from Figure 4.32 that the energy spectrum of the mantle is not well reproduced by the streaming Maxwellian distributions of Akasofu et al. (1973) and Lundin et al. (1981). In particular, the significant decrease in f at energies below a few tenths of a keV is not reproduced at all. Since this decrease in f at the lower energies is present in the vast majority of the spectra presented by Lundin et al. (1981) distribution 2 in Figure 4.32 was used in addition to distribution 1. This distribution has a significantly higher flow velocity (350 km s^{-1}) of the order of the solar wind velocity and a lower temperature (4×10^6 K) than does distribution 1. The high energy tail was taken to be $f = 9.34 \times 10^{-13} [K(\text{keV})]^{-3}$ for $K > 1.45$ keV. Although distribution 2 may not be the best fitting Maxwellian for the data, it gives a peak in the distribution as observed at 0.3–1 keV and the observed decrease in f at lower energies. The density 4.73×10^4 m^{-3} for distribution 2 was again adjusted so that f maximizes at the same value as that observed at the boundary of the plasma sheet. The angular distribution was taken as for distribution 1, except the streaming Maxwellian was applied below, and the factor-of-2 reduction above, 1.45 keV. Possible evolution of the mantle ion distribution between the regions of observation and the location of incidence upon the current sheet was not considered.

Using the initial distributions 1 and 2 of Figure 4.32 and the energy increase from relation (4.6) for the current sheet acceleration, we can obtain f versus energy for the accelerated particle distributions ejected from the current sheet with pitch angles near the field line direction. The results for $E_y = 2.5 \times 10^{-4}$ V m^{-1} and $B_z = 1$ nT ($E_y/B_z = 250$ km s^{-1}) are shown in Figure 4.33. Also shown in Figure 4.33 is an energy spectrum of the earthward-streaming ions at the outer boundary of the plasma sheet. The closed circles are observations at 0.05–45 keV obtained by DeCoster and Frank (1979) from the IMP–8 satellite on October 31, 1975, and the open circles are observations at energies from ~ 50 to 1000 keV obtained by Williams (1981) from the ISEE–1 satellite on May 3, 1978.

It can be seen that the calculated energy spectra agree well with the spectrum observed at the boundary of the plasma sheet. Over the entire range of output energies (1.5–1000 keV) for which the initial distribution was specified, the calculated values of f are within a factor of ~ 10 of the observations despite the ~ 9 order-of-magnitude change in the observed value of f over that energy range.

DeCoster and Frank (1979) also reported the distribution function in the ($v_\perp$, $v_\parallel$)-plane for the ions observed at the outer boundary of the plasma sheet. The distribution at the top of Figure 4.34 is a contour plot of the observed distribution function from which the energy spectrum in Figure 4.33 was obtained. Lyons and Speiser (1982) calculated the distribution function as a function of $v_\perp$ and $v_\parallel$ expected from the current sheet energization, and their results are shown in the middle of Figure 4.34 for the initial distribution 2 described earlier. The observed and calculated distributions are directly compared at the bottom of the figure.

The agreement between the overall shapes of the calculated and observed contours in Figure 4.34 is very good, especially considering that no attempt was made to adjust the parameters in the calculation to improve the agreement other than matching the peak values of f. The calculated distributions reproduce the observed peak in f at 500–1000 km s^{-1} along the field line direction, the more rapid fall in f from the peak toward lower energies than toward high energies, and the departure of the contours from circles

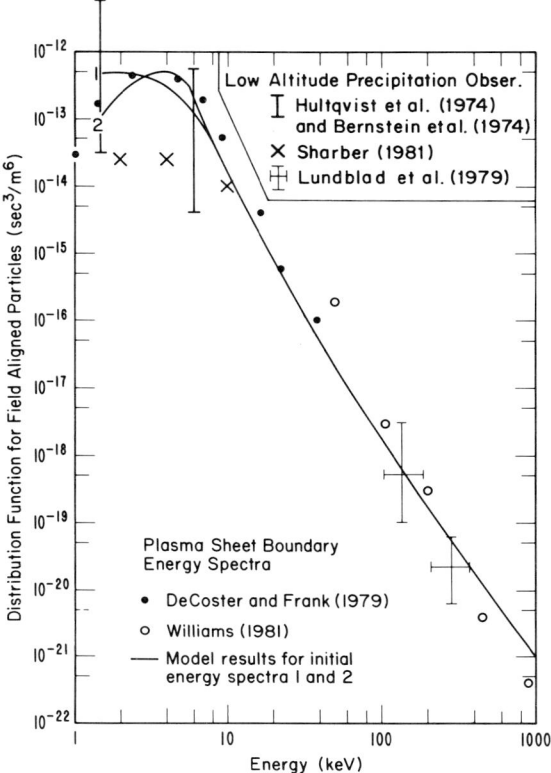

Fig. 4.33. Proton energy spectra calculated to result from the current sheet energization, compared with energy spectra observed at the outer boundary of the plasma sheet and with low altitude satellite observations of precipitating ions having isotropic pitch-angle distributions in the auroral zones (from Lyons and Speiser, 1982). (© by American Geophysical Union)

in velocity space. The departure of the contours from circles in the calculations results from the decrease in the ejection energy as the ejection pitch angle approaches 90° (see (Equation (4.9)), and we know of no other plasma acceleration processes that could cause the contours to display this feature.

The overall shape of the calculated contours of the accelerated particle distribution is not sensitive to the parameters of the tail model or to the details of the initial distribution function. However, changes in E_y/B_z will alter the $v_\parallel$ at which f peaks in the accelerated distribution, and changes in the values of the initial distribution function will change the value of the accelerated distribution function accordingly. Thus current sheet acceleration should generally result in a distribution function of the form shown in Figure 4.34, and observation of distribution functions of that form indicates that current sheet acceleration has formed the distributions. DeCoster and Frank reported that distributions of this type were frequently observed on the outer boundary of the plasma sheet, and such distributions have since been reported at that boundary on ISEE–1 (Frank et al., 1978) and on ISEE–2 (Forbes et al., 1981). These observations lead us to conclude that current

ELECTRIC FIELDS

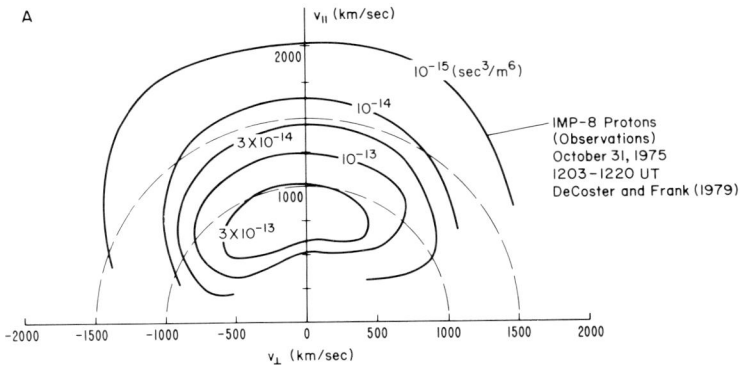
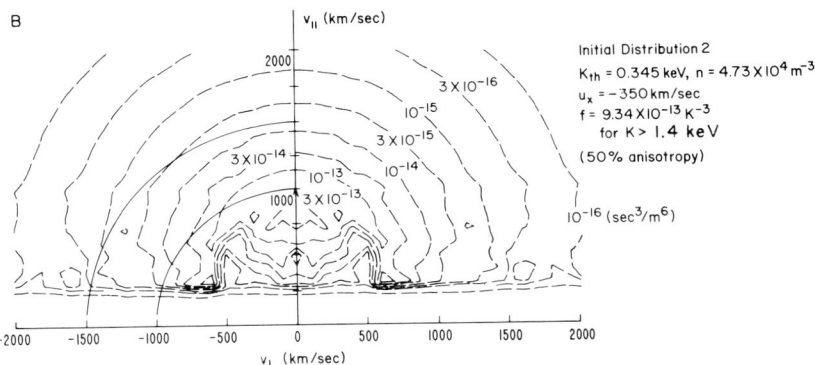
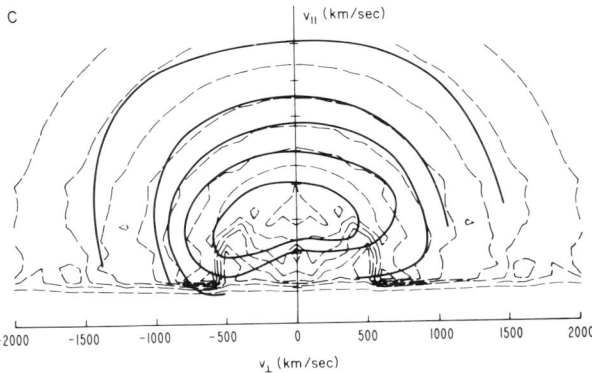

Fig. 4.34. Distribution function in $(v_\perp, v_\parallel)$-plane for earthward-streaming ions at the outer boundary of the plasma sheet. Shown are observations and the model results for the current sheet energization using initial model 2 of Figure 4.32 (from Speiser and Lyons, 1982). (© by American Geophysical Union)

sheet acceleration occurs frequently, and perhaps nearly continuously, within the earth's magnetosphere.

Accelerated ions ejected from the current sheet should be found streaming toward the earth along the entire plasma sheet boundary for energies $\leqslant 100$ keV. Higher energy ions, on the other hand, gain an appreciable fraction of the cross-tail, electric potential energy and should thus be seen more often on the dusk side of the tail than on the dawn side. In agreement with this prediction, such an asymmetry has been found in the occurrence frequency of 50–500 keV ions in the magnetotail (Meng et al., 1981).

The ions ejected from the current sheet outside the loss cone mirror and return to the equatorial plane at a location much closer to the earth than the location from which they were ejected, and the energy of the peak in the distribution of ejected ions is approximately the same as the mean energy of plasma sheet ions. Thus these ions are probably an important source for plasma sheet ions.

In addition, since the accelerated ion distributions include particles within the loss cone, the current sheet acceleration is a possibly important source of auroral ion precipitation as suggested by Jaeger and Speiser (1974). From Figures 4.33 and 4.34, the current sheet acceleration gives $f \approx 3 \times 10^{-13}$ s^3 m^{-6} for 3 keV protons within the loss cone, which gives a differential flux of 1.6×10^6 protons/cm^2-s-ster-keV. This is a factor of ~ 10 greater than the proton flux observed from rockets at several keV in an early evening proton aurora (Whalen et al., 1971) and in a diffuse midnight aurora (Moore and Evans, 1979).

The calculations show that the ion distribution ejected from the current sheet should be nearly isotropic within the loss cone for all energies from ~ 1 keV to nearly 1 MeV (see the model distribution near $v_\perp = 0$, Figure 4.34). Thus the resulting auroral ion precipitation should be isotropic in pitch angle over this entire energy range. Satellite measurements of auroral ion precipitation have shown such isotropy at energies from 1 to 10 keV (Hultqvist et al., 1974; Bernstein et al., 1974; Sharber, 1981) and from 115 to 360 keV (Lundblad et al., 1979). Ranges of fluxes of precipitating ions reported in the preceding references are compared in Figure 4.33 with the energy spectra calculated to result from the current sheet acceleration and the observations at the outer boundary of the plasma sheet. The agreement between the calculated and measured precipitation fluxes in Figure 4.33, and the observed isotropy of the auroral ion precipitation up to energies of several hundred keV, indicate that the isotropic auroral ion precipitation results directly from the accelerated ions ejected from the current sheet. The observations presented in the above references indicate that such isotropic precipitation is a major component of the total auroral ion precipitation.

Finally, we determine whether the earthward-streaming particles along the outer boundary of the plasma sheet carry sufficient energy to account for the energization due to the cross-tail current and electric field. When the estimates in Section 4.3.1 are used, the energization $\mathbf{I}_t \cdot \mathbf{E}$ per unit distance along the tail in the x-direction per unit distance across the tail in the y-direction is 8.0×10^{-6} W m^{-2}. If this energization is the result of current sheet particle acceleration, then $\frac{1}{2}(8.0 \times 10^{-6})$ W m^{-2} should be ejected from both sides of the current sheet. The unit of area along the current sheet maps to a unit of area normal to the lobe magnetic field that is smaller than the area along the current sheet by a factor of $1/\sin \xi$. Thus for $B_x = 20$ nT in the lobe, and $B_z = 1$ nT, an energy flux of 8.0×10^{-5} W m^{-2} is expected along the outer boundary of the plasma sheet. An integral over the observed distribution function in Figure 4.34 gives an energy flux of 3×10^{-4} W m^{-2}, in reasonable agreement with the estimate above of 8.0×10^{-5} W m^{-2} obtained from the value of $\mathbf{I}_t \cdot \mathbf{E}$ in the current sheet. We thus conclude that the current sheet acceleration from single particle motion can account for the particle energization that must occur within the tail current sheet.

4.4. Auroras and Parallel Electric Fields

4.4.1. EVIDENCE FOR THE ACCELERATION OF AURORAL ELECTRONS BY PARALLEL ELECTRIC FIELDS

Auroras are atmospheric emissions of visible light emanating from altitudes of ~ 100–300 km in the polar ($\sim 65°$–$75°$ geomagnetic latitude) atmosphere. They often display extended ray-like features which are approximately aligned with the geomagnetic field. This observation by Chapman and Bartels (1940) has led to the conclusion that auroras must be associated with charged particles. Identification of the responsible particles was later obtained from instruments onboard rockets launched directly into auroral displays. Specifically, McIlwain (1960) and Davis et al. (1960) found that visible auroras result from the precipitation of ~ 1–10 keV electrons into the auroral atmosphere.

McIlwain (1960) further concluded from his measurements over a bright active auroral arc that the electrons responsible for the aurora were distributed over a rather narrow energy range near 6 keV. He referred to the energy distribution of these electrons as being 'monoenergetic' and suggested that they may have been formed by electric field acceleration. This 'monoenergetic' electron distribution associated with discrete auroral features was considerably different from that which McIlwain observed within a relatively low intensity 'diffuse' auroral glow. The electron precipitation over the diffuse aurora was found to be less intense than that over the discrete aurora, and the electrons were distributed over a wide range of energies. Such diffuse auroral electron precipitation can result from the direct precipitation of geomagnetically trapped electrons (see Chapter 5) without additional acceleration by electric fields.

Several years later detailed measurements of the energy spectra of the electrons responsible for discrete auroras were obtained from rocket flights over aurora (Evans, 1967, 1968; Albert, 1967a, b). These observations confirmed McIlwain's (1960) result that the energy flux of the precipitating auroral electrons is confined to a narrow energy range centered near 1–10 keV. Examples of the detector count rate versus electron energy obtained by Evans (1968) are shown in Figure 4.35. Note that the count rate, or relative intensity, of the electron fluxes is proportional to the electron energy flux $\mathscr{E}$ per unit energy. The differential particle flux $j \sim \mathscr{E}/K$. Thus if the data in Figure 4.35 were plotted as j versus K, the energy spectra would be far less peaked. In fact the distribution function $f \sim \mathscr{E}/K^2$ would be a monotonically decreasing function of energy. Both Evans and Albert suggested that the auroral energy spectra indicated acceleration by electric fields; however they primarily considered electric fields perpendicular to, and not parallel to, the ambient magnetic field.

Measurements of the pitch-angle distributions of auroral electrons soon became available from satellites (Hoffman and Evans, 1968; Holmgren et al., 1970; Paschmann et al., 1972; Mizera et al., 1976) and rockets (O'Brien and Reasoner, 1971; Whalen and

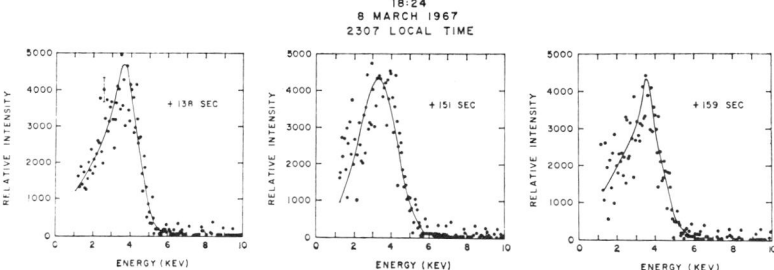

Fig. 4.35. Three representative examples of the detector count rate versus electron energy obtained from a rocket above an auroral display (from Evans, 1968). (© by American Geophysical Union)

McDiarmid, 1972; Maehlum and Moestue, 1973; Arnoldy et al., 1974; Lundin, 1976). These observations showed that the angular distributions of auroral electrons were occasionally peaked in the downward direction along magnetic field lines, and the authors suggested that acceleration by electric fields aligned parallel to the auroral magnetic field lines could account for these observations. Figure 4.36 shows an example of the energy spectra of precipitating electrons within three pitch-angle ranges as obtained by Arnoldy et al. (1974) over an active auroral display. Note the peaked energy spectrum and the field alignment in the vicinity of 5 keV. Field-aligned pitch-angle distributions such as

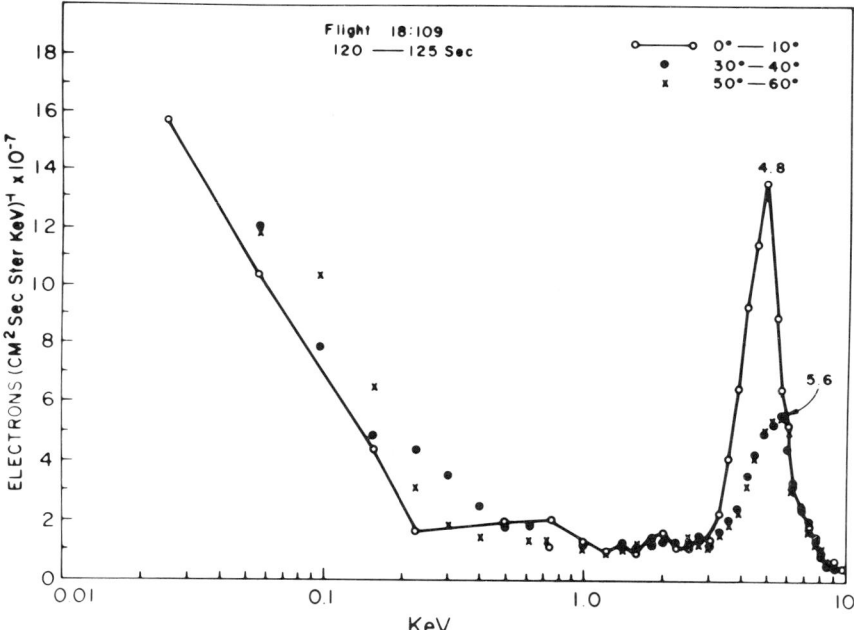

Fig. 4.36. An example of the energy spectra of precipitating electrons within three pitch-angle ranges obtained from a rocket above an auroral display (from Arnoldy et al., 1974). (© by American Geophysical Union)

those shown in Figure 4.36 are not always seen over auroras, however. This is because the field-aligned portion of the electron distribution is unstable to the generation of plasma waves. Such wave turbulence will interact with the electrons, and the resulting pitch-angle diffusion will drive the electron pitch-angle distribution towards isotropy.

In addition to the peak in the energy spectra of auroral electrons at 1–10 keV, the energy spectra show large fluxes of electrons at much lower energies. This can be seen in Figure 4.36 between 30 and 200 eV where the electron fluxes monotonically increase with decreasing electron energy. Similar distributions of low energy auroral electrons were observed earlier by Westerlund (1969). This led Westerlund (1969) and O'Brien (1970) to argue that a parallel electric field could not be responsible for the auroral electrons at higher energies, since all electrons would be accelerated by the total field-aligned potential difference.

A resolution to this difficulty was suggested by Evans (1974). He noted that a parallel electric field that accelerates electrons downward toward the atmosphere will act as a barrier for upgoing electrons. Thus upgoing electrons with a parallel energy too low to surmount the total potential difference will be reflected to appear as precipitating electrons. The primary precipitating auroral electrons, i.e., those from directly above the parallel potential difference, create a significant flux of secondary and backscattered electrons from the atmosphere. Evans noted that the reflection of those electrons must occur if the primary auroral electrons are accelerated by a parallel electric field, and he suggested that reflected electrons could account for the significant flux of low energy auroral electrons.

Model energy spectra of precipitating electrons at 0° and 45° pitch as calculated by Evans (1974) are shown in Figure 4.37. The electrons were assumed to have originated from an 800 eV plasma of density 1.5 cm^{-3}, and were assumed to have entered a region

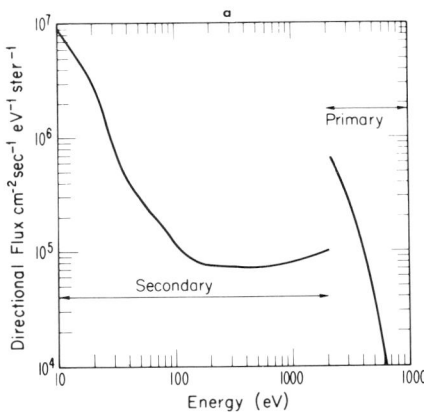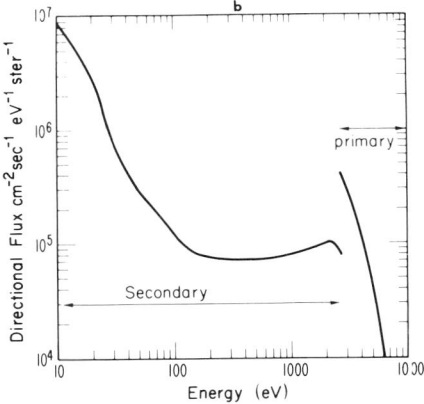

Fig. 4.37. Model energy spectra of precipitating electrons at (a) 0° pitch angle and (b) 45° pitch angle. The electrons were assumed to have originated from an 800 eV plasma of density 1.5 cm^{-3} and were assumed to have been accelerated by a total field-aligned potential difference of 2000 V located at 2000 km altitude. The discontinuities in the energy spectra separate the primary auroral electrons of magnetospheric origin from the lower energy backscattered and secondary electrons of atmospheric origin (from Evans, 1974). (© by American Geophysical Union)

of parallel electric field with a total potential difference of 2 kV. The discontinuity in the energy spectra in Figure 4.37 clearly separates the primary auroral electrons of magnetospheric origin at higher energies from the backscattered and secondary electrons of atmospheric origin. The discontinuity in the energy spectra could not be discerned by a real particle detector, and may well be smoothed by wave-particle interactions. However, the calculated energy spectra reproduce the peak in the auroral energy spectra and the enhanced fluxes of low energy electrons.

Evans fit his model of the auroral energy spectra to an observed auroral electron energy spectrum obtained on a satellite by Frank and Ackerson (1971) as shown in Figure 4.38. He assumed a 400 V parallel potential difference and an unaccelerated electron distribution of temperature 800 eV and density 5 cm^{-3}. The excellent agreement between the modeled and the observed energy spectra gives strong support to Evans's suggestion that the low energy electrons resulted from the electric field reflection of backscattered and secondary electrons originating from the atmosphere. Similar agreements between Evans's model and auroral electron observations were also obtained by Winningham et al. (1977) and Pulliam et al. (1981).

Compelling evidence that auroral electrons are accelerated by parallel electric fields has been obtained from S3-3 satellite observations over the auroral zone at ~ 5000–10 000 km altitude. Shelley et al. (1976) found ionospheric ions streaming upward from the

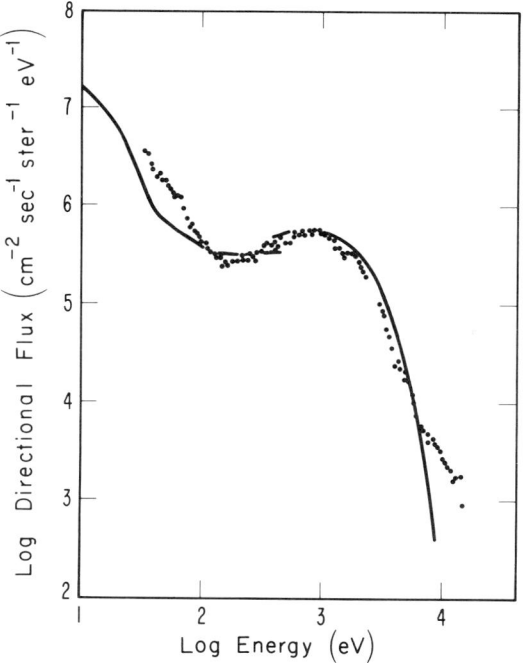

Fig. 4.38. Comparison of Evans's (1974) model auroral energy spectrum with an observed energy spectrum from Frank and Ackerson (1971). The electrons were assumed to have originated from an 800 eV plasma of density 5 cm^{-3}, and the field-aligned potential difference was taken to be 400 V (from Evans, 1974). (© by American Geophysical Union)

ionosphere, having been accelerated to energies of ~1 keV, and Mizera and Fennell (1977) found ions being accelerated upward simultaneously with electrons being accelerated downward. Figure 4.39 shows a representative pair of simultaneously measured electron and ion distributions as presented by Mizera and Fennell (1977). Contour plots in the ($v_\perp$, $v_\parallel$)-plane of the ion and electron distribution functions are shown in the upper and lower portions of the figure.

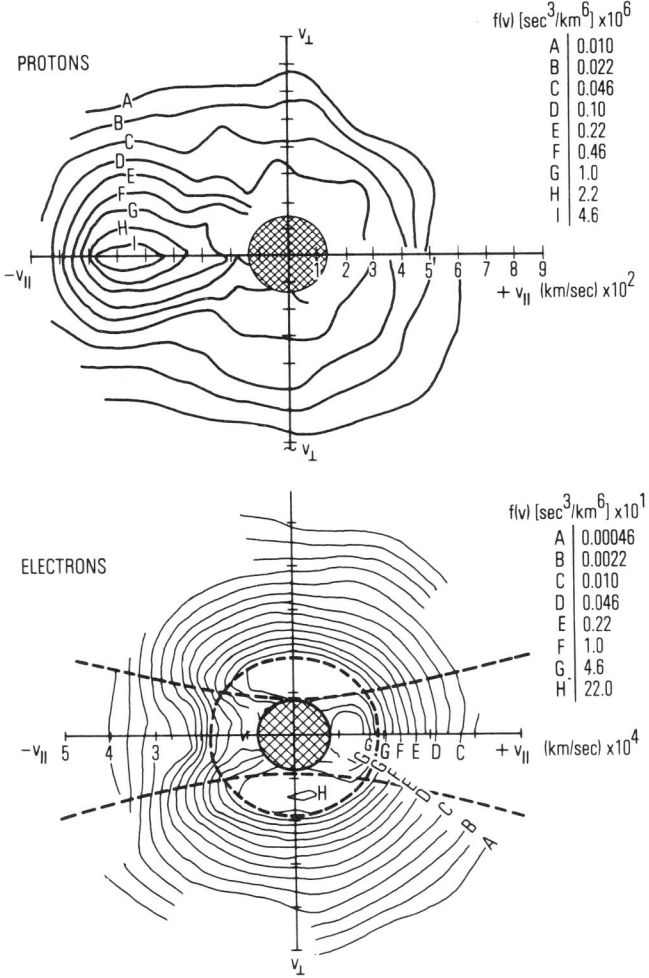

Fig. 4.39. Contour plots in the ($v_\perp$, $v_\parallel$)-plane of a representative pair of simultaneously measured electron and ion distribution functions from the S3–3 satellite at 7300 km altitude and ~ 1845 LT. These distributions indicate that the ions were accelerated upward simultaneously with the downward acceleration of electrons. The dashed ellipse gives the boundary between accelerated magnetospheric electrons and electrons that cannot surmount the parallel electric field. The hyperbola gives the boundary of the atmospheric loss cone (from Mizera and Fennell, 1977). (© by American Geophysical Union)

The ion distribution function in Figure 4.39 can be seen to peak in the upward direction at $v_\parallel \approx -600$ km s^{-1}, which corresponds to a proton energy ≈ 1.9 keV. (The ions at this time were identified as predominantly protons.) Mizera and Fennell estimated the angular half-width of the upward-directed ion beam to be $\sim 10°$, so that the beam was coming from well within the normal atmospheric loss cone. The electron distribution in the downward direction ($+v_\parallel$ axis) peaks at $v_\parallel = 1.8 \times 10^4$ km s^{-1} (≈ 1 keV) between the two contours labeled G. The dashed curves overlaid on the electron distribution are boundaries that delineate different regions in phase space that result from a parallel electric field (Whipple, 1977; Chiu and Schulz, 1978). The ellipse gives the boundary between accelerated magnetospheric electrons and electrons that cannot surmount the parallel electric field above the satellite, and the hyperbola gives the boundary of the atmospheric loss cone. The peak in the electron distribution function within the downgoing loss cone can be seen to be aligned with the dashed ellipse, as expected from electron acceleration by a parallel electric field combined with the magnetic mirroring force. The distributions in Figure 4.39 indicate the satellite was within the region of parallel electric field acceleration, a total potential difference of 2 kV being above the satellite and of 1 kV being below the satellite.

4.4.2. ASSOCIATION OF AURORAS WITH FIELD-ALIGNED CURRENTS

The precipitating electrons associated with auroras carry to the atmosphere a net flux of electrons that will give rise to an upward field-aligned current if the downward electron flux is not counteracted by a field-aligned flux of additional particles. It has now become established from rocket magnetometer and electron measurements that discrete auroras are associated with field-aligned currents of magnitude 10^{-6} A m^{-2} to a few times 10^{-5} A m^{-2} (Cloutier *et al.*, 1970; Vondrak *et al.*, 1971; Park and Cloutier, 1971; Choy *et al.*, 1971; Evans *et al.*, 1977; Anderson, 1978). This association has been verified by correlations of satellite and ground auroral photographs with satellite magnetometer observations (Kamide and Akasofu, 1976; Kamide and Rostoker, 1977). The field-aligned currents associated with discrete auroras represent structure on a larger scale field-aligned current system in auroral regions that has been identified from satellite magnetic field observations (Cummings and Dessler, 1967; Böstrom, 1967; Zmuda *et al.*, 1970; Fairfield, 1973; Arnoldy, 1974; Zmuda and Armstrong, 1974; Iijima and Potemra, 1976a, b). A summary of the distribution and directions of the large-scale field-aligned currents determined by Iijima and Potemra (1976a, b) is shown in Figure 4.40. This figure shows two dominant field-aligned current regions at most local times. The poleward currents are upward for noon-to-midnight local times and downward for midnight-to-noon local times. These directions are reversed for the equatorward currents.

4.4.3. LATITUDINAL VARIATIONS OVER AURORAS

The energy flux of electrons incident upon the auroral ionosphere depends upon the number flux and the energy of the electrons, the energy being dependent upon the amount of acceleration by parallel electric fields. Lyons *et al.* (1979) addressed the question of what quantity varies over auroras so as to give the enhanced electron energy deposition into the atmosphere responsible for the auroras — the magnitude of the

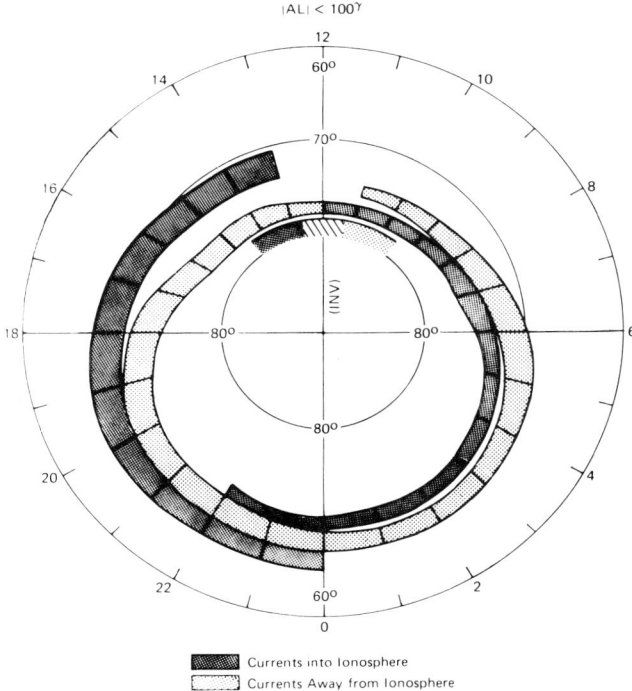

Fig. 4.40. Summary of the distribution and directions of the large-scale field-aligned currents as determined from Triad satellite magnetic field observations. The hatched area near noon indicates confused current directions (from Iijima and Potemra, 1976b). (© by American Geophysical Union)

parallel potential difference or the number flux of electrons that deposit their energy in the atmosphere?

To choose between these alternatives, it is necessary to determine simultaneously the potential difference along magnetic field lines and the net flux of electrons from the magnetosphere to the ionosphere in regions nearby and over auroral forms. The energy of the peak in the differential flux-versus-energy spectrum that typically occurs at 1–15 keV in electron distributions over auroras can be used to estimate the total field-aligned potential difference $V_\parallel$. In the absence of collisions, acceleration by a parallel potential difference $V_\parallel$ will create an electron energy spectrum with a discontinuous increase at a parallel energy $K_\parallel = |eV_\parallel|$. Assuming, for example, that a Maxwellian distribution of electrons of thermal energy K_{th} from the magnetosphere is incident upon a parallel potential difference, then the differential energy spectra at 0° pitch angle will have the form $j \sim K \exp[-(K - |eV_\parallel|)/K_{th}]$ for $K \geqslant |eV_\parallel|$ and $j = 0$ for $K < |eV_\parallel|$. Such an energy spectrum has a local maximum at $K = K_{th}$ if $|eV_\parallel| < K_{th}$ and at $K = |eV_\parallel|$ if $|eV_\parallel| \geqslant K_{th}$. Lyons et al. (1979) assumed that the incident electron distribution can be represented by a thermal energy K_{th} with $|eV_\parallel| \geqslant K_{th}$ so that the peak in the energy spectrum gives $|eV_\parallel|$.

Obtaining the net electron number flux from the magnetosphere to the ionosphere is not as straightforward as obtaining the magnitude of the field-aligned potential difference, since it is not sufficient simply to integrate energetic electron observations to obtain the net downward electron number flux. This is because a significant contribution to such an integration can come from ionospheric electrons at energies well below $|eV_\parallel|$ which were not accelerated by the parallel potential difference (Evans et al., 1977). These electrons result from the acceleration of ionospheric electrons to detectable energies by Coulomb collisions with the higher energy accelerated electrons below the parallel potential difference (Banks et al., 1974) and perhaps also by interactions with waves generated by the accelerated electrons and ions. The current carried by these accelerated topside ionospheric electrons may be compensated for by thermal electrons from the ionosphere (Evans et al., 1977), but thermal electron energies are generally too low for detection by instruments used for measuring auroral electrons.

However, the net number flux can be inferred by using the inferred field-aligned potential difference and the measured net electron energy flux precipitating into the ionosphere $\mathscr{E}_p$. This has a significant advantage over attempting to integrate the net number flux directly, since electrons with energies well below that of the peak in the energy spectrum contribute little to the measured net electron energy flux. Thus neither the low energy, accelerated electrons of ionospheric origin nor the lower energy ionospheric thermal electrons contribute significantly to the calculation of the net energy flux.

To infer the net flux of electrons from the magnetosphere to the ionosphere, $J_\parallel$, we divide $\mathscr{E}_p$ into two parts:

$$\mathscr{E}_p = \mathscr{E}_{p0} + |eV_\parallel| J_\parallel,$$

where $\mathscr{E}_{p0}$ is the downward energy flux of the precipitating electrons before acceleration by the parallel potential difference and $|eV_\parallel| J_\parallel$ is the increment in energy flux due to the acceleration. Letting $j(E, \alpha)$ be the differential, directional electron number flux as a function of pitch angle α and electron energy K,

$$J_\parallel = 2\pi \int_0^\infty \int_0^\pi j(K, \alpha) \sin \alpha \cos \alpha \, d\alpha \, dK$$

and

$$\mathscr{E}_p = 2\pi \int_0^\infty \int_0^\pi j(K, \alpha) K \sin \alpha \cos \alpha \, d\alpha \, dK.$$

The accelerated electron distribution may be unstable; however, as long as the net growth of wave energy by the electrons is well below $\mathscr{E}_p$, the energy lost by the electrons to the waves can be neglected.

If the average energy of the unaccelerated electron distribution is less than $|eV_\parallel|$, then $|eV_\parallel| J_\parallel$ is the dominant contributor to $\mathscr{E}_p$ and we have

$$\mathscr{E}_p \approx |eV_\parallel| J_\parallel \tag{4.10}$$

so that $J_\parallel$ can be obtained directly from $\mathscr{E}_p$ and $|V_\parallel|$. This assumption may introduce errors for small potential differences; however, as long as the average energy of the unaccelerated electrons is less than $|eV_\parallel|$, the error will be less than a factor of 2.

Lyons et al. (1979) applied (4.10) to rocket observations over auroras. They directly compared $|V_\||$ with $\mathscr{E}_p$, since enhancements in $\mathscr{E}_p$ clearly expose the regions above auroral forms, and the estimate for $J_\|$ comes simply from $\mathscr{E}_p/|eV_\||$.

The net downward electron energy flux (essentially equal to $\mathscr{E}_p$) throughout the Polar-3 rocket flight is given by the solid line in Figure 4.41. Electrostatic analyzers on this rocket obtained the complete pitch-angle distribution of upgoing and downgoing electrons from 20 to 16 keV south of, over, and north of a stable auroral arc. Details of this flight are given by Maynard et al. (1977) and Evans et al. (1977). One second of flight time corresponds to approximately 1 km of horizontal distance along the rocket trajectory, which was 16° to the west of geomagnetic north. The main arc was encountered between $\sim$ 150 and $\sim$ 210 s, and a weaker form was encountered near the end of the flight. South of the main arc, prior to 135 s, the observed electron energy spectra showed no peaks, which indicates that there was no significant parallel potential difference along field lines south of the arc. From 135 s to the end of the flight, the spectra displayed peaks at energies from 1 and 9 keV, indicating the precipitating electrons were accelerated by a parallel electric field both over and north of the main arc.

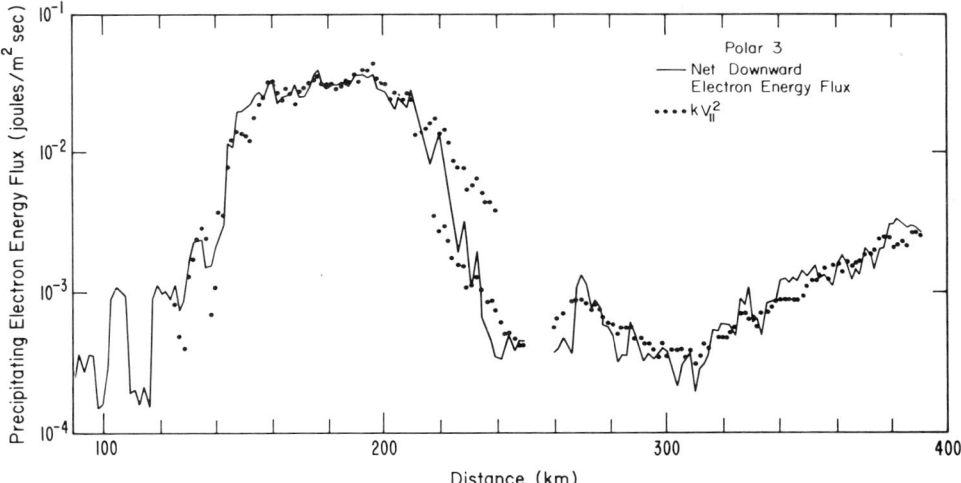

Fig. 4.41. Net downward electron energy flux throughout the Polar-3 flight (Maynard et al., 1977; Evans et al., 1977) and the functional form $kV_\|^2$. Values for $kV_\|^2$ are shown from 120 km until the end of the flight. $V_\|$'s $\gtrsim$ 1 kV were inferred from the peaks in the energy spectra of the precipitating electrons throughout this period. Two values of the constant k have been used, the higher value $(4.7 \times 10^{-10}$ J m^{-2}-s-kV$^2)$ giving a good fit before 220 km and the lower value $(1.0 \times 10^{-10}$ J m^{-2}-s-kV$^2)$ giving a good fit after 225 km (from Lyons et al., 1979). (© by American Geophysical Union)

The parallel potential difference $V_\|$ was inferred from the peaks in the energy spectra from 135 s until the end of the flight, and $\mathscr{E}_p$ was found to vary directly with $V_\|$. However, rather than finding $\mathscr{E}_p$ proportional to $V_\|$ as would be expected if variations in $V_\|$ were totally responsible for the variations in $\mathscr{E}_p$, it was found that $\mathscr{E}_p$ was proportional to $V_\|^2$ everywhere except for a small region (230–235 s) at the north edge of the arc. The dots in Figure 4.41 give $kV_\|^2$, where k has been chosen so as to give the best fit to $\mathcal{E}_p$. Two values of k were used; the higher value (0.47 ergs cm^{-2}·s^{-1}·kV^{-2} = 4.7 × 10^{-10} J m^{-2}·s^{-1}·V^{-2}) gives a good fit to $\mathcal{E}_p$ before 230 s, and the lower value (1.0 × 10^{-10} J m^{-2}·s^{-1}·V^{-2}) gives a good fit after 235 s.

Figure 4.41 shows that the relation $\mathcal{E}_p = kV_\parallel^2$ holds remarkably well at the southern edge, over, and north of the stable auroral arc traversed by Polar-3. The same relationship between $\mathcal{E}_p$ and $V_\parallel^2$ was found throughout two additional rocket flights; values of k were 1.8 × 10^{-10} and 9.6 × 10^{-10} J m^{-2}·s^{-1}·V^{-2}. This relation, together with Equation (4.10), implies that $V_\parallel$ is proportional to the net flux $J_\parallel$ of electrons from the magnetosphere to the ionosphere. This result implies that increases in $V_\parallel$ and increases in $J_\parallel$ are equally responsible for the enhanced electron energy deposition over auroral arcs. That $J_\parallel$ is generally portional to $V_\parallel$ indicates a direct physical connection between these two quantities.

In Figure 4.41, the latitudinal width of the discrete auroral arc traversed by Polar-3 can be seen to be ~ 60 km. This is somewhat broad for a discrete arc, ~ 10 km being more typical. Similar structures as shown in Figure 4.41 have also been observed on satellites (Frank and Ackerson, 1971; Lin and Hoffman, 1979; Hoffman and Lin, 1981; Fennell et al., 1981). The width of $\lesssim 10$ km auroral forms is generally below the spatial resolution of the satellite observations, but larger scale features have been resolved with mean latitudinal widths of ~ 20–50 km and maximum widths of ~ 200 km. Their characteristic appearance in satellite data displays, where peak fluxes move from lower energies to higher energies and then return to lower energies as a function of latitude, led Frank and Ackerson (1971) to refer to satellite observations of auroral precipitation regions as 'inverted-V' events. This term has generally referred to the larger scale, 20–200 km auroral features. The weaker auroral form in Figure 4.41, encountered by Polar-3 between 320 and 400 s, appears to be approximately half of a typical, large-scale inverted-V event as observed on satellites. In general, the latitudinal widths of discrete auroras range from ~ 1 to 200 km, though smaller scale structure is also observable.

In addition to the association with upward field-aligned currents, auroral precipitation regions are observed to be associated with changes in the ionospheric electric field where $\nabla \cdot \mathbf{E} < 0$. A commonly observed example of such an electric field divergence exists along the evening boundary between anti-sunward convection over the polar caps and sunward convection at lower latitudes (see Figures 4.9 and 4.10), and inverted-V precipitation regions are often observed to be coincident with this boundary (Frank and Gurnett, 1971; Gurnett and Frank, 1973). The auroral precipitation and electric fields are often highly structured in the vicinity of the reversal of the convection electric field; however, individual regions of auroral electron precipitation with latitudinal widths of the order of the 10 km width of discrete auroral arcs are also associated with electric field changes where $\nabla \cdot \mathbf{E} < 0$ (Swift and Gurnett, 1973; Maynard et al., 1977; Heelis et al., 1981; Burke, 1981; Temerin et al., 1981, Vondrak, 1981). Such a change in the electric field across auroral arcs can result from a poleward-directed electric field equatorward of an arc and an equatorward-directed electric field poleward of the arc. However, the electric field need not reverse across an arc, but can just change in magnitude across the arc so that $\nabla \cdot \mathbf{E} < 0$ in association with the arc.

4.4.4. THE CURRENT-VOLTAGE RELATION ALONG AURORAL FIELD LINES

In order to understand why parallel electric fields that accelerate auroral electrons exist along auroral field lines, it is necessary first to understand the physics of the relation between the field-aligned current density $j_\parallel$ and $V_\parallel$. The upward-directed field-aligned current associated with auroras can be carried only by ions flowing upward out of the ionosphere and electrons flowing downward from the magnetosphere.

Of critical importance to the generation of significant potential differences along auroral magnetic field lines is the field-aligned current that can be supplied by ionospheric ions. Measurements of the ionosphere thermal plasma density near 1000 km show the existence of high latitude troughs at auroral latitudes (Taylor *et al.*, 1975), and the boundary between sunward and anti-sunward convection is generally located within the troughs (Grebowsky *et al.*, 1976). The data presented in the above two references indicate that H$^+$ densities in the troughs are $\sim 10^1 - 10^3$ cm^{-3} and that O$^+$ densities are $\sim 10^3 - 10^4$ cm^{-3}, and it is plausible that these troughs result from the upward acceleration of ions from the ionosphere (Chiu and Schulz, 1978).

The maximum field-aligned current that can be supplied by the ionospheric plasma is that obtained by counting all particles of a given charge with a component of velocity upward along the magnetic field and neglecting all particles with a downward component. For a plasma species with a Maxwellian distribution of density N, thermal energy K_{th}, charge q, and mass m, we obtain the current from the upgoing particles as

$$j_{max} = Nq \left(\frac{K_{th}}{2\pi m} \right)^{1/2}.$$

It is difficult to determine precisely at what altitude to apply the relation. However, for the high latitude trough densities mentioned above, assuming a 2000 K temperature (0.172 eV), $j_{max} = 2.6 \times 10^{-9} - 2.6 \times 10^{-7}$ A m^{-2} for the protons and $6.5 \times 10^{-8} - 6.5 \times 10^{-7}$ A m^{-2} for the O$^+$. On the other hand, the ionospheric electrons can contribute $j_{max} = 1.1 \times 10^{-5} - 1.1 \times 10^{-4}$ A m^{-2} to a downward-going current for an electron density of $10^3 - 10^4$ cm^{-3}.

Typical intensities of the large-scale, field-aligned currents observed on low altitude satellites are 10^{-6} A m^{-2} (Iijima and Potemra, 1976a) for both the upward and downward currents, and the upward currents associated with discrete auroras extend up to a few times 10^{-5} A m^{-2}. Such currents can easily be supplied by the ionosphere for the case of the downward field-aligned currents (by parallel potential differences $V_\parallel$ such that $eV_\parallel$ is the order of the ionospheric thermal energy). However, the observed upward currents cannot be supplied by the ionospheric ions within the high latitude troughs (Lennartsson, 1980; Lyons, 1980), so that electrons from high altitudes (e.g., the magnetosheath and/or the plasma sheet) are required.

The magnetic mirroring of magnetospheric electrons, however, restricts the electron flux that can flow to the ionosphere to that which is carried by electrons within the loss cone. As we shall show shortly, the maximum current that can be carried by magnetospheric electrons within the nominal atmospheric loss cone is too low to give the current associated with discrete auroras. However acceleration by electric fields along magnetic field lines will increase the electrons' parallel velocity without affecting their perpendicular velocity. Thus parallel electric fields can increase the field-aligned current that can be carried by magnetospheric electrons by increasing the number of electrons within the loss cone.

The field-aligned current density from the high altitude electron population, assuming an isotropic Maxwellian distribution of density N and thermal energy K_{th} and single-particle motion, is related to the total field-aligned potential difference $V_\parallel$ by (Knight, 1973; Lemaire and Scherer, 1974; Antonova and Tverskoy, 1975)

$$j_\parallel = eN \left(\frac{K_{th}}{2\pi m_e}\right)^{1/2} \frac{B_i}{B_{V_\parallel}} \left[1 - \left(1 - \frac{B_{V_\parallel}}{B_i}\right) \exp\left\{-\frac{eV_\parallel}{K_{th}\left[(B_i/B_{V_\parallel}) - 1\right]}\right\}\right]. \quad (4.11)$$

Here B_i is the ionospheric magnetic field strength, and $B_{V_\parallel}$ is the magnetic field strength at the top of the potential variation along field lines. This relation is independent of the distribution of the potential along field lines, except for the assumption that no particles incident upon the top of the parallel potential variation mirror before falling through the entire potential difference. However, this assumption is not a significant restriction, since $B_i/B_{V_\parallel}$ can be varied in (4.11) and $j_\parallel$ has little dependence upon $B_i/B_{V_\parallel}$ as long as $B_i/B_{V_\parallel} > eV_\parallel/K_{th}$, and this inequality is generally satisfied along auroral field lines.

Figure 4.42 shows $j_\parallel$ versus $V_\parallel$ as obtained from Equation (4.4) for $N = 1$ cm^{-3} and $K_{th} = 1$ keV, which are reasonable values for the plasma sheet (Vasyliunas, 1968; Schield and Frank, 1970) and the high latitude magnetosheath (Bahnsen, 1978). The results for other values of N and K_{th} can easily be obtained from Figure 4.42 by multiplying the current densities on the vertical axis by N(cm^{-3})$K_{th}^{1/2}$ (keV) and the potential differences on the horizontal axis by K_{th} (keV). Curves are shown for values of $B_i/B_{V_\parallel}$ ranging from 3, corresponding to a parallel potential variation entirely below ~ 2000 km in altitude, to 1000, corresponding to a parallel potential variation extending to many earth radii in altitude. S3-3 satellite observations of accelerated particles show that the potential variation generally occurs at altitudes between 5000 and 10 000 km (Gorney et al., 1981), so that $B_i/B_{V_\parallel} \approx 30$.

Figure 4.42 shows that the maximum field-aligned current density that can be carried from the ionosphere by precipitating electrons from high altitudes is $\sim 10^{-6}$ A m^{-2} for $V_\parallel = 0$, which is approximately equal to the lower limit observed in association with discrete auroras. However $j_\parallel$ increases with increasing $V_\parallel$, reaching a maximum value dependent upon $B_i/B_{V_\parallel}$. This maximum is attained when the magnitude of $V_\parallel$ is sufficiently large that nearly all electrons incident upon the top of the region of potential variation along field lines are accelerated so as to be within the atmospheric loss cone. The maximum value of $j_\parallel$ increases with increasing $B_i/B_{V_\parallel}$ since increasing numbers of electrons are available to be accelerated into the loss cone as $B_i/B_{V_\parallel}$ is increased. For $B_i/B_{V_\parallel} = 30$, the maximum $j_\parallel$ is $\sim 3 \times 10^{-5}$ A m^{-2}, comparable with the maximum value of $j_\parallel$ typically observed in association with discrete auroras.

Figure 4.42 shows that single-particle motion in the presence of the observed auroral $V_\parallel$'s of $\sim 1-10$ keV gives $j_\parallel$'s of the observed magnitudes of $\sim 10^{-6} - 10^{-5}$ A m^{-2}. Thus the observed potential differences are required for the observed field-aligned current densities to exist (Knight, 1973).

It is important to note the consistency between the $j_\parallel$ versus $V_\parallel$ relation given by Equation (4.11) and shown in Figure 4.42 and the observation of Lyons et al. (1979)

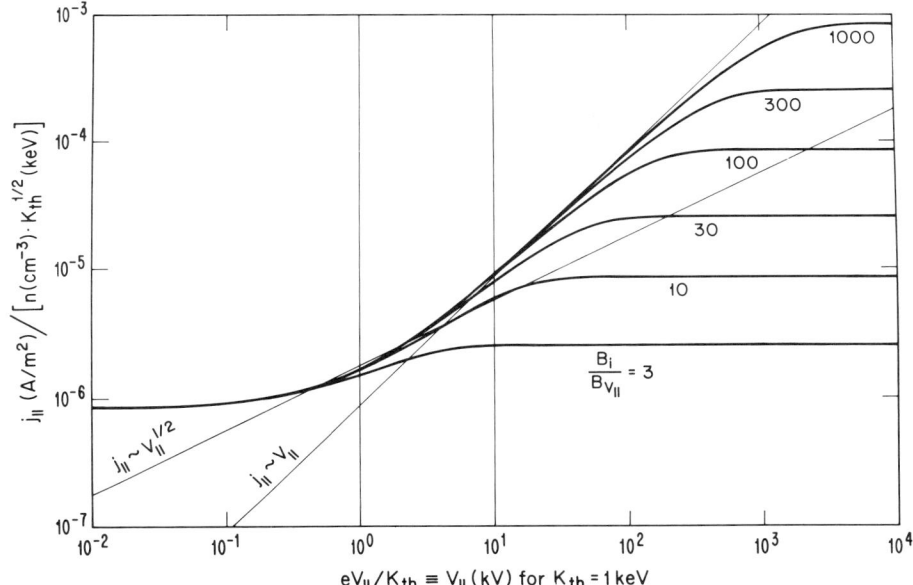

Fig. 4.42. The $j_\parallel$ versus $V_\parallel$ relation for single-particle motion along field lines for a high-altitude electron plasma with density $n = 1$ cm^{-3} and $K_{th} = 1$ keV. Results for other values of n and K_{th} can be obtained by multiplying the current densities on the vertical axis by n (cm^{-3}) $K_{th}^{1/2}$ (keV) and the potential differences on the horizontal axis by K_{th} (keV). Curves are shown for $B_i/B_{V_\parallel}$ from 3 to 1000. Lines for $j_\parallel \sim V_\parallel^{1/2}$ and $V_\parallel^1$ and for $eV_\parallel/K_{th} = 1$ and 10 are shown for reference (from Lyons, 1981b). (© by American Geophysical Union)

that $j_\parallel \sim V_\parallel^2$ (Fridman and Lemaire, 1980). Figure 4.42 shows that $j_\parallel \sim V_\parallel^{0.5-1.0}$ over the range of typical auroral values of $j_\parallel$ and $V_\parallel$. For $eV_\parallel/K_{th} > 1$, most electrons precipitating into the ionosphere will have an energy $\approx eV_\parallel$, so that the total precipitating electron energy flux $\mathscr{E}_p$ will be proportional to $j_\parallel V_\parallel$. Thus Equation (4.11) gives $\mathscr{E}_p \sim V_\parallel^{1.5-2.0}$ for $eV_\parallel/K_{th} > 1$, in agreement with the observed relation that $\mathscr{E}_p \sim V_\parallel^2$. Thus the observed magnitudes of $V_\parallel$ and $j_\parallel$ and the observed relationship between them are easily explained by considering nothing more than the single-particle motion along auroral field lines.

The $j_\parallel - V_\parallel$ relation for single-particle motion ignores the effects of electron scattering driven by plasma waves. It is well established that the accelerated auroral electron distribution is unstable to the generation of plasma waves, and the detailed energy and angular distribution of auroral electrons often cannot be quantitatively explained by parallel electric field accelerations without invoking electron scattering by plasma waves (Evans, 1976). However, the energy that goes into the generated waves is at least two to three orders of magnitude less than that precipitated into the auroral ionosphere by the electrons, so that energy scattering by waves does not significantly affect the net energy deposited into the ionosphere.

Pitch-angle scattering could modify the $j_\parallel - V_\parallel$ relation if a significant flux of precipitating electrons were scattered into the upgoing loss cone, allowing such electrons to surmount the total field-aligned potential difference and thus to decrease $j_\parallel$. However, for such an effect to be significant, the upgoing flux of electrons within the upgoing loss cone in auroral regions would need to be comparable with the flux of precipitating auroral electrons. Observations such as those in Figure 4.39 have given no evidence that this occurs. Electrons can also be scattered into the pitch-angle range where they become trapped between the parallel electric field above and the magnetic mirror below. However, such scattering can be significant for only an insignificantly short time before the flux of the trapped electrons becomes much greater than the flux of precipitating electrons. No evidence exists for such a pile-up of trapped electrons, and wave-particle scattering generally smooths and does not create peaks in particle distribution functions. Thus while electron scattering may modify the distribution of precipitating auroral electrons, it does not significantly affect the $j_\| - V_\|$ relation along auroral field lines.

The above discussion gives no information about the distribution of the potential along field lines. Calculations of this potential distribution using single-particle motion show that both smoothly varying distributions (Lemaire and Scherer, 1973, 1974; Chiu and Schulz, 1978) and distributions with a large variation over a short distance, which occur in double layers (Block, 1972, 1978; Lemaire and Scherer, 1978; Lennartsson, 1980; Stern, 1981), are possible. A definitive conclusion as to how the potential is distributed along field lines has not yet been reached. In addition, it is possible that the effects of particle scattering by plasma waves will have to be included to evaluate this potential distribution quantitatively, even though the scattering by plasma waves does not affect the relation between $j_\|$ and $V_\|$.

4.4.5. GENERATION OF LARGE-SCALE INVERTED-V PRECIPITATION REGIONS

In this section we investigate the generation of the large-scale inverted-V precipitation regions by spatial variations in the magnetospheric convection electric field where $\nabla \cdot \mathbf{E} < 0$ (Lennartsson, 1977; Lyons, 1980; Chiu and Cornwall, 1980). A commonly observed example of such an electric field divergence exists along the evening boundary between anti-sunward and sunward convection, a boundary along which inverted-V precipitation regions are observed. Other variations in the electric field, where $\nabla \cdot \mathbf{E} < 0$, also exist in association with discrete auroras and over the polar caps in association with polar-cap auroras. These polar-cap electric field variations might result from spatial variations in the magnetosheath flow over open, polar-cap magnetic field lines.

The convection boundary and polar-cap regions where $\nabla \cdot \mathbf{E} < 0$ are schematically illustrated in Figure 4.43, where magnetic field lines emanating from the dusk-dawn meridian plane are shown as seen from the geomagnetic tail. In Figure 4.43, the boundary between open and closed field lines is shown coincident with the convection reversal for illustration. This gives an extended line, where $\nabla \cdot \mathbf{E} < 0$, running roughly along the evening auroral oval. Since $\nabla \cdot \mathbf{E} < 0$, this discontinuity must be negatively charged, which can result in a parallel current out of the ionosphere (Lennartsson, 1977). On the dawn side, the boundary produces an extended line which is positively charged and can cause a parallel current into the ionosphere. It is likely that this pair of currents corresponds to the poleward large-scale parallel currents generally observed in the auroral zone as shown in Figure 4.40. Similarly, flow-aligned discontinuities in the magnetosheath plasma velocity will produce extended, flow-aligned, narrow regions of upward (where $\nabla \cdot \mathbf{E} < 0$) and downward (where $\nabla \cdot \mathbf{E} > 0$) currents over the polar cap.

ELECTRIC FIELDS

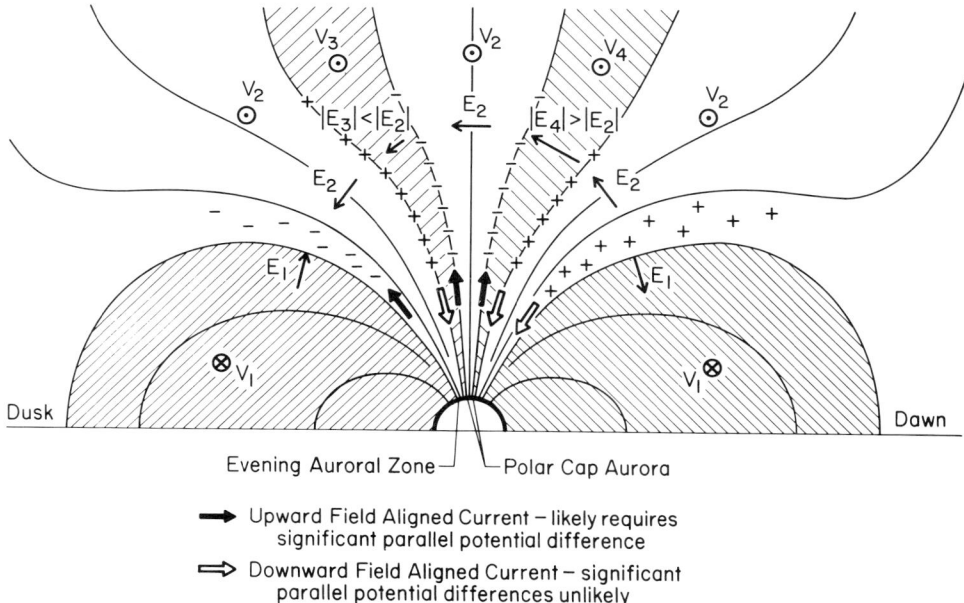

Fig. 4.43. Magnetic field lines emanating from the dusk-dawn meridian plane as viewed from the geomagnetic tail looking toward the sun. A converging electric field exists along the evening boundary between sunward and anti-sunward convection and along any flow-aligned discontinuity in the magnetosheath flow speed increasing from left to right in the figure. Such converging electric fields are associated with negatively charged regions which drive upward field-aligned currents likely to require significant field-aligned electric potential differences. A diverging electric field exists along the morning boundary between sunward and anti-sunward convection and along magnetosheath flow discontinuities where flow speed decreases from left to right in the figure. Such diverging electric fields are associated with positively charged regions that drive downward field-aligned currents which generally will not require significant field-aligned potential differences (from Lyons, 1980).

The equatorward, field-aligned currents in Figure 4.40 which flow into the ionosphere on the dusk side and out of the ionosphere on the dawn side, are readily accounted for by the charge separation of trapped electrons and ions convecting inward from the tail, i.e., the Alfvén layer (Schield et al., 1969; Wolf, 1974; Lennartsson, 1977). Such space charges will also result in $\nabla \cdot \mathbf{E} \neq 0$, so that the following analysis should be applicable to these equatorward currents as well as to those illustrated in Figure 4.43.

The initial assumption is that a discontinuity in the magnetospheric convection electric field, where $\nabla \cdot \mathbf{E} < 0$, exists well above the ionosphere. This upper boundary condition is then applied to obtain the ionospheric potential, assuming a steady state so that $\nabla \cdot \mathbf{j} = 0$ where $\mathbf{j}$ is current density. The problem is reduced to two dimensions (one horizontal dimension x) by assuming that the electric field is solely along the $\pm x$-axis and that there is no divergence in the ionospheric current perpendicular to x (the Hall current). The Hall current along the x direction from an electric field along the other horizontal direction could be included in a two-dimensional model but is not included here.

In the following analysis, the problem is first solved analytically in an approximate form, and the results are found to agree with the typically observed parallel potential differences and currents in the large-scale precipitation regions. The problem is then solved numerically in a more general form, and the results are compared with a specific observed example.

Figure 4.44 illustrates the upper boundary electric potential ϕ as a function of x, assumed to exist at altitudes above any significant parallel potential differences, and the solution to be obtained for the ionospheric electric potential ϕ_i as a function of distance in the ionosphere x_i. The distance x along the upper boundary maps along magnetic field lines to the ionospheric distance x_i, and all quantities can be discussed as a function of x_i. Thus ϕ is shown in Figure 4.44, and will later be specified, as a function of x_i. The electric field discontinuity is taken to be at $x_i = 0$. In Figure 4.44 the upper boundary electric field E_2 for $x_i > 0$ is equal to $-E_1$, the electric field for $x_i < 0$; however the following analysis allows for any E_1 and E_2. (In fact, any variation of ϕ with x_i could be assumed when obtaining a numerical solution for ϕ_i.) Once the ionospheric potential $\phi_i(x_i)$ is obtained, the parallel potential difference $V_\parallel(x_i)$ as a function of x_i is given by $\phi_i(x_i) - \phi(x_i)$ which is assumed to approach zero as $x_i \to \pm \infty$.

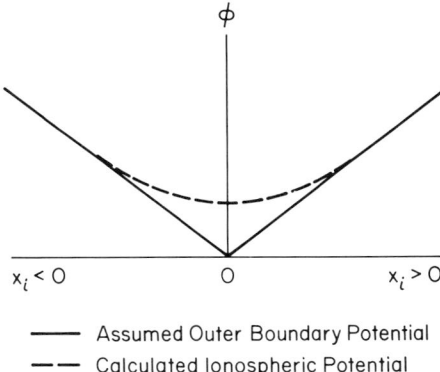

Fig. 4.44. Assumed upper boundary potential at an altitude along field lines above any significant parallel potential differences (solid line) and the ionospheric potential to be calculated (dashed line) as a function of horizontal distance across the field lines in the ionosphere x_i (from Lyons, 1980).

The height-integrated Pedersen current in the ionosphere $I_p = -\Sigma_p \partial \phi_i/\partial x_i$, where Σ_p is the height-integrated, ionospheric Pedersen conductivity. Thus, $\nabla \cdot \mathbf{j} = 0$ gives (Atkinson, 1970; Coroniti and Kennel, 1972; Chiu and Cornwall, 1980)

$$j_\parallel = -\frac{\partial}{\partial x_i} I_p = \frac{\partial}{\partial x_i} \left(\Sigma_p \frac{\partial \phi_i}{\partial x_i} \right), \tag{4.12}$$

where $j_\parallel$ is the upward field-aligned current density exiting the ionosphere. Equation (4.12) is a soluble, second-order differential equation in the ionospheric potential ϕ_i for a specified $\phi(x_i)$ provided $j_\parallel$ and Σ_p can be expressed as a function of ϕ_i and ϕ.

A useful, yet very simple, relation is easily obtainable by integrating (4.12) from $x_i = -\infty$ to $x_i = +\infty$:

$$I_\parallel = \Sigma_{p,\infty}(E_{1,i} - E_{2,i}),$$

where $\Sigma_{p,\infty}$ is the height-integrated Pedersen conductivity for $V_\| = 0$, and $E_{1,i}$ and $E_{2,i}$ are the upper boundary electric fields as mapped into the ionosphere. This relates the total, upward, field-aligned current $I_\|$ per unit length normal to x_i to the imposed electric field change. This relation is independent of the dependence of $j_\|$ and Σ_p upon ϕ and ϕ_i, is valid for any electric field change, and can be applied to the downward current from $E_2 - E_1 > 0$ as well as the upward current from $E_2 - E_1 < 0$.

The relation (4.11) between $j_\|$ and $V_\|$ must be simplified to solve (4.12) analytically. Considerable simplification is obtained by assuming $eV_\|/K_{th} \ll B_i/B_{V_\|}$, so that (4.11) reduces to

$$j_\| = eN \left(\frac{K_{th}}{2\pi m_e}\right)^{1/2} \left(1 + \frac{eV_\|}{K_{th}}\right). \tag{4.13}$$

This assumes the parallel potential difference is located above the altitude where $B_i/B_{V_\|} = eV_\|/K_{th}$. For $eV_\|/K_{th} \gg 1$, (4.13) reduces to simply

$$j_\| = kV_\| = k(\phi_i - \phi) \tag{4.14}$$

where $k = e^2 N/(2\pi m_e K_{th})^{1/2}$. This latter assumption makes $j_\|$ invalid by a factor of more than 2 whenever $eV_\|/K_{th} < 1$. However, as long as the majority of the current contributing to the total parallel current $I_\|$ is from regions where $eV_\|/K_{th} > 1$, the error in making this assumption is not significant. Note that $j_\| = eJ_\|$, where $J_\|$ is the precipitating electron flux. Thus if $eV_\|/K_{th} > 1$ (essentially the same assumption as that used in obtaining Equation (4.10)), most precipitating electrons will have an energy $eV_\|$. Thus Equation (4.14) gives the observed relationship $\mathcal{E}_p = kV_\|^2$ and allows the observed value of k to be related to N and K_{th}.

Letting Σ_p be constant as a function of x_i, using Equation (4.14), and applying the upper boundary condition $\phi(x_i) = -E_{1,i}x_i$ for $x_i \leq 0$ and $\phi(x_i) = -E_{2,i}x_i$ for $x_i \geq 0$, allows Equation (4.12) to be written as

$$\frac{d^2\phi_i}{dx_i^2} = \frac{k}{\Sigma_p}\left[\phi_i + \begin{Bmatrix} E_{1,i} \\ E_{2,i} \end{Bmatrix} x_i\right], \quad \text{for} \quad x_i \begin{Bmatrix} \leq 0 \\ \geq 0 \end{Bmatrix}. \tag{4.15}$$

The general solution to (4.15) is

$$\phi_i(x_i) = \begin{cases} C_1 e^{x_i/x_w} + C_2 e^{-x_i/x_w} - E_{1,i}x_i, & \text{for } x_i \leq 0 \\ C_1' e^{x_i/x_w} + C_2' e^{-x_i/x_w} - E_{2,i}x_i, & \text{for } x_i \geq 0 \end{cases}. \tag{4.16}$$

where the half-width

$$x_w = (\Sigma_p/k)^{1/2} = (\Sigma_p/e^2 N)^{1/2} (2\pi m_e K_{th})^{1/4}.$$

For $\phi_i(x_i) \to \phi(x_i)$ as $x_i \to \pm\infty$, $C_2 = C_1' = 0$. Continuity of ϕ_i at $x_i = 0$ gives $C_1 = C_2'$, and continuity of the ionospheric electric field at $x_i = 0$ gives $C_1 = \frac{1}{2}x_w(E_{1,i} - E_{2,i})$. (The assumption of a discontinuity in the upper boundary electric field is valid as long as the upper boundary electric field changes over a distance in x_i much smaller than does the ionospheric electric field). Thus the parallel potential difference as a function of x_i is

$$V_\|(x_i) = V_{\|,0} e^{\pm x_i/x_w}, \quad \text{for} \quad x_i \lessgtr 0 \tag{4.17}$$

where the maximum parallel potential difference is

$$V_{\|,0} = \frac{x_w}{2} (E_{1,i} - E_{2,i}).$$

Values for the constants in (4.16) can also be obtained for the boundary condition of any specified values of $V_\|$ at any two x_i, and Chiu et al. (1981) present a solution to Equation (4.12) for more general upper boundary conditions.

As an example, taking Σ_p = 5 mhos, and assuming an incident electron population where N = 1 cm^{-3} and K_{th} = 0.25 keV, the above solution gives x_w = 54 km. This gives a total width of ~ 100 km, which is comparable with the typical latitudinal width of inverted V's. This width is not sensitive to the choice of N and K_{th}, since $x_w \sim K_{th}^{1/4}/N^{1/2}$. The total width remains on the order of 100 km for densities of 0.1–10 cm^2 and thermal energies of $2.5 \times 10^{-3} - 2.5 \times 10^1$ keV. Typical changes in the observed low altitude electric field across the boundary from sunward to anti-sunward convection are 75 mV m^{-1}. This gives $V_{\|,0}$ = 2 kV which agrees well with inferred parallel potential differences within inverted V's. When the above parameters are used, the field-aligned current strength at x_i = 0 is 2.4×10^{-6} A m^{-2}; this is also in reasonable agreement with observations.

Thus the simple analytical solution (4.17) to the current continuity Equation (4.12) shows that the typical latitudinal width and magnitudes of the observed field-aligned potential differences and currents associated with large-scale, inverted-V precipitation regions can result from the observed electric-field discontinuity along the evening auroral zone. However, to make definitive comparisons with specific observations, it is desirable to let Σ_p vary with $V_\|$ and to use the full expression (4.11) for $j_\|$. This avoids the requirement that $eV_\|/K_{th} \gg 1$, which becomes violated as $V_\|$ decreases with increasing $|x_i|$, and allows for the variation in Σ_p with the energy flux of the precipitating electrons. However, it becomes no longer possible to solve (4.12) analytically so that a numerical solution is necessary.

Any functional form can be chosen for $\phi(x_i), j_\|(\phi, \phi_i)$, and $\Sigma_p(\phi, \phi_i)$ when the current continuity Equation (4.12) is solved numerically. For the present, the upper boundary potential $\phi(x_i)$ is chosen as before, $j_\|$ is given by Equation (4.11), and Σ_p varies with the precipitating electron energy flux $\mathscr{E}_p$ as (Harel et al., 1977; 1981)

$$\Sigma_p = 0.5 + 1.6 \times 10^2 [\mathscr{E}_p (\text{J m}^{-2}\cdot\text{s}^{-1})]^{1/2} \text{ mhos}. \qquad (4.18)$$

(As a check on the above relation, it adequately reproduces the values of Σ_p computed by Evans et al. [1977] from the classical expressions for the Pedersen conductivity using rocket observations of precipitating auroral electrons.) Under the same conditions used in deriving (4.11), the energy flux of the precipitating electrons is related to $V_\|$ by (Lundin and Sandahl, 1978)

$$\mathscr{E}_p = NK_{th} \left(\frac{K_{th}}{2\pi m_e}\right)^{1/2} \frac{B_i}{BV_\|} \left\{\left[2 + \frac{eV_\|}{K_{th}}\right] - \left[\frac{eV_\|}{K_{th}} + 2\left(1 - \frac{BV_\|}{B_i}\right)\right]\right.$$
$$\left. \cdot \exp\left[-\frac{eV_\|/K_{th}}{(B_i/BV_\|) - 1}\right]\right\}. \qquad (4.19)$$

There are five variable parameters in the numerical solution of Equation (4.12): the density N and thermal energy K_{th} of the incident electron population, the ionospheric electric fields $E_{1,i}$ as $x_i \to -\infty$ and $E_{2,i}$ as $x_i \to \infty$, and the magnetic field ratio $B_i/B_{V_\parallel}$ at the altitude of the parallel potential difference. To approximate the evening convection boundary, $E_{2,i} = -E_{1,i}$ was assumed (corresponding to the schematic illustration in Figure 4.44) and $B_i/B_{V_\parallel}$ was taken to be 10. Then N and K_{th} were varied in an attempt to obtain reasonable agreement with the observation of a specific inverted-V region at ~ 1800 LT (Gurnett and Frank, 1973).

Gurnett and Frank show, as a function of latitude, the simultaneously measured differential electron energy flux as a function of energy (from which an approximation for the parallel potential difference has been inferred from the peak in the energy spectrum), the total precipitating energy flux, and the horizontal electric field. The measured electric field change across the precipitation region was $\sim 8.5 \times 10^{-2}$ V m^{-1} in a direction $\sim 45°$ to the N–S direction. Assuming the actual horizontal electric field was in the N–S direction, corresponding to the idealized model used here, the actual electric field change was ~ 0.12 V m^{-1}. Thus $E_{1,i} = -E_{2,i} = 0.06$ V m^{-1} were taken as the upper boundary electric fields.

A reasonable fit to the magnitudes and spatial extent of the observed parallel potential differences $V_\parallel(x_i)$ and total precipitating electron fluxes $\mathscr{E}_p(x_i)$ was obtained for $N = 1$ cm^3 and $K_{th} = 500$ eV. These values are typical of values observed in the high latitude magnetosheath and the plasma sheet. The comparisons of the model results with the observations are shown in Figure 4.45, where latitude has been converted to horizontal distance in the ionosphere x_i. The measured precipitating energy fluxes, presented per steradian by Gurnett and Frank, have been multiplied by π to estimate the total precipitation electron energy flux $\mathscr{E}_p$. This presumes the precipitating electrons were isotropic in pitch angle. Note in Figure 4.45 that neither the detailed structures within the inverted V, which have magnitudes and scale sizes comparable with those of discrete auroral structure, nor the steep sides of the inverted V are reproduced by the model. However, the model does reproduce the overall magnitudes and spatial extent of the observed precipitating energy fluxes and parallel potential differences, showing that the overall precipitation region can result from the observed electric field change. Remember that, although the electric field change was measured at low altitudes below any significant parallel potential difference, the inferred parallel potential difference approaches zero away from the center of the precipitation region. Thus the measured electric field change can be mapped along magnetic field lines to any desired altitude.

Figure 4.46 shows the complete results of the numerical solution and illustrates how the results depend upon the input parameters. Each of the four columns in Figure 4.46 shows the effects of varying one parameter, and five separate plots are stacked vertically in each column. The top panel shows both the assumed upper boundary potential $\phi(x_i)$ (the lines with a discontinuous slope at $x_i = 0$), and the calculated ionospheric potential $\phi_i(x_i)$ (the smooth curves). The next four panels, from top to bottom, give the parallel potential difference $V_\parallel(x_i) = \phi_i(x_i) - \phi(x_i)$, the precipitating electron energy flux $\mathscr{E}_p(x_i)$, the parallel current density $j_\parallel(x_i)$, and the height-integrated, ionospheric Pedersen conductivity $\Sigma_p(x_i)$. In each panel the solid curve gives the results for the parameters used in Figure 4.45 for comparison with observations ($E_{1,i} = -E_{2,i} = 0.06$ V m^{-1}, $K_{th} = 500$ eV, $N = 1$ cm^3, $B_i/B_{V_\parallel} = 10$), and the other curves give the results when one parameter

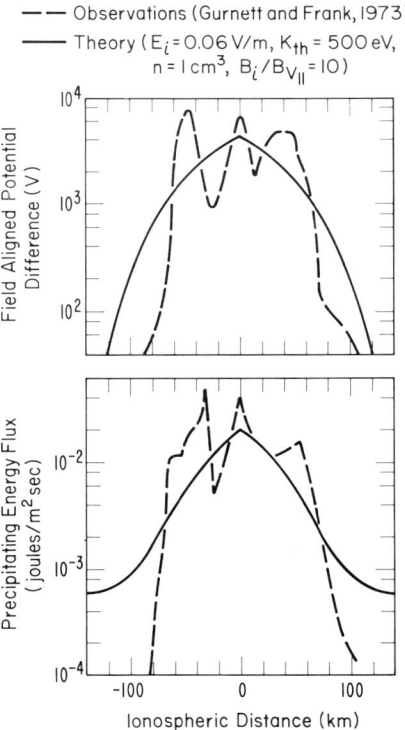

Fig. 4.45. Comparison of the theoretically calculated field-aligned potential differences and precipitating electron energy fluxes with observations within an 'inverted-V' precipitation region at ~ 1800 LT (from Gurnett and Frank, 1973). The calculations used the electric field change measured simultaneously with the precipitating electrons as described in the text, and the density N and thermal energy K_{th} of the electron population incident upon the parallel difference have been adjusted in an attempt to obtain reasonable agreement with the observations. The 'observed' parallel potential differences have been inferred from the peak in the electron energy spectra, and the measured precipitating electron fluxes, presented per steradian by Gurnett and Frank, have been multiplied by π to estimate the total precipitating electron energy fluxes (from Lyons, 1980).

is varied. Field-aligned potential differences and current densities that are not meaningful, owing to approximations made in the numerical solution for $V_{\|}(x_i) \leqslant 30$ V, are indicated by lighter lines.

The first column gives the effects of varying the magnitude of the electric field change by showing the numerical results for $E_{1,\,i} = -E_{2,\,i} = 0.03, 0.06$, and 0.12 V m^{-1}. It can be seen that the magnitude of the electric field change significantly affects the magnitude of the parallel potential differences, currents, and precipitation fluxes but has significantly less effect on the width of the precipitation region. This is as expected from the analytical solution (4.17), except that the maximum $V_{\|}$ increases more than linearly with the upper boundary electric field change because of the increase in Σ_p with $V_{\|}$. The second column gives the results for $K_{th} = 100, 250, 500, 1000$, and 2500 eV and shows that the magnitudes and latitudinal widths of the results vary directly with K_{th}. This result

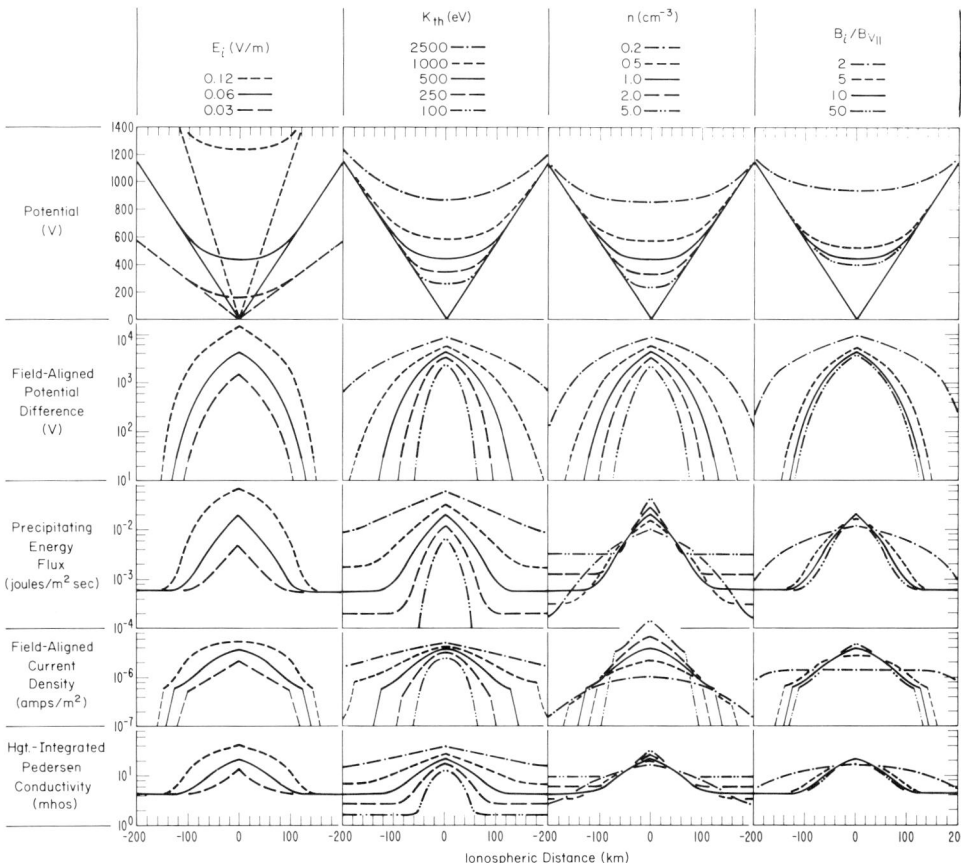

Fig. 4.46. Complete results of the numerical solution to the current continuity equation and the dependence of the results upon the input parameters. The solid line in each panel is for $E_{\perp,i} = -E_{2,i} = 0.06$ mV m^{-1}, $K_{th} = 500$ eV, $N = 1$ cm^{-3}, and $B_i/BV_\| = 10$; the other curves show the results when one parameter is varied as indicated above each column. The top panel in each column shows the assumed upper boundary potential (lines with a discontinuity at 0 km) and the calculated ionospheric potential (smooth curves) as a function of horizontal distance across the field lines in the ionosphere. The other panels, from top to bottom, give the parallel potential difference, the precipitating electron energy flux, the parallel current density, and the height-integrated ionospheric Pedersen conductivity. Lighter lines are used for field-aligned potential differences ≤ 30 eV and corresponding current densities where the numerical results are not meaningful (from Lyons, 1980).

could have been anticipated from Equation (4.17), since $\mathscr{E}_p \sim K_{th}^{3/2}$ for $V_\| = 0$ so that $\Sigma_p \sim K_{th}^{3/4}$ and $x_w \sim K_{th}^{3/8}$ for $x_i \to \pm\infty$. The effects of varying N are slightly more complicated. The widths of the precipitation region and the parallel potentials decrease, while the peak magnitudes of $j_\|$ and $\mathscr{E}_p$ increase, with increasing N. Finally, the last column shows that the results are insensitive to the altitude of the parallel potential difference (and thus also to the altitude distribution of the potential) unless $B_i/B_{V_\|}$

becomes well below $eV_{\parallel}/K_{th}$. The curves for $B_i/B_{V_{\parallel}} \geq 5 = eV_{\parallel,o}/K_{th}$ show little dependence on $B_i/B_{V_{\parallel}}$ but significant effects appear for $B_i/B_{V_{\parallel}} \leq 2 = 0.4 \, eV_{\parallel,o}/K_{th}$.

4.4.6. SMALLER SCALE DISCRETE AURORAL STRUCTURE

A density $N = 1$ cm^{-3} and a thermal energy $K_{th} = 500$ eV were taken in the previous section as typical values for the high altitude electron plasma, and it was found that the width of the precipitation region decreases as N is increased or K_{th} is decreased. Thus, with a sufficiently large value of N and a sufficiently small value of K_{th}, it is possible to obtain precipitation regions the size of discrete auroral forms. The left-hand panels of Figure 4.47 give the field-aligned potential difference and the precipitating electron energy flux versus x_i for $N = 5$ cm^{-3} and $K_{th} = 200$ eV and 100 eV, as compared with the results for $N = 1$ cm^{-3} and $K_{th} = 500$ eV. In all three cases, the magnitude of the high altitude electric field as mapped into the ionosphere E_i was taken to be 0.06 V m^{-1} on both sides of the electric field discontinuity, giving a total discontinuity of 0.12 V m^{-1}. It can be seen from Figure 4.47 that with the rather extreme values for N and K_{th}, precipitation regions result having $V_{\parallel} \gtrsim 1$ kV and precipitating electron fluxes $\gtrsim 10^{-2}$ J m^{-2}·s^{-1} ($\gtrsim 10$ ergs/cm^2·s^{-1}) over a 10–30 km latitudinal width, as is required to account for a discrete auroral form. The question then arises as to whether the parameters $N \sim 5$ cm^{-3} and $K_{th} \sim 100$ eV are typical of the high altitude plasma accelerated above discrete auroras.

It is possible to determine whether such extreme values for N and K_{th} occur over discrete auroras from the rocket observations of the constant k relating $\mathcal{E}_p$ and $V_{\parallel}^2$ (see Section 4.4.3). Four values of k were obtained from the auroral electron data examined by Lyons *et al.* (1979): 1.0×10^{-10}, 1.8×10^{-10}, 4.7×10^{-10}, and 9.6×10^{-10} J m^{-2}·s^{-1}·V^{-2}. From the expression for k used in Equation (4.14), we obtain $k = 1.07 \times 10^{-23}$ $N/K_{th}^{1/2}$ J m^{-2}·s^{-1}·V^{-2}. Expressing N in cm^{-3} and K_{th} in eV for convenience gives $k = 2.7 \times 10^{-8}$ N(cm^{-3})/$K_{th}^{1/2}$ (eV) J m^{-2}·s^{-1}·V^{-2}. For $N = 5$ cm^{-3} and $K_{th} = 100$ eV, $k = 1.4 \times 10^{-8}$ J m^{-2}·s^{-1}·V^{-2}, which is more than an order of magnitude greater than the highest value of k found during the three rocket flights. It is thus reasonable to conclude that a high value of k, corresponding to a high density and a low thermal energy, is not generally responsible for the narrowness of discrete auroral forms.

The high altitude potential distribution used in the previous section has a region, where $\nabla \cdot \mathbf{E} < 0$, which is negatively charged and results in a field-aligned current out of the ionosphere requiring $V_{\parallel} > 0$. On the other hand, a region in the high altitude potential distribution, where $\nabla \cdot \mathbf{E} > 0$, is positively charged. A positively charged region can drive a current into the ionosphere, which cannot occur where $V_{\parallel} > 0$. Thus it is reasonable to consider a high altitude potential distribution where $\nabla \cdot \mathbf{E} < 0$ in one region, say near $x_i = 0$, but where $\nabla \cdot \mathbf{E} > 0$ in the vicinity of an $x_1 < 0$ and an $x_2 > 0$. The region where $\nabla \cdot \mathbf{E} < 0$ will drive a current out of the ionosphere having $V_{\parallel} > 0$, but if $\nabla \cdot \mathbf{E}$ at x_1 and x_2 is sufficiently large, there will be a current into the ionosphere near x_1 and x_2. Such an upper boundary potential distribution will force $V_{\parallel}$ to be zero at an x_i where $x_1 < x_i < 0$ and an x_i where $0 < x_i < x_2$, so that such a distribution can force precipitation regions to be of any width less than that of the inverted-V regions.

The right-hand panels of Figure 4.47 show the field-aligned potential difference and the precipitating electron energy flux versus x_i under the boundary condition $V_{\parallel} = 0$ at

ELECTRIC FIELDS

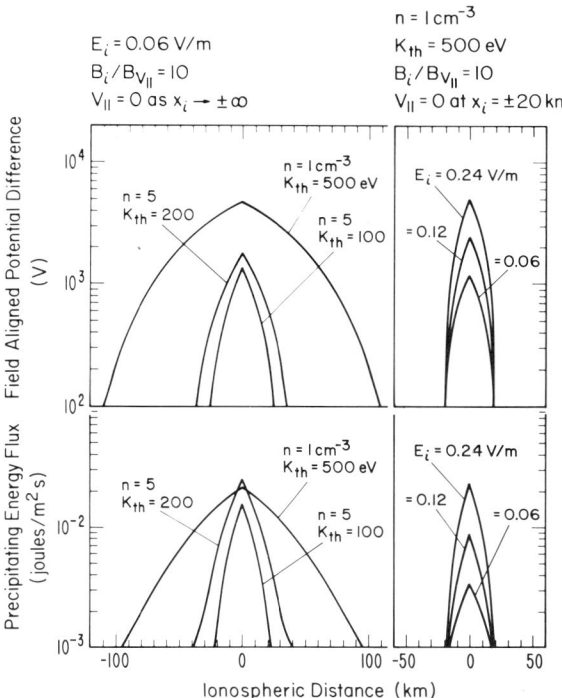

Fig. 4.47. Parameter variations that can give narrow precipitation regions. The left-hand panels give the field-aligned potential difference and the precipitating electron energy flux versus ionospheric distance for $N = 5$ cm^{-3} with $K_{th} = 200$ eV and 100 eV, compared with the results for $N = 1$ cm^{-3} and $K_{th} = 500$ eV. The magnitude of the high altitude electric field as mapped into the ionosphere, 0.06 V m^{-1}, and the magnetic field ratio $B_i/B_{V_\parallel} = 10$ are the same in all three cases. The boundary condition $V_\parallel = 0$ as $x_i \to \pm \infty$ is employed. The right-hand panels give the field-aligned potential difference and the precipitating electron energy flux versus ionospheric distance under the boundary condition $V_\parallel = 0$ at $x_i = \pm 20$ km. The results are shown for $E_i = 0.06$, 0.12, and 0.24 V m^{-1}. The other parameters were maintained at $N = 1$ cm^{-3}, $K_{th} = 500$ eV, and $B_i/B_{V_\parallel} = 10$ in all three cases (from Lyons, 1981a).

$x_i = \pm 20$ km. This boundary condition forces the precipitation region to have a latitudinal width < 40 km. The values $N = 1$ cm^{-3} and $K_{th} = 500$ eV used in the previous section have again been chosen. However, while an upper boundary electric field as mapped into the ionosphere E_i of 0.06 V m^{-1} adequately accounts for the inverted-V regions, the right-hand panels of Figure 4.46 show that $E_i = 0.06$ V m^{-1} does not give precipitating energy fluxes $\gtrsim 10^{-2}$ J m^{-2}·s^{-1} and field-aligned potential differences $\gtrsim 1$ kV as is required to account for discrete auroras. However, with a high altitude electric field that is four times larger ($E_i = 0.24$ V m^{-1}, which gives a total electric field discontinuity of 0.48 V m^{-1} at $x_i = 0$), precipitating electron energy fluxes and field-aligned potential differences are obtained that have the magnitudes and latitudinal widths associated with discrete auroras.

It is possible to determine the high altitude potential distribution above discrete auroras from rocket observations of precipitating electrons and electric fields. The observations should cover horizontal distances south of, over, and north of a discrete auroral form, and the aurora should not have varied significantly during the time over which the measurements were obtained.

An excellent set of such observations was obtained from the Polar-3 rocket flight (Maynard *et al.*, 1977; Evans *et al.*, 1977), one of the flights for which Lyons *et al.* (1979) found $\mathscr{E}_p = kV_\parallel^2$ as shown in Figure 4.41. The measured electric field was primarily normal to the arc, and the ionospheric electric potential obtained from integrating the measured normal electric field is shown versus horizontal distance along the rocket trajectory (which was approximately normal to the arc) in Figure 4.48. This electric potential is at ~ 200 km altitude and is thus assumed to be below any significant field-aligned potential difference. Given the total field-aligned potential difference, the high altitude potential distribution of the auroral generator can be obtained by subtracting the field-aligned potential difference from the measured ionospheric potential. The result of subtracting the field-aligned potential difference, inferred from the peak in the measured energy spectra of precipitating electrons, from the measured ionospheric potential gives the inferred upper boundary potential shown by the jagged, solid line in Figure 4.48.

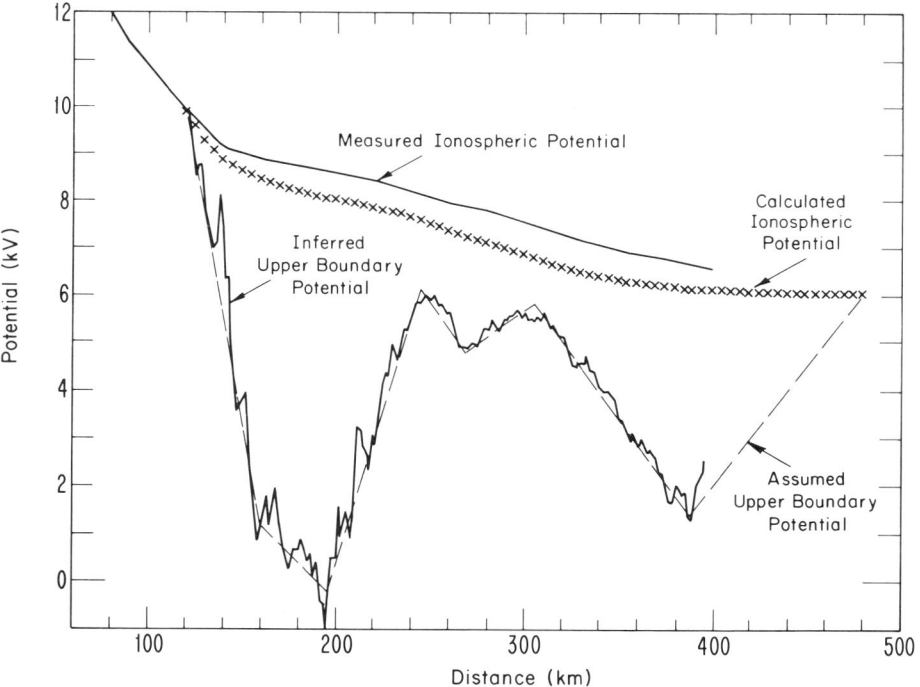

Fig. 4.48. Shown as a function of distance in the ionosphere, in a direction nearly normal to a large, discrete, auroral arc encountered between ~ 150 km and ~ 200 km, are the measured ionospheric potential from the Polar-3 auroral rocket flight (Maynard *et al.*, 1977; Evans *et al.*, 1977), the inferred and assumed upper boundary potentials above all field-aligned potential differences, and the ionospheric potential obtained by solving the current continuity equation in the ionosphere (from Lyons, 1981a).

The inferred upper boundary potential in Figure 4.48 shows an overall electric field change having $\nabla \cdot \mathbf{E} < 0$, consistent with that used in the previous section. However, the inferred upper boundary potential associated with the discrete aurora has considerable significant structure that was not included in the earlier analysis. This structure gives a total electric field change as mapped into the ionosphere of several hundred mV m^{-1} above the auroral arc, which is considerably greater than the electric field change of the convection boundary. However, this is just the magnitude of the electric field change that is required to obtain precipitating energy fluxes and $V_\parallel$'s of the magnitudes associated with discrete auroras under the $V_\parallel = 0$ boundary condition at the edges of the precipitation region (see right-hand panels of Figure 4.47). In addition, at the southern (equatorward) edge of the region of field-aligned potential differences (120 km), the upper boundary electric field had $\nabla \cdot \mathbf{E} > 0$. Lyons (1981a) inferred that this edge of the arc was sufficiently positively charged to drive a downward field-aligned current requiring $V_\parallel \approx 0$, and $V_\parallel = 0$ was inferred at this edge of the arc. The inferred high altitude electric field also had $\nabla \cdot \mathbf{E} > 0$ at the northern edge of the large arc. However, another region with $\nabla \cdot \mathbf{E} < 0$, associated with the weaker auroral form, was located ~ 200 km north of the center of the main arc, and $V_\parallel$ did not reduce to zero at the northern edge of the main arc. The electric field variation north of the weaker arc could not be obtained since the rocket re-entered the atmosphere.

Nevertheless, the overall inferred, high altitude, electric potential distribution in Figure 4.48 qualitatively agrees with that used to obtain the results in Figure 4.47. This suggests that structure in the high altitude, electric field distribution may be responsible for the latitudinal size of discrete auroral forms.

To determine whether the inferred structure in the upper boundary potential distribution can indeed account for the observed discrete auroral structure, Lyons (1981a) solved the current continuity Equation (4.12) for the ionospheric potential as in the previous section using the high altitude potential distribution inferred from the Polar-3 flight. The boundary condition $V_\parallel = 0$ at 120 km was employed, and the upper boundary potential was extrapolated from 390 to 480 km, as shown in Figure 4.48, to give a $V_\parallel = 0$ boundary condition at 480 km. Fortunately, the solution for the ionospheric potential at distances < 390 km was found not to depend significantly upon the assumed upper boundary potential beyond 390 km when a reasonable extrapolation is used. Rather than including every data point for the inferred upper boundary potential, the dashed line drawn through the data in Figure 4.47 was used for simplicity. It should be noted that, although the measured ionospheric potential was used for inferring the upper boundary potential, the measured ionospheric potential was not directly used for solving the current continuity equation.

The measured value of k, from Section 4.4.3, was used in specifying N and K_{th}. The value of k gives the ratio $N/K_{th}^{1/2}$ and not the absolute values of N and K_{th}. However, the ratio $N/K_{th}^{1/2}$ is more important than the absolute values in the solution of the continuity equation. (In fact, the analytical solution in Section 4.4.5 depends only upon the ratio $N/K_{th}^{1/2}$). The measurements showed $k = 4.7 \times 10^{-10}$ J m^{-2}·s^{-1}·V^{-2} at distances $\lesssim 225$ km, and k decreased to 1.0×10^{-10} J m^{-2}·s^{-1}·V^{-2} in the vicinity of 225 km and remained near that value for the remainder of the flight. In the calculations, k was reduced from the higher value to the lower value at 225 km. More specifically $N = 5$ cm^{-3}, $K_{th} = 865$ eV was used at distances < 225 km, and $N = 0.106$ cm^{-3}, $K_{th} = 865$ eV was used at 225 km and beyond. The values of N and K_{th} before 225 km were chosen to give the observed value of k for $V_{\parallel} \gtrsim 1$ kV and to give precipitating electron energy fluxes for $V_{\parallel} = 0$ in agreement with those observed south of main arc. The choice of changing N and not changing K_{th} at 225 km was completely arbitrary.

The calculated ionospheric potential is given by the x's in Figure 4.48. The results can be seen to agree with the measured ionospheric potential to within ~ 0.5 kV throughout the flight, and the non-zero $V_{\parallel}$'s at the northern edge of the main arc are reproduced by the model. That $V_{\parallel}$ did not reach zero between the two auroral forms appears to be a result of the detailed structure of the upper boundary potential distribution, and it is not appropriate to infer that $V_{\parallel}$ generally does not reach zero between auroral arcs.

Figure 4.49 shows a comparison between the measured electric field normal to the arc and the electric field obtained from differentiating the calculated ionospheric potential. The results agree to within a factor of 2. Even the overall structure of the increases in the ionospheric electric field poleward of the main arc, which result from a marked decrease in the ionospheric Pedersen conductivity in the calculations, is reproduced by the calculations. The measured and calculated precipitating electron energy fluxes are compared in Figure 4.50, and the agreement is good. However, the agreement must be good whenever the field-aligned potential difference is in agreement with the observations and the measured value for k is used.

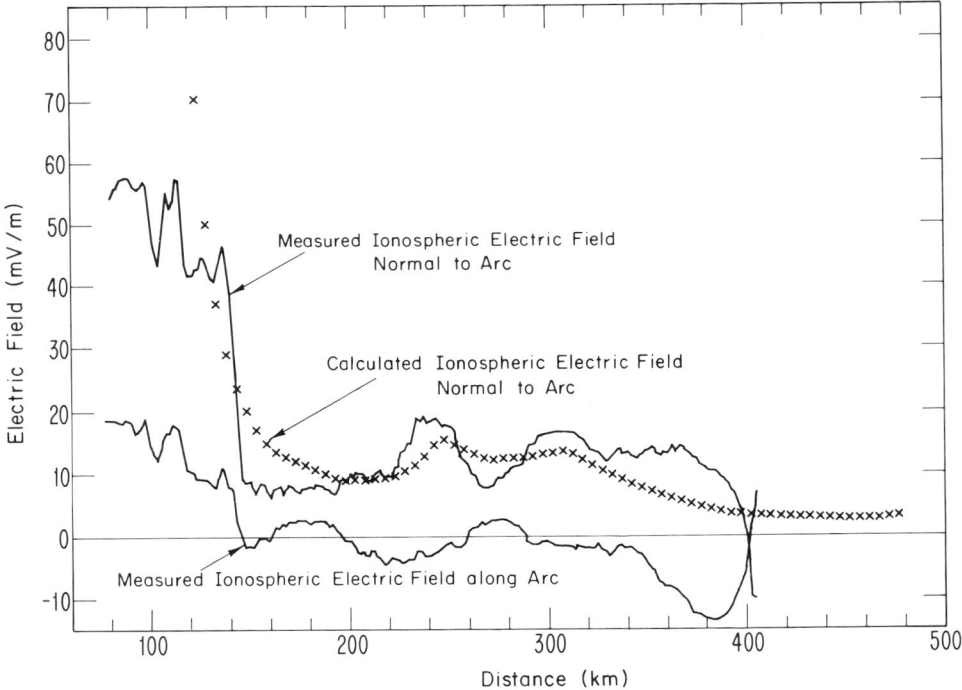

Fig. 4.49. Measured electric fields normal and parallel to the large, discrete auroral arc from the Polar-3 rocket flight and the normal electric field calculated from the solution to the ionospheric current continuity equation (from Lyons, 1981a).

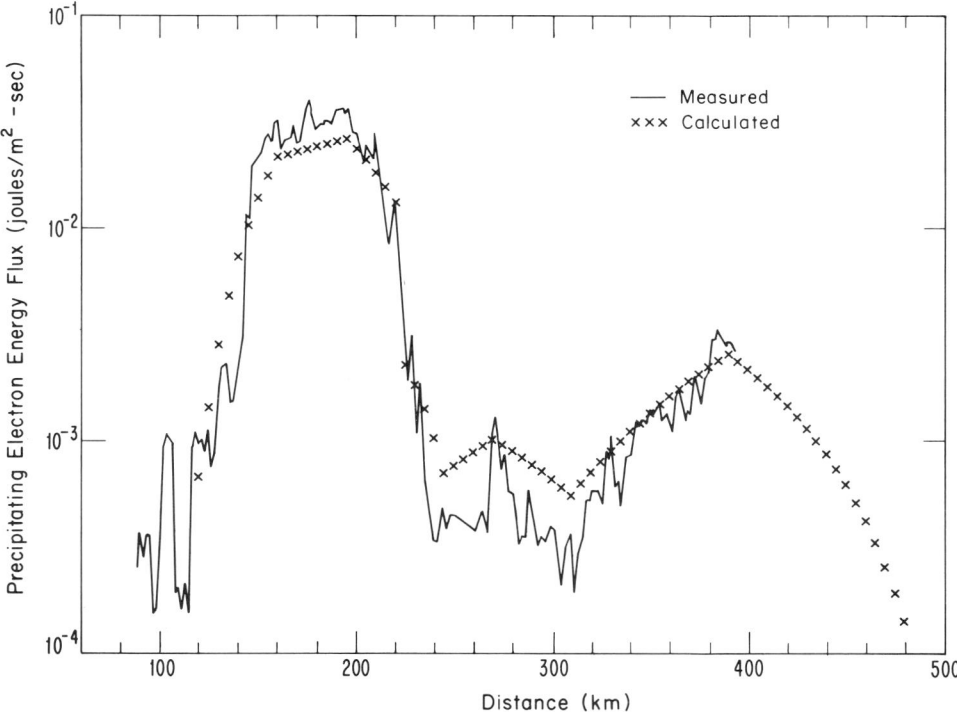

Fig. 4.50. Measured precipitating electron energy flux from the Polar-3 flight and the calculated precipitating electron energy flux from the solution to the ionospheric current continuity equation (from Lyons, 1981a).

The results in the last two sections show the width of large-scale, inverted-V precipitation regions to be a natural result of the current versus electric potential relations along auroral field lines and in the ionosphere. A negative divergence of the high altitude, magnetospheric electric field is required; however the inverted-V scale size need not be imposed by structure in the high altitude electric field.

On the other hand, the width of smaller scale, discrete auroral precipitation structure appears not to be a natural result of these current versus electric potential relations, except in the case of extremely high densities ($N \gtrsim 5$ cm^{-3}) and low thermal energies ($K_{th} \lesssim 200$ eV) for the high altitude electron population incident upon the field-aligned potential variation. However, using the same current versus electric potential relations, it has been shown that discrete auroral structure can result from structure in the high altitude magnetospheric electric field. The analysis shows discrete auroral forms to be approximately centered along field lines that connect to regions where the high altitude electric field has $\nabla \cdot \mathbf{E} < 0$, as is the case for the inverted-V precipitation regions. However the high altitude electric field must have $\nabla \cdot \mathbf{E}$ sufficiently greater than zero at the edges of discrete auroras to force their width to be narrow. In addition, the magnitude of the total, high altitude, electric field change giving $\nabla \cdot \mathbf{E} < 0$ must be several hundred mV m^{-1} as mapped into the ionosphere, which is several times greater than that required for the inverted-V precipitation regions.

References

Akasofu, S.-I., *Polar and Magnetospheric Substorms*, D. Reidel Publ. Co., Dordrecht, Holland (1968).

Akasofu, S.-I., E. W. Hones, Jr., S. J. Bame, J. R. Asbridge, and A. T. Y. Lui, 'Magnetotail and boundary layer plasma at a geocentric distance of $\sim 18 R_e$: Vela 5 and 6 observations', *J. Geophys. Res.* 78, 7257 (1973).

Albert, R. D., 'Nearly monoenergetic electron fluxes detected during a visible aurora', *Phys. Rev. Lett.* 18, 369 (1967a).

Albert, R. D., 'Energy and flux variations of nearly monoenergetic auroral electrons', *J. Geophys. Res.* 72, 5811 (1967b).

Alfvén, H., 'Some properties of magnetospheric neutral surfaces', *J. Geophys. Res.* 73, 4379 (1968).

Alfvén, H. and C.-G. Fälthammar, 'A new approach to the theory of the magnetosphere', *Cosmic Electrodynamics* 2, 78 (1971).

Anderson, H. R., 'Birkeland currents and auroral structure', *J. Geomag. Geoelec.* 30, 381 (1978).

Angerami, J. J. and D. L. Carpenter, 'Whistler studies of the plasmapause in the magnetosphere. 2. Electron density and total tube contact near the knee in magnetospheric ionization', *J. Geophys. Res.* 71, 711 (1966).

Antonova, E. E. and B. A. Tverskoy, 'Nature of the electron precipitation band of the "inverted V" type and of the Harang discontinuity in the evening sector of the auroral ionosphere', *Geomagnetism and Aeronomy* 15, 85 (1975).

Arnoldy, R. L., 'Auroral particle precipitation and Birkeland currents', *Rev. Geophys. and Space Phys.* 12, 217 (1974).

Arnoldy, R. L., P. B. Lewis, and P. O. Isaacson, 'Field-aligned auroral electron fluxes', *J. Geophys. Res.* 79, 4208 (1974).

Atkinson, G., 'Auroral arcs: result of the interaction of a dynamic magnetosphere with the ionosphere', *J. Geophys. Res.* 75, 4796 (1970).

Axford, W. I. and C. O. Hines, 'A unifying theory of high-latitude geophysical phenomena and geomagnetic storms', *Can. J. Phys.* 30, 1433 (1961).

Bahnsen, A., 'Recent techniques of observations and results from the magnetopause regions', *J. Atm. Terr. Phys.* 40, 235 (1978).

Banks, P. M., C. R. Chappell, and A. F. Nagy, 'A new model for the interaction of auroral electrons with the atmosphere: spectral degradation, backscatter, optical emission, and ionization', *J. Geophys. Res.* 79, 1459 (1974).

Bernstein, W., B. Hultqvist, and H. Borg, 'Some implications of low altitude observations of isotropic precipitation of ring current protons beyond the plasmapause', *Planet. Space Sci.* 22, 767 (1974).

Block, L. P., 'Potential double layers in the ionosphere', *Cosmic Electrodynamics* 3, 349 (1972).

Block, L. P., 'A double layer review', *Astrophys. Space Sci.* 55, 59 (1978).

Böstrom, R., 'Auroral electric fields', in *Aurora and Airglow*, B. M. McCormac (ed.), Reinhold Publ. Corp., New York, 293 (1967).

Burke, W. J., 'Electric fields, Birkeland currents, and electron precipitation in the vicinity of discrete auroral arcs', in *Physics of Auroral Arc Formation*, S.-I. Akasofu and J. R. Kan (eds.), Amer. Geophys. Union, Washington, D.C., 169 (1981).

Carpenter, D. L., 'Whistler evidence of a "knee" in the magnetospheric ionization density profile', *J. Geophys. Res.* 68, 1675 (1963).

Carpenter, D. L., 'Whistler studies of the plasmapause in the magnetosphere, 1. Temporal variations in the position of the knee and some evidence on plasma motions near the knee', *J. Geophys. Res.* 71, 693 (1966).

Carpenter, D. L., 'Whistler evidence of the dynamic behavior of the duskside bulge in the plasmasphere', *J. Geophys. Res.* 75, 3837 (1970).

Cauffman, D. P. and D. A. Gurnett, 'Double-probe measurements of convection electric fields with the Injun–5 satellite', *J. Geophys. Res.* 76, 6014 (1971).

Chapman, S., 'The electric current-systems of magnetic storms', *Terr. Magn. and Atmos. Elec.* 40, 349 (1935).

Chapman, S. and J. Bartels, *Geomagnetism*, Oxford University Press, New York, 302 (1940).

Chappell, C. R., K. K. Harris, and G. W. Sharp, 'A study of the influence of magnetic activity on the location of the plasmapause as measured by OGO 5', *J. Geophys. Res.* **75**, 50 (1970).

Chappell, C. R., K. K. Harris, and G. W. Sharp, 'The dayside of the plasmasphere', *J. Geophys. Res.* **76**, 7632 (1971).

Chiu, Y. T. and J. M. Cornwall, 'Electrostatic model of a quiet auroral arc', *J. Geophys. Res.* **85**, 543 (1980).

Chiu, Y. T. and M. Schulz, 'Self-consistent particle and parallel electrostatic field distributions in the magnetospheric-ionospheric auroral regions', *J. Geophys. Res.* **83**, 629 (1978).

Chiu, Y. T., A. L. Newman, and J. M. Cornwall, 'On the structures and mapping of auroral electrostatic potentials', *J. Geophys. Res.* **86**, 10029 (1981).

Choy, L. W., R. L. Arnoldy, W. Potter, P. Kintner, and L. J. Cahill, Jr., 'Field-aligned currents near an auroral arc', *J. Geophys. Res.* **76**, 8279 (1971).

Cloutier, P. A., H. R. Anderson, R. J. Park, R. R. Vondrak, R. J. Spiger, and B. R. Sandel, 'Detection of geomagnetically aligned currents associated with an auroral arc', *J. Geophys. Res.* **75**, 2595 (1970).

Coroniti, F. V. and C. F. Kennel, 'Polarization of the auroral electrojet', *J. Geophys. Res.* **77**, 2835 (1972).

Cowley, S. W. H., 'Plasma populations in a simple open model magnetosphere', *Space Sci. Rev.* **26**, 217 (1980).

Cowley, S. W. H. and M. Ashour-Abdalla, 'Adiabatic plasma convection in a dipole field: variation of plasma bulk parameters with L', *Planet. Space Sci.* **23**, 1527 (1975).

Cowley, S. W. H. and M. Ashour-Abdalla, 'Adiabatic plasma convection in a dipole field: electron forbidden-zone effects for a simple electric field model', *Planet. Space Sci.* **24**, 805 (1976a).

Cowley, S. W. H. and M. Ashour-Abdalla, 'Adiabatic plasma convection in a dipole field: proton forbidden-zone effects for a simple electric field model', *Planet. Space Sci.* **24**, 821 (1976b).

Cowley, S. W. H. and D. J. Southwood, 'Some properties of a steady-state geomagnetic tail', *Geophys. Res. Lett.* **7**, 833 (1980).

Cummings, W. D. and A. J. Dessler, 'Field-aligned currents in the magnetosphere', *J. Geophys. Res.* **72**, 1007 (1967).

Davis, L. R., O. E. Berg, and L. H. Meredith, 'Direct measurements of particle fluxes in and near auroras', *Space Res.*, *I*, North Holland Publ. Co., Amsterdam, 721 (1960).

Davis, T. N., 'The morphology of the polar aurora', *J. Geophys. Res.* **65**, 3497 (1960).

Davis, T. N., 'The morphology of the auroral displays of 1957–1958, 2. Detailed analyses of Alaska data and analyses of high-latitude data', *J. Geophys. Res.* **67**, 75 (1962).

DeCoster, R. J. and L. A. Frank, 'Observations pertaining to the dynamics of the plasma sheet', *J. Geophys. Res.* **84**, 5099 (1979).

Dungey, J. W., 'Conditions for the occurrence of electrical discharges in astrophysical systems', *Phil. Mag.* **44**, 725 (1953).

Dungey, J. W., 'Interplanetary magnetic field and the auroral zones', *Phys. Rev. Letters* **6**, 47 (1961).

Dungey, J. W., 'The structure of the exosphere or adventures in velocity space', in *Geophysics, The Earth's Environment*, C. DeWitt, J. Hieblot, and A. Lebeau (eds.), Gordon and Breach, New York, 505 (1963).

Dungey, J. W., 'The length of the magnetospheric tail', *J. Geophys. Res.* **70**, 1753 (1965).

Evans, D. S., 'A 10-cps periodicity in the precipitation of auroral zone electrons', *J. Geophys. Res.* **72**, 4281 (1967).

Evans, D. S., 'The observations of a near monoenergetic flux of auroral electrons', *J. Geophys. Res.* **73**, 2315 (1968).

Evans, D. S., 'Precipitating electron fluxes formed by a magnetic field aligned potential difference', *J. Geophys. Res.* **79**, 2853 (1974).

Evans, D. S., 'The acceleration of charged particles at low altitudes', in *Physics of Solar Planetary Environments*, D. J. Williams (ed.), Amer. Geophys. Union, Washington, D.C., 730 (1976).

Evans, D. S., N. C. Maynard, J. Troim, T. Jacobsen, and A. Egeland, 'Auroral vector electric field and particle comparisons, 2. Electrodynamics of an arc', *J. Geophys. Res.* **82**, 2235 (1977).

Fairfield, D. H., 'Magnetic field signatures of substorms on high-latitude field lines in the nighttime magnetosphere', *J. Geophys.* **78**, 1553 (1973).

Feldman, W. C., J. R. Asbridge, S. J. Bame, M. D. Montgomery, and S. P. Gray, 'Solar wind electrons', *J. Geophys. Res.* 80, 4181 (1975).
Fennell, J. F., 'Access of solar protons to the earth's polar caps', *J. Geophys. Res.* 78, 1036 (1973).
Fennell, J. F., P. F. Mizera, and D. R. Croley, 'Low energy polar cap electrons during quiet times', *Proc. Int. Conf. Cosmic Rays* 14, MG8–3, 1267 (1975).
Fennell, J. F., D. J. Gorney, and P. F. Mizera, 'Aurora particle distribution functions and their relationship to inverted V's and auroral arcs', in *Physics of Auroral Arc Formation*, S.-I. Akasofu and J. R. Kan (eds.), Amer. Geophys. Union, Washington, D.C., 91 (1981).
Forbes, T. G., E. W. Hones, S. J. Bame, J. R. Asbridge, G. Paschmann, N. Sckopte, and C. T. Russell, 'Evidence for the tailward retreat of a magnetic neutral line in the magnetotail during substorm recovery', *Geophys. Res. Lett.* 8, 261 (1981).
Frank, L. A. and K. L. Ackerson, 'Observations of charged particle precipitation into the auroral zone', *J. Geophys. Res.* 76, 3612 (1971).
Frank, L. A. and D. A. Gurnett, 'Distributions of plasmas and electric fields over the auroral zones and polar caps', *J. Geophys. Res.* 76, 6829 (1971).
Frank, L. A. and D. A. Gurnett, 'Direct observations of low-energy solar electrons associated with a type-III solar radio burst', *Solar Physics* 27, 446 (1972).
Frank, L. A., K. L. Ackerson, R. J. DeCoster, and B. G. Burek, 'Three-dimensional plasma measurements within the earth's magnetosphere', *Space Sci. Rev.* 22, 739 (1978).
Fridman, M. and J. Lemaire, 'Relationship between auroral electron fluxes and field-aligned electric potential difference', *J. Geophys. Res.* 85, 664 (1980).
Fukushima, N. and Oguti, T., 'Polar magnetic storms and geomagnetic bays, Appendix 1. A theory of S_D-field', *Rep. Ionos. Space Res. Japan* 7, 137 (1953).
Gold, T., 'Motions in the magnetosphere of the earth', *J. Geophys. Res.* 64, 1219 (1959).
Gorney, D. J., A. Clarke, D. Croley, J. Fennell, J. Luhmann, and P. Mizera, 'The distribution of ion beams and conics below 8000 km', *J. Geophys. Res.* 86, 83 (1981).
Grebowsky, J. M., A. J. Chen, and H. A. Taylor, Jr., 'High-latitude troughs and the polar cap boundary', *J. Geophys. Res.* 81, 690 (1976).
Gurnett, D. A. and L. A. Frank, 'Observed relationships between electric fields and auroral particle precipitation', *J. Geophys. Res.* 78, 145 (1973).
Harel, M., R. A. Wolf, P. H. Reiff, and H. K. Hillis, 'Study of plasma flow near the earth's plasmapause', U.S. Air Force Geophysics Lab. report, AFGL–TR–77–0286 (1977).
Harel, M., R. A. Wolf, P. H. Reiff, and R. W. Spiro, 'Quantitative simulation of a magnetospheric substorm, 1. Model logic and overview', *J. Geophys. Res.* 86, 2217 (1981).
Heelis, R. A., W. B. Hanson, and J. L. Burch, 'AE–C observations of electric fields around auroral arcs', in *Physics of Auroral Arc Formation*, S.-I. Akasofu and J. R. Kan (eds.), Amer. Geophys. Union, Washington, D.C., 154 (1981).
Heikkila, W. J., 'Impulsive penetration and viscous interaction', *Proceedings of Magnetospheric Boundary Layer Conference*, Alpach, ESAASP–148, 375 (1979).
Heikkila, W. J., R. J. Pellinen, C.-G. Fälthammar, and L. P. Block, 'Potential and inductive electric fields in the magnetosphere during auroras', *Planet. Space Sci.* 27, 1383 (1979).
Heppner, J. P., 'Electric field variations during substorms: OGO–6 measurements', *Planet. Space Sci.* 20, 1475 (1972a).
Heppner, J. P., 'Electric fields in the magnetosphere', in *Critical Problems of Magnetospheric Physics*, E. R. Dyer (ed.), IUCSTP Secretariat, c/o National Academy of Sciences, Washington, D.C., 107 (1972b).
Heppner, J. P., 'Empirical models of high-latitude electric fields', *J. Geophys. Res.* 82, 115 (1977).
Hoffman, R. A. and D. S. Evans, 'Field-aligned electron bursts at high latitudes observed by OGO–4', *J. Geophys. Res.* 73, 6201 (1968).
Hoffman, R. A. and C. S. Lin, 'Study of inverted V auroral precipitation events', in *Physics of Auroral Arc Formation*, S.-I. Akasofu and J. R. Kan (eds.), Amer. Geophys. Union, Washington, D.C., 80 (1981).
Holmgren, L.-Å, P. Christophersen, and W. Riedler, 'On the pitch-angle dependence of auroral electron fluxes in the keV range', *Physica Norvegica* 4, 85 (1970).

Hones, E. W., Jr., J. R. Asbridge, S. J. Bame, M. D. Montgomery, S. Singer, and S.-I. Akasofu, 'Measurements of magnetotail plasma flow with Vela 4B', *J. Geophys. Res.* 77, 5503 (1972).

Hultqvist, B., H. Borg, P. Christophersen, W. Riedler, and W. Bernstein, 'Energetic protons in the keV energy range and associated keV electrons observed at various local times and disturbance levels in the upper ionosphere', NOAA Technical Report ERL 305–SEL 29, U.S. Dept. of Commerce, Boulder, CO (1974).

Iijima, T. and T. A. Potemra, 'The amplitude distribution of field-aligned currents at northern high latitudes observed by Triad', *J. Geophys. Res.* 81, 2165 (1976a).

Iijima, T. and T. A. Potemra, 'Field-aligned currents in the dayside cusp observed by Triad', *J. Geophys. Res.* 81, 5971 (1976b).

Jaeger, E. F. and T. W. Speiser, 'Energy and pitch angle distributions for auroral ions using the current sheet acceleration model', *Astrophys. and Space Sci.* 28, 129 (1974).

Kamide, Y. and S.-I. Akasofu, 'The location of the field-aligned currents with respect to discrete auroral arcs', *J. Geophys. Res.* 81, 3999 (1976).

Kamide, Y. and G. Rostoker, 'The spatial relationships of field-aligned currents and auroral electrojets to the distribution of nightside auroras', *J. Geophys. Res.* 82, 5589 (1977).

Knight, L., 'Parallel electric fields', *Planet. Space Sci.* 21, 741 (1973).

Lemaire, J., 'Impulsive penetration of filamentary plasma elements into the magnetospheres of Earth and Jupiter', *Planet. Space Sci.* 27, 47 (1977).

Lemaire, J. and M. Scherer, 'Plasma sheet particle precipitation: a kinetic model', *Planet. Space Sci.* 21, 281 (1973).

Lemaire, J. and M. Scherer, 'Ionosphere-plasma sheet field-aligned currents and parallel electric fields', *Planet. Space Sci.* 22, 1485 (1974).

Lemaire, J. and M. Scherer, 'Field aligned distribution of plasma mantle and ionospheric plasma', *J. Atmos. Terr. Phys.* 40, 337 (1978).

Lennartsson, W., 'On high latitude convection field inhomogeneities, parallel electric fields and inverted V precipitation events', *Planet. Space Sci.* 25, 89 (1977).

Lennartsson, W., 'On the consequences of the interaction between the auroral plasma and the geomagnetic field', *Planet. Space Sci.* 28, 135 (1980).

Levy, R. H., H. E. Petschek, and G. L. Siscoe, 'Aerodynamic aspects of the magnetospheric flow', *AIAA Journal* 2, 2065 (1964).

Lin, C. S. and R. A. Hoffman, 'Characteristics of the inverted-V event', *J. Geophys. Res.* 84, 1514 (1979).

Lui, A. T. Y., E. W. Hones, Jr., F. Yasuhara, S.-I. Akasofu, and S. J. Bame, 'Magnetotail plasma flow during plasma sheet expansions: Vela 5 and 6 and IMP 6 observations', *J. Geophys. Res.* 82, 1235 (1977).

Lundblad, J. Å., F. Soraas, and K. Aarsnes, 'Substorm morphology of > 100 keV protons', *Planet. Space Sci.* 27, 841 (1979).

Lundin, R., 'Rocket observations of electron spectral and angular characteristics in an "inverted V" event', *Planet. Space Sci.* 24, 499 (1976).

Lundin, R. and I. Sandahl, 'Some characteristics of the parallel electric field acceleration of electrons over discrete auroral arcs as observed from two rocket flights', Symposium on European Rocket Research, Ajaccio, Corsica, 1978, ESAMSP–135, 125 (1978).

Lundin, R., B. Hultqvist, N. Pissarenko, and A. Zackarov, 'The plasma mantle: composition and other characteristics observed by means of the PROGNOZ–7 satellite', Kiruna Geophysical Institute preprint #81:1, Kiruna, Sweden (1981).

Lyons, L. R., 'Generation of large-scale regions of auroral currents, electric potentials, and precipitation by the divergence of the convection electric field', *J. Geophys. Res.* 85, 17 (1980).

Lyons, L. R., 'Discrete aurora as the direct result of an inferred, high-altitude generating potential distribution', *J. Geophys. Res.* 86, 1 (1981a).

Lyons, L. R., 'The field-aligned current versus electric potential relation and auroral electrodynamics', in *Physics of Auroral Arc Formation*, S.-I. Akasofu and J. R. Kan (eds.), Amer. Geophys. Union, Washington, D.C., 252 (1981b).

Lyons, L. R. and T. W. Speiser, 'Evidence for current-sheet acceleration in the geomagnetic tail', *J. Geophys. Res.* 87, 2276 (1982).

Lyons, L. R., D. S. Evans, and R. Lundin, 'An observed relation between magnetic field aligned electric fields and downward electron energy fluxes in the vicinity of auroral forms', *J. Geophys. Res.* 84, 457 (1979).

Maehlum, B. N. and H. Moestue, 'High temporal and spatial resolution observations of low energy electrons by a mother-daughter rocket in the vicinity of two quiescent auroral arcs', *Planet. Space Sci.* 21, 1957 (1973).

Maynard, N. C., D. S. Evans, B. Maehlum, and A. Egeland, 'Auroral vector electric field and particle comparisons, 1, Premidnight convection topology', *J. Geophys. Res.* 82, 2227 (1977).

McIlwain, C. E., 'Direct measurements of particles producing visible auroras', *J. Geophys. Res.* 65, 2727 (1960).

Meng, C.-I., A. T. Y. Lui, S. M. Krimigis, S. Ismail, and D. J. Williams, 'Spatial distribution of energetic particles in the distant magnetotail', *J. Geophys. Res.* 86, 5682 (1981).

Mizera, P. F. and J. F. Fennell, 'Signatures of electric fields from high and low altitude particle distributions', *Geophys. Res. Lett.* 4, 311 (1977).

Mizera, P. F. and J. F. Fennell, 'Satellite observations of polar, magnetotail lobe and interplanetary electrons at low energies', *Rev. Geophys. Space Phys.* 16, 147 (1978).

Mizera, P. F., D. R. Croley, Jr., and J. F. Fennell, 'Electron pitch-angle distributions in an inverted "V" structure', *Geophys. Res. Lett.* 3, 149 (1976).

Moore, T. E. and D. S. Evans, Distribution of energetic positive ion species above a diffuse midnight aurora, *J. Geophys. Res.* 84, 6443 (1979).

Nagata, T. and S. Kokubun, 'An additional geomagnetic daily variation field (S_q^p) in the polar region on a geomagnetically quiet day', *Rep. Ionos. Space Res. Japan* 16, 256 (1962).

Ness, N. F., 'The earth's magnetic tail', *J. Geophys. Res.* 70, 2989 (1965).

Nishida, A., 'Formation of the plasmapause, or magnetospheric plasma knee, by the combined action of magnetospheric convection and plasma escape from the tail', *J. Geophys. Res.* 71, 5669 (1966).

Nishida, A., 'Coherence of geomagnetic DP2 fluctuations with interplanetary magnetic variations', *J. Geophys. Res.* 73, 5549 (1968).

Nishida, A. and S. Kokubun, 'New polar magnetic disturbances: S_q^p, SP, DPC, and DP2', *Rev. Geophys. Space Phys.* 9, 417 (1971).

Nishida, A., N. Iwasaki, and T. Nagata, 'The origin of fluctuations in the equatorial electrojet; a new type of geomagnetic variation', *Ann. Geophys.* 22, 478 (1966).

O'Brien, B. J., 'Consideration that the source of auroral energetic particles is not a parallel electrostatic field', *Planet. Space Sci.* 18, 1821 (1970).

O'Brien, B. J. and D. L. Reasoner, 'Measurements of highly collimated short-duration bursts of auroral electrons and comparisons with existing auroral models', *J. Geophys. Res.* 76, 8258 (1971).

Park, C. G., 'Whistler observations of the interchange of ionization between the ionosphere and the protonosphere', *J. Geophys. Res.* 75, 4249 (1970).

Park, R. J. and P. A. Cloutier, 'Rocket-based measurements of Birkeland currents related to an auroral arc and electrojet', *J. Geophys. Res.* 76, 7714 (1971).

Paschmann, G., R. G. Johnson, R. D. Sharp, and E. G. Shelley, 'Angular distributions of auroral electrons in the energy range 0.8–16 keV', *J. Geophys. Res.* 77, 6111 (1972).

Pulliam, D. M., H. R. Anderson, K. Stamnes, and M. H. Rees, 'Auroral electron acceleration and atmospheric interactions: (1) Rocket-borne observations and (2) scattering calculations', *J. Geophys. Res.* 86, 2397 (1981).

Reid, G. C. and H. Leinbach, 'Low-energy cosmic-ray events associated with solar flares', *J. Geophys. Res.* 64, 1801 (1959).

Reid, G. C. and H. H. Sauer, 'Evidence for nonuniformity of solar proton precipitation over the polar caps', *J. Geophys. Res.* 72, 4383 (1967).

Rishbeth, H. and O. K. Garriott, *Introduction to Ionospheric Physics*, Academic Press, New York, pp. 132–139, p. 227 (1969).

Roederer, J. G., *Dynamics of Geomagnetically Trapped Radiation*, Springer-Verlag, New York (1970).

Rosenbauer, H. A., A. Grunwaldt, M. D. Montgomery, G. Paschmann, and N. Sckopke, 'HEOS 2 plasma observations in the distant polar magnetosphere: the plasma mantle', *J. Geophys. Res.* 80, 2723 (1975).

Schield, M. A. and L. A. Frank, 'Electron observations between the inner edge of the plasma sheet and the plasmapause', *J. Geophys. Res.* 75, 5401 (1970).

Schield, M. A., J. W. Freeman, and A. J. Dessler, 'A source for field-aligned currents at auroral latitudes', *J. Geophys. Res.* 74, 247 (1969).

Sharber, J. R., 'The continuous (diffuse) aurora and auroral-E ionization', in *Physics of Space Plasmas*, T. S. Chang, B. Coppi and J. R. Jasperse (eds.), Scientific Publishers, Cambridge, Mass. (1981) (in press).

Shelley, E. G., R. D. Sharp, and R. G. Johnson, 'Satellite observations of an ionospheric acceleration mechanism', *Geophys. Res. Lett.* 3, 654 (1976).

Silsbee, H. C. and E. H. Vestine, 'Geomagnetic bays, their frequency and current systems', *Terr. Magn. Atmos. Elec.* 47, 195 (1942).

Speiser, T. W., 'Particle trajectories in model current sheets, 1. Analytical solutions', *J. Geophys. Res.* 70, 4219 (1965).

Speiser, T. W., 'Particle trajectories in model current sheets, 2, applications to auroras using a geomagnetic tail model', *J. Geophys. Res.* 72, 3919 (1967).

Speiser, T. W., 'On the uncoupling of parallel and perpendicular particle motion in a neutral sheet', *J. Geophys. Res.* 73, 1112 (1968).

Speiser, T. W. and L. R. Lyons, 'Comparison of an analytical approximation for particle motion in a current sheet with precise numerical calculation', *J. Geophys. Res.* 1983 (in press).

Speiser, T. W. and N. F. Ness, 'The neutral sheet in the geomagnetic tail: its motion, equivalent currents, and field line connection through it', *J. Geophys. Res.* 72, 131 (1967).

Stern, D. P., 'A study of the electric field in an open magnetospheric model', *J. Geophys. Res.* 78, 7292 (1973).

Stern, D. P., 'The motion of a proton in the equatorial magnetosphere', *J. Geophys. Res.* 80, 595 (1975).

Stern, D. P., *Energetics of the magnetosphere*, presented at 1980 AGU Fall Meeting, NASA Technical Memorandum 82039, Goddard Space Flight Center, Greenbelt, Md. (1980).

Stern, D. P., 'One-dimensional models of quasi-neutral parallel electric fields', *J. Geophys. Res.* 86, 5839 (1981).

Swift, D. W., 'Possible mechanisms for formation of the ring current', *J. Geophys. Res.* 76, 2276 (1971).

Swift, D. W., 'The effects of the neutral sheet on magnetospheric dynamics', *J. Geophys. Res.* 82, 1288 (1977).

Swift, D. W. and D. A. Gurnett, 'Direct comparison between satellite electric field measurements and the visual aurora', *J. Geophys. Res.* 78, 7306 (1973).

Taylor, H. A., Jr., H. C. Brinton, and M. W. Pharo III, 'Contraction of the plasmasphere during geomagnetically disturbed periods', *J. Geophys. Res.* 73, 961 (1968).

Taylor, H. A., Jr., J. M. Grebowsky, and A. J. Chen, 'Ion composition irregularities and ionosphere-plasmasphere coupling: Observations of a high latitude ion trough', *J. Atmos. Terr. Phys.* 37, 613 (1975).

Temerin, M., M. H. Boehm, and F. S. Mozer, 'Paired electrostatic shocks', *Geophys. Res. Lett.* 8, 799 (1981).

Van Allen, J. A., J. F. Fennell, and N. F. Ness, 'Asymmetric access of energetic solar protons to the earth's north and south polar caps', *J. Geophys. Res.* 76, 4262 (1971).

Vasyliunas, V. M., 'A survey of low energy electrons in the evening sector of the magnetosphere with OGO−1 and OGO−3', *J. Geophys. Res.* 73, 2839 (1968).

Vasyliunas, V. M., 'The interrelationship of magnetospheric processes', in *Earth's Magnetospheric Processes*, B. M. McCormac (ed.), D. Reidel Publ. Co., Dordrecht, Holland, 29 (1972).

Volland, H., 'Models of global electric fields within the magnetosphere', *Ann. Geophys.* 31, 154 (1975).

Vondrak, R. R., 'Chatanika radar measurements of the electrical properties of auroral arcs', in *Physics of Auroral Arc Formation*, S.-I. Akasofu and J. R. Kan (eds.), Amer. Geophys. Union, Washington, D.C., 185 (1981).

Vondrak, R. R., H. R. Anderson, and R. J. Spiger, 'Rocket-based measurements of particle fluxes and currents in an auroral arc', *J. Geophys. Res.* 76, 7701 (1971).

West, H. I., Jr. and A. L. Vampola, 'Simultaneous observations of solar-flare electron spectra in interplanetary space and within earth's magnetosphere', *Phys. Rev. Lett.* **26**, 458 (1971).

Westerlund, L. H., 'The auroral electron energy spectrum extended to 45 eV', *J. Geophys. Res.* **74**, 351 (1969).

Whalen, B. A. and I. B. McDiarmid, 'Observations of magnetic-field-aligned auroral-electron precipitation', *J. Geophys. Res.* **77**, 191 (1972).

Whalen, B. A., J. R. Miller, and I. B. McDiarmid, 'Evidence for a solar wind origin of auroral ions from low-energy ion measurements', *J. Geophys. Res.* **76**, 2406 (1971).

Whipple, E. C., Jr., 'The signature of parallel electric fields in a collisionless plasma', *J. Geophys. Res.* **82**, 1525 (1977).

Winningham, J. D. and W. J. Heikkila, 'Polar cap auroral electron fluxes observed with Isis 1', *J. Geophys. Res.* **79**, 949 (1974).

Winningham, J. D., T. W. Speiser, E. W. Hones, Jr., R. A. Jeffries, W. H. Roach, D. S. Evans, and H. C. Stenbaek-Nielsen, 'Rocket-borne measurements of the dayside cleft plasma: The Tordo experiments', *J. Geophys. Res.* **82**, 1876 (1977).

Williams, D. J., 'Observations of significant magnetosheath antisolar energy flow', *J. Geophys. Res.* **84**, 2105 (1979).

Williams, D. J., 'Energetic ion beams at the edge of the plasma sheet: ISEE 1 observations plus a simple explanatory model', *J. Geophys. Res.* **86**, 5507 (1981).

Wolf, R. A., 'Calculations of magnetospheric electric fields', in *Magnetospheric Physics*, B. M. McCormac (ed.), D. Reidel Publ. Co., Dordrecht, Holland, 167 (1974).

Wolf, R. A., M. Harel, R. W. Spiro, G.-H. Voigt, P. H. Reiff, and C.-K. Chen, 'Computer simulation of inner magnetospheric dynamics for the magnetic storm of July 29, 1977', *J. Geophys. Res.* **87**, 5949 (1982).

Yeager, D. M. and L. A. Frank, 'Low-energy electron intensities at large distances over the earth's polar cap', *J. Geophys. Res.* **81**, 3966 (1976).

Zmuda, A. J. and J. C. Armstrong, 'The diurnal variation of the region with vector magnetic field charges associated with field-aligned currents', *J. Geophys. Res.* **79**, 2501 (1974).

Zmuda, A. J., J. C. Armstrong, and F. T. Heuring, 'Characteristics of transverse magnetic disturbances observed at 1100 km in the auroral oval', *J. Geophys. Res.* **75**, 4757 (1970).

CHAPTER 5

WAVE-PARTICLE INTERACTIONS

Note: Magnetospheric studies of the interactions between plasma waves and particles have generally been performed using Gaussian units. To allow the material presented here to be conveniently used in conjunction with other published material on magnetospheric wave-particle interactions, we have not converted equations from the literature into MKS units. Thus all equations in this chapter are in Gaussian units.

5.1. General Relations for Wave Growth and Particle Diffusion

Magnetospheric particle distributions are often unstable to the generation of plasma waves, and the waves can have important effects on the distribution of the particles. Kennel and Englemann (1966) and Kennel and Wong (1967a, b) have provided a convenient framework for studying magnetospheric wave-particle interactions. They developed general, and physically understandable, expressions for wave growth resulting from interactions with plasma particles and for particle diffusion in velocity space driven by plasma waves. These expressions are valid for arbitrary plasma waves propagating at arbitrary angles to a uniform magnetic field in a spatially homogeneous plasma. Great simplification is obtained in this approach, since it separates the problems of wave growth and of particle diffusion. Thus the cumbersome problems associated with treating both simultaneously are avoided. The crucial assumption made by Kennel and Englemann, one that is generally valid within the magnetosphere, is that the wave growth rate γ is much less than the real wave frequency ω. An additional assumption is that the wave amplitudes are sufficiently small that the time scale for changes in the particle distribution function due to interactions with the waves is much longer than the periods of the waves under consideration.

Taking the limit that $\gamma/\omega \ll 1$ gives what is referred to as the limit of resonant diffusion. That is, only particles of species 'j' satisfying the resonance condition

$$\omega - k_{\|}v_{\|} + n\Omega_j = 0 \qquad (5.1)$$

for some integer value of n interact with the waves. Here $k_{\|}$ is the component of the wave vector **k** along the ambient magnetic field direction, and $\Omega_j = |(q_j B)/(m_j c)|$ is the gyrofrequency of a plasma particle with charge q_j and mass m_j. The wave frequency ω is not an independent variable but is a unique function of **k**. In (5.1), $n = 0$ gives the Landau resonance where the parallel wave phase velocity equals the parallel particle velocity. All other values of n are cyclotron harmonic resonances where the wave frequency in the frame of reference moving along field lines with a particle is some harmonic of the particle's gyrofrequency. The principal cyclotron harmonic resonance, given by $n = -1$, occurs for a wave frequency in the particle's frame of reference equal to its gyrofrequency.

The resonance condition (5.1) depends upon the wave frequency ω and the parallel wave number $k_\parallel$. The dispersion relation for any particular wave mode relates ω to the wave vector $\mathbf{k}$ and must be known to solve (5.1) for the parallel particle velocity $v_{\parallel,\text{res}}$ for resonance with a particular wave. Resonance does not occur for all $v_\parallel$ for a given value of n, but only at values of $v_\parallel$ for which a wave mode exists that can propagate at a value of ω and $k_\parallel$ required for resonance.

The resonance condition defines a surface in velocity space for which resonance occurs at each value of n with a wave of a given ω and $k_\parallel$. Non-relativistically, this surface is a straight line in the $(v_\perp, v_\parallel)$-plane at constant $v_\parallel$. As the relativistic correction to the mass factor in the particle gyrofrequency becomes important, the line becomes an ellipse for the $n \neq 0$ cyclotron resonances.

Non-resonant wave-particle interactions and other non-linear effects are excluded by taking the limit of resonant diffusion. However such effects should dominate the linear, resonant effects only when the assumption $\gamma/\omega \ll 1$ is violated or where wave amplitudes are sufficiently large that significant diffusion occurs in a time less than a resonant particle's gyroperiod. Such situations generally do not occur in association with the natural generation of plasma waves in the magnetosphere or with the diffusion of particles interacting with natural magnetospheric plasma waves. Interesting non-linear phenomena do occur when artificial, coherent waves are injected into the magnetosphere (e.g., Helliwell and Katsufrakes, 1974); however such man-made waves do not generally have significant effects on the structure of magnetospheric particle distributions (Lyons and Williams, 1978; Thorne and Tsurutani, 1979, 1981).

To determine the effects of a wave with a particular ω and $k_\parallel$ on particles, and to determine whether such a wave is amplified or damped by the particles, it is necessary to determine how particles move in velocity space as they interact with the wave. Kennel and Englemann (1966) presented a simple, physical argument for determining how such diffusion occurs. If a particle gains a quantum of energy ΔK from a wave, then $\Delta K = \hbar \omega$. Similarly, the gain in parallel momentum $m \Delta v_\parallel = \hbar k_\parallel$. Assuming the energy gain is small compared with the total particle energy, then $\Delta K = m(v_\parallel \Delta v_\parallel + v_\perp \Delta v_\perp)$. This gives $v_\parallel \Delta v_\parallel + v_\perp \Delta v_\perp = \hbar \omega / m = \Delta v_\parallel \omega / k_\parallel$, which can be integrated to give

$$v_\perp^2 + (v_\parallel - \omega/k_\parallel)^2 = \text{const}. \tag{5.2}$$

Equation (5.2) defines surfaces in velocity space, referred to as 'single-wave characteristics', on which particles are constrained to move during their interaction with small amplitude waves. At each point in the $(v_\perp, v_\parallel)$-plane for which resonance occurs with a wave of specified ω and $k_\parallel$, particles are equally likely to move in either direction along these single-wave characteristics. Thus if each interaction with a wave causes a small increment in a particle's velocity, then particle motion in velocity space may be viewed as a diffusion process along the single-wave characteristics.

At the Landau resonance, where all resonant particles have $v_{\parallel,\text{res}} = \omega/k_\parallel$, the single-wave characteristics become $v_\perp = $ constant so that particle diffusion is solely in the $\pm v_\parallel$ direction. At the cyclotron resonances, where $v_{\parallel,\text{res}} = (\omega + n\Omega)/k_\parallel$, resonant particles diffuse in the directions tangent to the circles defined by (5.2). These circles are centered at $\omega/k_\parallel$ along the $v_\parallel$ axis and have a radius that increases with increasing $v_\perp$ for a given $v_{\parallel,\text{res}}$.

The net flux of particles along the single-wave characteristics at any point in velocity space will always be towards decreasing values of the distribution function f, under the assumption that particle motion in velocity space may be treated as a diffusion process. If this net flux is towards increasing particle energy, then there will be a net transfer of energy from the resonant wave distribution to the particles at that point in velocity space. Thus, under such conditions, resonant waves will be damped by their interaction with that portion of the particle distribution function. On the other hand, if the net flux of particles along the single-wave characteristics is towards decreasing particle energy, then that portion of the particle distribution function will cause wave growth.

The net particle diffusion along single-wave characteristics is illustrated in Figure 5.1 for particle distribution functions representative of some forms that naturally occur within the magnetosphere. The solid curves in each panel of the figure are contours of the particle distribution function f in the $(v_\perp\ v_\parallel)$-plane, the distribution function values for the contours satisfying $f_a > f_b > f_c$. The dashed circles and vertical lines are sample single-wave characteristics, and the thin, solid horizontal lines are sample values of $v_\parallel$ where resonance is assumed to occur with a propagating wave mode. Arrows tangent to the single-wave characteristics indicate the directions of net particle diffusion for the sample values of $v_{\parallel,\,\text{res}}$. Relativistic corrections to the gyrofrequency are assumed to be unimportant to the resonance condition.

The left-hand panels illustrate the net particle diffusion for the cyclotron ($n \neq 0$) resonances. The single-wave characteristics are thus circles, which are shown in the figure for an assumed value of $\omega/k_\parallel$. In the upper panel, a distribution function is illustrated that is isotropic in pitch angle and has f monotonically decreasing with increasing energy. It can be seen that the net particle diffusion is towards increasing particle energy throughout velocity space for such a distribution function. Thus any wave in cyclotron resonance with any part of such a particle distribution will be damped by the particles.

The lower left-hand panel shows a particle distribution representative of what is commonly observed in the radiation belts as a result of the atmospheric loss cone. This distribution function is anisotropic as a function of pitch angle and has a minimum in both directions along the magnetic field lines. It can be seen that if the pitch-angle anisotropy is sufficiently large, the net particle diffusion is towards decreasing energy at $v_\parallel$'s $< \omega/k_\parallel$ over a range of pitch angles surrounding the field line direction. This pitch-angle range is indicated by the shading in the figure, and this region in velocity space will amplify any wave resonant with that portion of the particle distribution. However, for each value of $v_{\parallel,\,\text{res}}$ for which wave growth occurs because of resonance with the shaded portion of the particle distribution function, wave damping also occurs because of resonance at higher values of $v_\perp$. Thus an integral along the contour in velocity space for which resonance occurs must be evaluated to determine whether the total particle distribution causes net wave growth of net wave damping. This integral for the net wave growth rate is given by Equation (5.3).

The right-hand panels of Figure 5.1 illustrate the net particle diffusion for the $n = 0$ Landau resonance. For this resonance, the single-wave characteristics become straight lines parallel to the $v_\parallel$ axis. Thus whenever $\partial f/\partial |v_\parallel| < 0$, the net particle diffusion is towards higher energies so that such regions in velocity space extract energy from any resonant waves. This is illustrated in the upper panel for a particle distribution isotropic in pitch angle with f monotonically decreasing with increasing energy. However, regions in

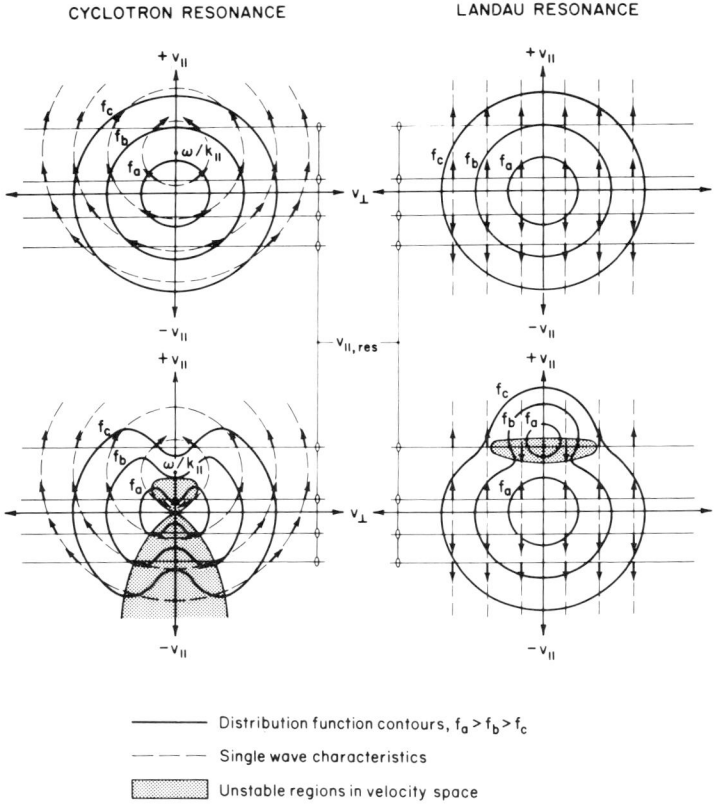

Fig. 5.1. Illustration of the net particle diffusion along single-wave characteristics for particle distribution functions representative of some forms that naturally occur within the magnetosphere. Arrows tangent to the single-wave characteristics indicate the direction of net particle diffusion for the sample values of $v_{\parallel,\,\text{res}}$ shown in the figure. The isotropic distribution in the upper panels causes wave damping for both the cyclotron and Landau resonances. The distribution in the lower left panel can be unstable to cyclotron wave growth, and the distribution in the lower right-hand panel can be unstable to Landau wave growth.

velocity space where $\partial f/\partial |v_\parallel| > 0$ can contribute to wave growth as illustrated in the lower right-hand panel of the figure. As in the case of the cyclotron resonances, an integral along the resonant contour in velocity space must be evaluated to determine whether or not the Landau resonance gives net wave growth or damping.

The above discussion indicates that the contribution to the growth rate of a wave from a particular point in velocity space at which resonance occurs should be proportional to the gradient of the particle distribution in the direction of increasing energy along the single-wave characteristics at that point in velocity space. Letting ζ be the angle between the tangent to a single-wave characteristic and the $v_\perp$-axis given by $\tan \zeta = dv_\parallel/dv_\perp$, letting ℓ_c be distance along the single-wave characteristic in the direction of increasing energy,

WAVE-PARTICLE INTERACTIONS

and letting positive $v_\parallel$ be in the direction of the parallel phase velocity $\omega/k_\parallel$ of the wave, we have

$$\frac{\partial f}{\partial \ell_c} = \frac{\partial f}{\partial v_\perp} \cdot \frac{\partial v_\perp}{\partial \ell_c} + \frac{\partial f}{\partial v_\parallel} \cdot \frac{\partial v_\parallel}{\partial \ell_c} = \frac{\partial f}{\partial v_\perp} \cos \zeta + \frac{\partial f}{\partial v_\parallel} \sin \zeta.$$

The relation between $d\ell$, $dv_\perp$, $dv_\parallel$, and ζ is illustrated in Figure 5.2. From Equation (5.2), we obtain

$$\tan \zeta = \frac{v_\perp}{(\omega/k_\parallel) - v_\parallel}$$

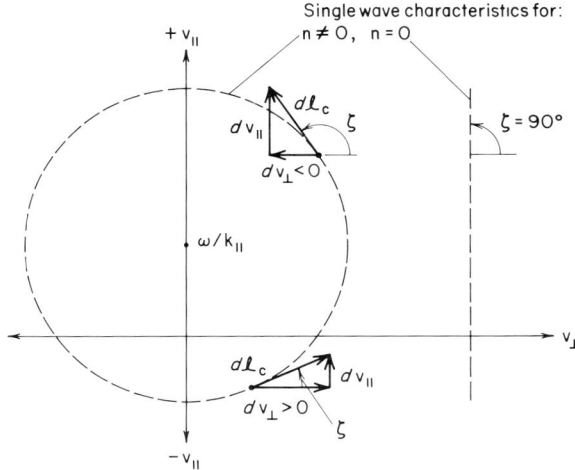

Fig. 5.2. Illustration of the relationships between the angle ζ, the angle that the tangent to a single-wave characteristic makes with the $v_\perp$-axis, and the elements of length $d\ell_c$, $dv_\perp$, and $dv_\parallel$ along the single-wave characteristics, the $v_\perp$-axis, and the $v_\parallel$-axis, respectively.

which gives

$$\frac{\partial f}{\partial \ell_c} = \frac{\cos \zeta}{(\omega/k_\parallel) - v_\parallel} \left[\left(\frac{\omega}{k_\parallel} - v_\parallel \right) \frac{\partial f}{\partial v_\perp} + v_\perp \frac{\partial f}{\partial v_\parallel} \right].$$

Noting that $\cos \zeta / [(\omega/k_\parallel) - v_\parallel]$ is positive for all $\mathbf{v}$, we obtain

$$\frac{\partial f}{\partial \ell_c} \propto \hat{G} f,$$

where the operator $\hat{G}$ is given by

$$\hat{G} = \left(\frac{\omega}{k_\parallel} - v_\parallel \right) \frac{\partial}{\partial v_\perp} + v_\perp \frac{\partial}{\partial v_\parallel}.$$

Since Equation (5.5) describes particle diffusion in velocity space, and magnetospheric particle distributions are often viewed as a function of pitch angle and energy, it is convenient to write (5.5) as a diffusion equation in particle speed v and pitch angle α. Following Lyons (1974a), we transform from $(v_\perp, v_\parallel)$-space to (v, α)-space using

$$\frac{\partial}{\partial v_\perp} = \sin\alpha \frac{\partial}{\partial v} + \frac{\cos\alpha}{v} \frac{\partial}{\partial \alpha}, \qquad \frac{\partial}{\partial v_\parallel} = \cos\alpha \frac{\partial}{\partial v} - \frac{\sin\alpha}{v} \frac{\partial}{\partial \alpha}.$$

This transformation gives

$$\hat{G} = \left(\frac{\cos\alpha}{v} - \frac{k_\parallel}{\omega}\right) \frac{\partial}{\partial \alpha} + \sin\alpha \frac{\partial}{\partial v}$$

and

$$\hat{G}_k + \frac{\omega - k_\parallel v_\parallel}{\omega v_\perp} = \frac{1}{\sin\alpha} \frac{\partial}{\partial \alpha} \sin\alpha \left(\frac{\cos\alpha}{v} - \frac{k_\parallel}{\omega}\right) + \frac{1}{v^2} \frac{\partial}{\partial v} v^2 \sin\alpha,$$

which allows (5.5) to be written in the form of a diffusion equation in spherical coordinates:

$$\frac{\partial f_j}{\partial t} = \nabla \cdot (\mathbf{D} \cdot \nabla f) = \frac{1}{v \sin\alpha} \frac{\partial}{\partial \alpha} \sin\alpha \left(D_{\alpha\alpha} \frac{1}{v} \frac{\partial f}{\partial \alpha} + D_{\alpha v} \frac{\partial f}{\partial v}\right) +$$
$$+ \frac{1}{v^2} \frac{\partial}{\partial v} v^2 \left(D_{v\alpha} \frac{1}{v} \frac{\partial f}{\partial \alpha} + D_{vv} \frac{\partial f}{\partial v}\right). \tag{5.7}$$

The elements of the diffusion matrix $\mathbf{D}$ are given by

$$D_{\alpha\alpha} = \sum_{n=-\infty}^{\infty} \int_0^\infty k_\perp \, dk_\perp D_{\alpha\alpha}^{nk_\perp}$$

$$D_{\alpha v} = D_{v\alpha} = \sum_{n=-\infty}^{\infty} \int_0^\infty k_\perp \, dk_\perp D_{\alpha v}^{nk_\perp} \tag{5.8}$$

$$D_{vv} = \sum_{n=-\infty}^{\infty} \int_0^\infty k_\perp \, dk_\perp D_{vv}^{nk_\perp},$$

where the pitch-angle diffusion coefficient for a given n and $k_\perp$ is given by

$$D_{\alpha\alpha}^{nk_\perp} = \lim_{\mathscr{V} \to \infty} \frac{\pi q_j^2}{(2\pi)^2 \mathscr{V} m_j^2} \left[\frac{-\sin^2\alpha - n\Omega_j/\omega}{\cos\alpha}\right]^2 \frac{\Theta_{nk}}{|v_\parallel - \partial\omega/\partial k_\parallel|}\bigg|_{k_\parallel = k_{\parallel,\text{res}}}, \tag{5.9}$$

and the mixed and speed diffusion coefficients for each n and $k_\perp$ are simply related to $D_{\alpha\alpha}^{nk_\perp}$ by

$$D_{\alpha v}^{nk_\perp} = D_{\alpha\alpha}^{nk_\perp} \left[\frac{\sin\alpha \cos\alpha}{-\sin^2\alpha - n\Omega_j/\omega} \right]\bigg|_{k_\| = k_{\|,\text{res}}}$$

$$D_{vv}^{nk_\perp} = D_{\alpha\alpha}^{nk_\perp} \left[\frac{\sin\alpha \cos\alpha}{-\sin^2\alpha - n\Omega_j/\omega} \right]^2 \bigg|_{k_\| = k_{\|,\text{res}}}$$

(5.10)

The resonant value of $k_\|$ is given by

$$k_{\|,\text{res}} = (\omega + n\Omega_j)/v_\|, \qquad (5.11)$$

and the term $|v_\| - \partial\omega/\partial k_\||^{-1}$ in (5.9) comes from transforming the integral over $k_\|$ in (5.5) to an integral over the argument, $k_\| v_\| - n\Omega_j - \omega$, of the delta function. Since ω is a function of $k_\|$, the argument of the delta function can equal zero at more than one value of $k_\|$ for a given n and $k_\perp$. When this occurs, dividing the $k_\|$ integration into intervals, each containing a zero of the delta function argument, shows that it is simply necessary to add the value of $D_{\alpha\alpha}^{nk_\perp}$ corresponding to each value of $k_{\|,\text{res}}$.

The above results for the quasi-linear diffusion matrix can be generalized to include relativistic effects by using Lerche's (1968) relativistic analogue of Kennel and Englemann's quasi-linear diffusion equation. Taking the limit of resonant diffusion, and transforming from $(p_\perp, p_\|)$-space to (p, α)-space, where $\mathbf{p} = \gamma_r m_j \mathbf{v}$ is the particle momentum and $\gamma_r = (1 - v^2/c^2)^{-1/2}$, yields the same relations for the relativistic diffusion coefficients $D_{\alpha\alpha}$, $D_{p\alpha}$, $D_{\alpha p}$, and D_{pp} as those given by (5.8)–(5.11) except for the following: The gyrofrequency Ω_j must everywhere be replaced by its relativistic value Ω_j/γ_r, and $D_{\alpha\alpha}^{nk_\perp}$ must be multiplied by m_j^2.

That such a simple relationship as (5.10) should exist between the elements of the diffusion matrix is easily understood from the geometry of the single-wave characteristics (5.2). Differentiating (5.2), using the resonance condition (5.10), and transforming to (v, α)-space, yields

$$\frac{\Delta v}{\Delta \alpha} = \frac{v \sin\alpha \cos\alpha}{-\sin^2\alpha - n\Omega_j/\omega}. \qquad (5.12)$$

Since

$$D_{\alpha\alpha} \sim \frac{(v\Delta\alpha)^2}{2\Delta t}, \qquad D_{\alpha v} = D_{v\alpha} \sim \frac{(v\Delta\alpha\Delta v)}{2\Delta t}, \quad \text{and} \quad D_{vv} \sim \frac{(\Delta v)^2}{2\Delta t},$$

the ratio of $\Delta v/\Delta \alpha$ given by (5.12) reproduces the relationship between the diffusion coefficients given by (5.10).

5.2. General Results from Cold Plasma Theory

To apply the results of the previous section to resonant interactions with a specific mode of plasma waves, it is necessary to specify the wave dispersion relation and the components of the wave electric field. Such a specification is quite straightforward for wave modes that can be described by cold plasma theory. The theory of cold plasma wave propagation as formulated by Stix (1962) allows us to evaluate the quasi-linear diffusion equation and the linear growth rate equation for resonant particle interactions with any wave mode accurately described by cold plasma theory.

We first define

$$\left.\begin{matrix}\mathscr{R}\\ \mathscr{L}\end{matrix}\right\} = 1 - \sum_j \frac{\omega_{pj}^2}{\omega} \frac{1}{(\omega \pm \epsilon_j \Omega_j)}$$

$$\left.\begin{matrix}\mathscr{S}\\ \mathscr{D}\end{matrix}\right\} = \frac{1}{2}(\mathscr{R} \pm \mathscr{L})$$
(5.13)

and

$$\mathscr{P} = 1 - \sum_j \frac{\omega_{pj}^2}{\omega^2}.$$

The frequency ω is then related to the wave vector $\mathbf{k}$ by means of the dispersion relation for cold plasma waves:

$$\tilde{\mu}^2 = \frac{(\mathscr{R}\mathscr{L} - \mathscr{P}\mathscr{S})\sin^2\theta + 2\mathscr{P}\mathscr{S} \pm [(\mathscr{R}\mathscr{L} - \mathscr{P}\mathscr{S})^2 \sin^4\theta + 4\mathscr{P}^2\mathscr{D}^2 \cos^2\theta]^{1/2}}{2(\mathscr{S}\sin^2\theta + \mathscr{P}\cos^2\theta)} \quad (5.14)$$

or equivalently

$$\tan^2\theta = -\frac{\mathscr{P}(\tilde{\mu}^2 - \mathscr{R})(\tilde{\mu}^2 - \mathscr{L})}{(\mathscr{S}\tilde{\mu}^2 - \mathscr{R}\mathscr{L})(\tilde{\mu}^2 - \mathscr{P})},$$

where $\tilde{\mu} = c|k|/\omega$ is the refractive index and θ is the wave normal angle (the angle between $\mathbf{B}$ and $\mathbf{k}$). The parallel kinetic energy $K_{\|,\,res} = \frac{1}{2} m_j v_{\|,\,res}^2$ for particle resonance can be written in terms of $\tilde{\mu}$ as

$$K_{\|,\,res} = \frac{1}{2} m_j c^2 \frac{(1 + n\Omega_j/\omega)^2}{\tilde{\mu}^2 \cos^2\theta}.$$

Using the wave polarizations for a cold plasma, Kennel and Wong (1967b) obtained the following expression for the weighting function $\Theta_{n\mathbf{k}}/W_{\mathbf{k}}$ in the linear growth rate expression (5.3):

$$\frac{\Theta_{n\mathbf{k}}}{W_{\mathbf{k}}} = \frac{1}{2\rho(\mathbf{k}, \omega)} \left| (\tilde{\mu}^2 - \mathscr{R})J_{n+\epsilon_j} + (\tilde{\mu}^2 - \mathscr{L})J_{n-\epsilon_j} - 2\epsilon_j M(\mathbf{k}, \omega) \frac{v_\|}{v_\perp} J_n \right|^2, \quad (5.15)$$

where

$$M(\mathbf{k}, \omega) = \frac{\tilde{\mu}^2 \cos\theta \sin\theta}{\mathcal{P} - \tilde{\mu}^2 \sin^2\theta} (\tilde{\mu}^2 - \mathcal{S}),$$

and

$$16\pi\rho(\mathbf{k}, \omega) = (\tilde{\mu}^2 - \mathcal{L})^2 \frac{\partial}{\partial \omega} \omega\mathcal{R} + (\tilde{\mu}^2 - \mathcal{R})^2 \frac{\partial}{\partial \omega} \omega\mathcal{L} +$$

$$+ 2M^2(\mathbf{k}, \omega) \frac{\partial}{\partial \omega} \omega\mathcal{P} + 2\left[\left(\frac{\mathcal{P}(\tilde{\mu}^2 - \mathcal{S})\cos\theta}{\mathcal{P} - \tilde{\mu}^2 \sin^2\theta}\right)^2 + \mathcal{D}^2\right].$$

The weighting function $\Theta_{n\mathbf{k}}$ can be expressed solely in terms of the magnitude of the Fourier transform of the wave magnetic field at each $\mathbf{k}$, $\mathbf{B}_\mathbf{k}$, by using the cold plasma relations (Lyons, 1974b):

$$|E_{\mathbf{k},\parallel}|^2 = \frac{|\mathbf{B}_k|^2}{\tilde{\mu}^2} \left[\left(\frac{\mathcal{D}}{(\tilde{\mu}^2 - \mathcal{S})}\right)^2 \left(\frac{\tilde{\mu}^2 \sin^2\theta - \mathcal{P}}{\tilde{\mu}^2 \cos\theta \sin\theta}\right)^2 + \left(\frac{\mathcal{P}}{\tilde{\mu}^2 \sin\theta}\right)^2\right]^{-1},$$

$$\left.\begin{array}{c}E_{\mathbf{k},R}\\E_{\mathbf{k},L}\end{array}\right\} = \frac{\tilde{\mu}^2 \sin^2\theta - \mathcal{P}}{\tilde{\mu}^2 \cos\theta \sin\theta}\left(1 \pm \frac{\mathcal{D}}{\tilde{\mu}^2 - \mathcal{S}}\right)\frac{E_{\mathbf{k},\parallel}}{\sqrt{2}}.$$

This gives

$$\Theta_{n\mathbf{k}} = \frac{|\mathbf{B}_k|^2}{\tilde{\mu}^2}\left[\left(\frac{\mathcal{D}}{\tilde{\mu}^2 - \mathcal{S}}\right)^2 \left(\frac{\tilde{\mu}^2 \sin^2\theta - \mathcal{P}}{\tilde{\mu}^2}\right)^2 + \left(\frac{\mathcal{P}\cos\theta}{\tilde{\mu}^2}\right)^2\right]^{-1}.$$

$$\cdot \left[\frac{\tilde{\mu}^2 \sin^2\theta - \mathcal{P}}{2\tilde{\mu}^2}\left(1 + \frac{\mathcal{D}}{\tilde{\mu}^2 - \mathcal{S}}\right) J_{n-\epsilon_j} +\right.$$

$$+ \frac{\tilde{\mu}^2 \sin^2\theta - \mathcal{P}}{2\tilde{\mu}^2}\left(1 - \frac{\mathcal{D}}{\tilde{\mu}^2 - \mathcal{S}}\right) J_{n+\epsilon_j} +$$

$$\left. + \epsilon_j \cot\alpha \sin\theta \cos\theta\, J_n \right]^2. \tag{5.16}$$

The above allows the quasi-linear diffusion equation (5.5, 5.7) to be evaluated once $|\mathbf{B}_k|^2$ as a function of $\mathbf{k}$ is specified for a wave distribution of interest.

5.3. $|\mathbf{B}_k|^2$ in Terms of the Measurable Wave Intensity

Wave measurements in space are generally of the real wave magnetic (or electric) field $\mathbf{B}_{\text{wave}}(\omega, \theta)$ as a function of frequency and wave normal angle. However, the Fourier transform of the wave magnetic field $\mathbf{B}_\mathbf{k}$ is required to evaluate the quasi-linear diffusion of resonant particles. In analogy with Equation (5.6), and assuming the wave distribution is uniform over $\mathscr{V}$ and cylindrically symmetric with respect to the ambient magnetic field, the total wave magnetic field $\mathbf{B}_{\text{wave}}(\mathbf{x}, t)$ as a function of position $\mathbf{x}$ and time can be expressed as a function of $\mathbf{B_k}$ as

$$\mathscr{V}|\mathbf{B}_{\text{wave}}(\mathbf{x}, t)|^2 = \frac{1}{(2\pi)^2} \int_{-\infty}^{\infty} \int_0^{\infty} |\mathbf{B_k}|^2 \, k_\perp \, dk_\perp \, dk_\parallel. \tag{5.17}$$

Any distribution of waves can be written as

$$|\mathbf{B}_{\text{wave}}(\omega, \theta)|^2 = B^2(\omega) g_\omega(\theta), \tag{5.18}$$

where $B^2(\omega)$ is the wave magnetic field intensity squared per unit frequency, $g_\omega(\theta)$ gives the variation of wave magnetic field energy with wave normal angle for each frequency, and $B^2(\omega)$ and $g_\omega(\theta)$ are normalized such that the total wave magnetic field can be expressed as

$$|\mathbf{B}_{\text{wave}}(\mathbf{x}, t)|^2 = \int_0^{\infty} B^2(\omega) \, d\omega \tag{5.19}$$

and

$$\int_0^{\pi} g_\omega(\theta) = 1.$$

We next write $|\mathbf{B_k}|^2$ in the form

$$|\mathbf{B_k}|^2 = \frac{\mathscr{V}}{N(\omega)} B^2(\omega) g_\omega(\theta), \tag{5.20}$$

where $N(\omega)$ is a normalization factor which ensures that the wave amplitude squared per unit frequency is given by $B^2(\omega)$. Substituting (5.20) into (5.17), and transforming the integral over $k_\perp$ and $k_\parallel$ to an integral over ω and $\tan \theta$ yields

$$|\mathbf{B}_{\text{wave}}(\mathbf{x}, t)|^2 = \int_0^{\infty} B^2(\omega) \cdot$$

$$\cdot \left\{ \frac{1}{(2\pi)^2} \frac{1}{N(\omega)} \int_{-\infty}^{\infty} g_\omega(\theta) \left| J\left(\frac{k_\perp, k_\parallel}{\omega, \tan \theta}\right) \right| k_\perp \, d(\tan \theta) \right\} d\omega, \tag{5.21}$$

where $\int_{-\infty}^{\infty} d(\tan \theta)$ covers $180° \leq \theta \leq 90°$ for $\tan \theta \leq 0$ and $0° \leq \theta \leq 90°$ for $\tan \theta \geq 0$, and $J([k_\perp, k_\parallel]/[\omega, \tan \theta])$ is the Jacobian of the transformation which can be evaluated from the wave dispersion relation. Comparing (5.21) with (5.19), we see that the bracketed quantity in (5.21) must be equal to 1. Therefore,

$$N(\omega) = \frac{1}{(2\pi)^2} \int_{-\infty}^{\infty} g_\omega(\theta) \left| J\left(\frac{k_\perp, k_\parallel}{\omega, \tan \theta}\right) \right| k_\perp \, d(\tan \theta). \tag{5.22}$$

Thus for $|\mathbf{B}_{\text{wave}}|^2$ in the form given by (5.18), $|\mathbf{B_k}|^2$ is given by (5.20), and $N(\omega)$ is given by (5.22).

The Jacobian $N(\omega)$ involves the evaluation of four derivatives:

$$J\left(\frac{k_\perp, k_\parallel}{\omega, \tan\theta}\right) = \frac{\partial k_\perp}{\partial \omega}\bigg|_{\tan\theta} \cdot \frac{\partial k_\parallel}{\partial \tan\theta}\bigg|_\omega - \frac{\partial k_\perp}{\partial \tan\theta}\bigg|_\omega \cdot \frac{\partial k_\parallel}{\partial \omega}\bigg|_{\tan\theta}.$$

However, $(k_\perp, k_\parallel)$-space can be transformed to $(\omega, \tan\theta)$-space in two steps by transforming one variable at a time. This can be done in four different ways to yield the following equivalent expressions for the Jacobian which involve only two derivatives:

$$\begin{aligned}J\left(\frac{k_\perp, k_\parallel}{\omega, \tan\theta}\right) &= \frac{\partial k_\perp}{\partial \omega}\bigg|_{k_\parallel} \cdot \frac{\partial k_\parallel}{\partial \tan\theta}\bigg|_\omega = \frac{\partial k_\perp}{\partial \tan\theta}\bigg|_\omega \cdot \frac{\partial k_\parallel}{\partial \omega}\bigg|_{k_\parallel \tan\theta} \\ &= \frac{\partial k_\parallel}{\partial \omega}\bigg|_{k_\perp} \cdot \frac{\partial k_\parallel}{\partial \tan\theta}\bigg|_\omega = \frac{\partial k_\parallel}{\partial \tan\theta}\bigg|_\omega \cdot \frac{\partial k_\perp}{\partial \omega}\bigg|_{k_\parallel \tan\theta}.\end{aligned} \quad (5.23)$$

5.4. Auroral Kilometric Radiation (AKR)

We now apply the general results of the previous sections to three modes of plasma waves that are important to magnetospheric physics and that have dispersion and polarization properties describable by cold plasma theory. The intense electromagnetic radiation referred to as auroral kilometric radiation (AKR) offers an excellent example of how the very general results of this chapter can be applied to understand the basic causes of wave generation without having to perform detailed growth rate calculations.

AKR was first observed as intense electromagnetic radiation propagating away from the earth in the frequency range 30–700 kHz, the maximum intensity observed being near 100–300 kHz (Dunkel et al., 1970; Brown, 1973; Gurnett, 1974). Gurnett (1974) found the total wave power to be as high as 10^9 W. Thus he concluded that the earth is an intense planetary radio source, the AKR intensity being comparable with that of the decametric radio emissions from Jupiter. The wavelengths of the radiation were found to be in the kilometric range, so that Gurnett referred to the emission as 'terrestrial kilometric radiation'. An example of the power spectra of the AKR electric field from IMP–8 at a radial distance of $25.2R_e$ in the local evening is shown in Figure 5.3. The many-order-of-magnitude enhancement of the radiation above the receiver noise level as shown in Figure 5.3 was found to be typical of AKR. Gurnett detected the wave magnetic field simultaneously with the wave electric field, thus showing that AKR is electromagnetic.

Gurnett (1974) and Kurth et al. (1975) noted that AKR appeared to be associated with discrete aurora and to emanate from the low altitude ($\lesssim 10\,000$ km) auroral region; hence Kurth et al. (1975) referred to this emission as 'auroral kilometric radiation'. Additional evidence for such an association of AKR with aurora has been presented by Alexander and Kaiser (1976), Green et al. (1977), and Gallagher and Gurnett (1979).

Direct measurements of AKR within the auroral source region have now been obtained from measurements on the low altitude (apogee 3550 km), polar-orbiting ISIS–1 satellite. Benson and Calvert (1979) found that AKR is generated within regions of local depletions of the electron density, where the ratio of $\omega_{pe}/\Omega_e < 0.2$. They and Calvert (1981a)

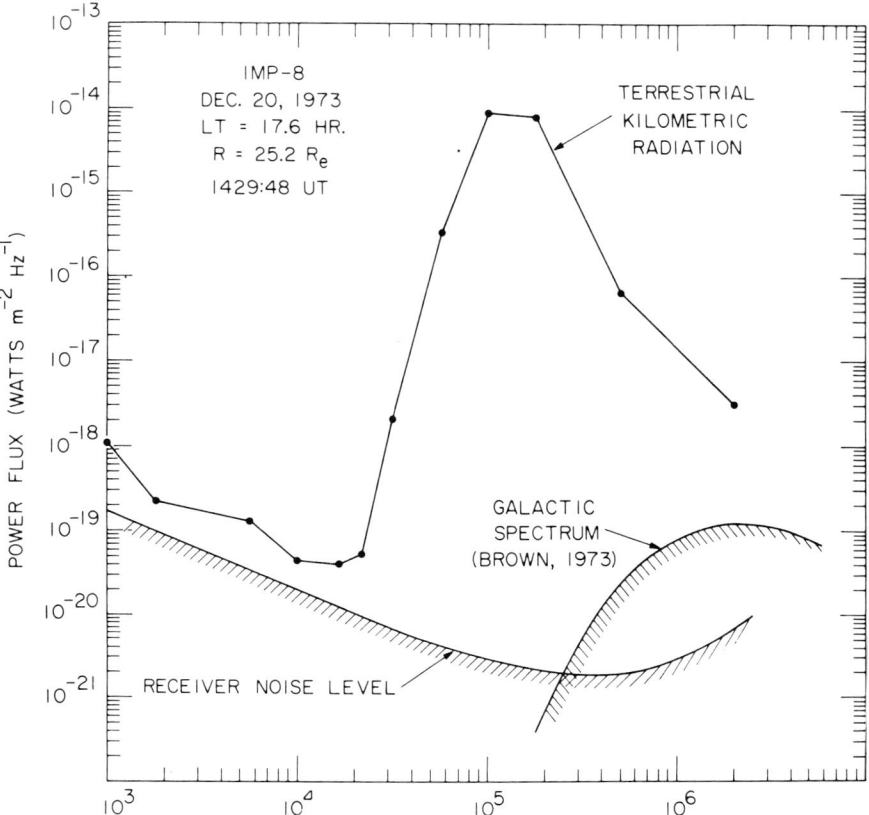

Fig. 5.3. An example of the power spectra of the auroral kilometric radiation electric field from IMP–8 at a radial distance of $25.2 R_e$ in the local evening (from Gurnett, 1974). (© by American Geophysical Union)

found that the radiation was generated just above the local electron gyrofrequency at wave normal angles nearly normal to the ambient magnetic field. In addition they found a small gap between Ω_e and the lowest frequency of the AKR emissions, indicating that AKR is generated above the cutoff frequency ω_x for the right-hand (extraordinary) mode. These observations are consistent with cold plasma theory, since wave propagation is not allowed between the upper hybrid frequency ω_{UH} and ω_x, where $\Omega_e < \omega_{UH} < \omega_x$ but ω_x is only slightly above Ω_e for $(\omega_{pe}/\Omega_e)^2 \ll 1$.

Benson et al. (1980) compared AKR observations with auroral electron observations from the ISIS–1 satellite. They concluded that it is necessary to have keV electron precipitation of the form associated with discrete aurora and inverted-V precipitation regions, in addition to the density depletion, in order to generate AKR. This analysis thus confirmed the earlier inferences of the association of AKR with discrete aurora.

Calvert (1981b) extended the analysis of the density depletions associated with AKR, further demonstrating the importance of these depletions in controlling AKR. He found values of ω_{pe}/Ω_e as low as ~ 0.03, this minimum occurring at a geocentric altitude between $+1.5-2.0R_e$. For AKR emissions just above Ω_e, the frequencies corresponding to the $1.5-2.0R_e$ altitude region are $\sim 200-300$ kHz. This frequency range corresponds to the peak intensity of AKR emissions, suggesting that the peak intensities are generated in the region of the minimum values of ω_{pe}/Ω_e. In addition, Calvert found that a ratio of $\omega_{pe}/\Omega_e \lesssim 0.1$ extends from 1.3 to $3R_e$ within the density depletion regions, corresponding to local electron gyrofrequencies of 50–600 kHz. This frequency range very nearly covers the range of large AKR intensities. These observations strongly suggest that the existence of the density depletion regions in the topside ionosphere, and in particular the ratio of ω_{pe}/Ω_e, plays a crucial role in the generation of AKR.

Figure 5.4 shows the solution of the cold plasma dispersion relation for wave normal angles near 90° and a ratio of $\omega_{pe}/\Omega_e = 0.1$, a typical value for the density depletion regions. The $L-O$ mode is left-hand, circularly polarized for parallel wave propagation, and the wave dispersion relation becomes independent of the ambient magnetic field; i.e., $\omega = |k|c$, for $\theta = 90°$ (ordinary mode). The $R-X$ mode is right-hand, circularly polarized for parallel wave propagation, and its dispersion relation is dependent upon the ambient magnetic field at $\theta = 90°$ (extraordinary mode). The $L-O$ mode has only one high frequency branch which is continuous in the vicinity of Ω_e. The $R-X$ mode, however, has a high and a low frequency branch separated by a stop band (shaded in Figure 5.4) which extends from ω_{UH} to ω_x. For $(\omega_{pe}/\Omega_e)^2 \ll 1$, this stop band is narrow and just above Ω_e as shown in the figure. The observed frequency distribution of AKR, having a low frequency cutoff just above Ω_e, is thus consistent with propagation in the high frequency branch of the $R-X$ mode.

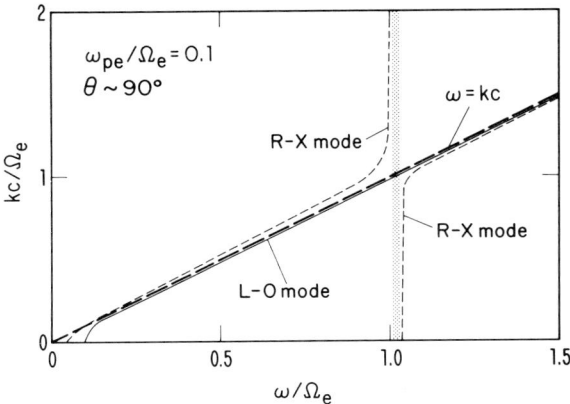

Fig. 5.4. Schematic illustration of the solution of the cold plasma dispersion relation for wave normal angles near 90° and $\omega_{pe}/\Omega_e = 0.1$. Notice that the $R-X$ mode has two branches separated by a stop band just above Ω_e (from Dusenbery and Lyons, 1982). (© by American Geophysical Union)

Owing to the association of AKR with discrete auroral electron precipitation, Wu and Lee (1979) examined the distribution of auroral electrons and suggested that this distribution is indeed unstable to the direct generation of AKR in the high frequency branch of the $R-X$ mode within the density depletion regions. This generation process was analyzed in more detail by Omidi and Gurnett (1982), Wu et al. (1982) and Dusenbery and Lyons (1982). We follow Dusenbery and Lyons's analysis of Wu and Lee's proposal in the following discussion.

Figure 5.5a shows an electron distribution function from the S3−3 satellite (from Croley et al., 1978) typical of that observed as a result of the parallel electric field acceleration along auroral field lines. This distribution was measured at 7330 km, an altitude well within the range from which AKR is generated. The nearly circular ellipse separates particles that are turned around by the parallel potential difference above the satellite and are thus trapped below the top of the parallel electric field region, from the particles at higher energies that can surmount the total parallel potential difference along the field line. The hyperbola demarcates the atmospheric loss cone, where the effects of electron acceleration by the field-aligned potential difference below the satellite have been included.

The rectangles in Figure 5.5a illustrate potential sources of free energy for wave growth within the distribution function. In the downward direction, shown as positive $v_\parallel$'s, $\partial f/\partial v_\parallel > 0$ just inside the ellipse. In addition, $\partial f/\partial v_\perp > 0$ throughout the rectangle in the downward direction. This region of $\partial f/\partial v_\perp > 0$ results from the acceleration of all downgoing electrons by the same parallel energy increment together with the particle propagation into a region of increased magnetic field strength which causes the particle pitch angles to approach 90°. In the upward direction, there is a large region near the edge of the loss cone where $\partial f/\partial v_\perp > 0$.

From Equation (5.2), we know that particles in cyclotron resonance with waves are constrained to diffuse along single-wave characteristics in velocity space which are circles centered at $\omega/k_\parallel$ along the $v_\parallel$-axis. From Figure 5.4, we see that $\omega/k_\parallel > c$ for the upper frequency branch of the $R-X$ mode so that $\omega/k_\parallel$ is much greater than the speeds of all relevant auroral electrons. Thus the single-wave characteristics for cyclotron resonance of AKR with auroral electrons are essentially straight lines parallel to the $v_\perp$-axis, and the $n = 0$ Landau resonance cannot occur between auroral electrons and AKR. Sample single-wave characteristics passing through potentially unstable regions of the auroral electron distribution function are shown in Figure 5.5b; arrows on these characteristics indicate the direction of net particle diffusion. Since the characteristics are normal to the $v_\parallel$-axis, it is apparent that regions of $\partial f/\partial v_\parallel > 0$ cannot affect wave growth. On the other hand, the regions of $\partial f/\partial v_\perp > 0$ can cause wave growth if wave propagation is possible for waves that can resonate with those regions of velocity space. AKR has $\omega \approx \Omega_e$, and from the analysis of wave growth in Section 5.1, we found that such waves have resonant parallel velocities $v_{\parallel, \text{res}} \ll \omega/k_\parallel$ for the cyclotron resonances and growth rates that are dependent upon $\partial f/\partial v_\perp$ and independent of $\partial f/\partial v_\parallel$. This suggests that the regions of $\partial f/\partial v_\perp > 0$ in the auroral electron distribution function may be important for the growth of AKR.

Before determining whether AKR can resonate with the region of the auroral electron distribution where $\partial f/\partial v_\perp > 0$, let us examine the surfaces in velocity space given by Equation (5.1) for which resonance can occur as a function of ω and $k_\parallel$. From (5.1), it is seen that for $|\omega - n\Omega_e| \ll \Omega_e$, the relativistic mass correction to the gyrofrequency can be important in determining the resonant $v_\parallel$'s, even for $v_\parallel \ll c$. Including the relativistic mass correction by replacing Ω by $\Omega(1 - v^2/c^2)^{1/2}$, the resonance condition (5.1) may be written in the form of an ellipse in velocity space:

$$\frac{(v_\parallel - v_{\parallel 0})^2}{a^2} + \frac{v_\perp^2}{(1 + h_\parallel^2)a^2} = 1, \tag{5.24}$$

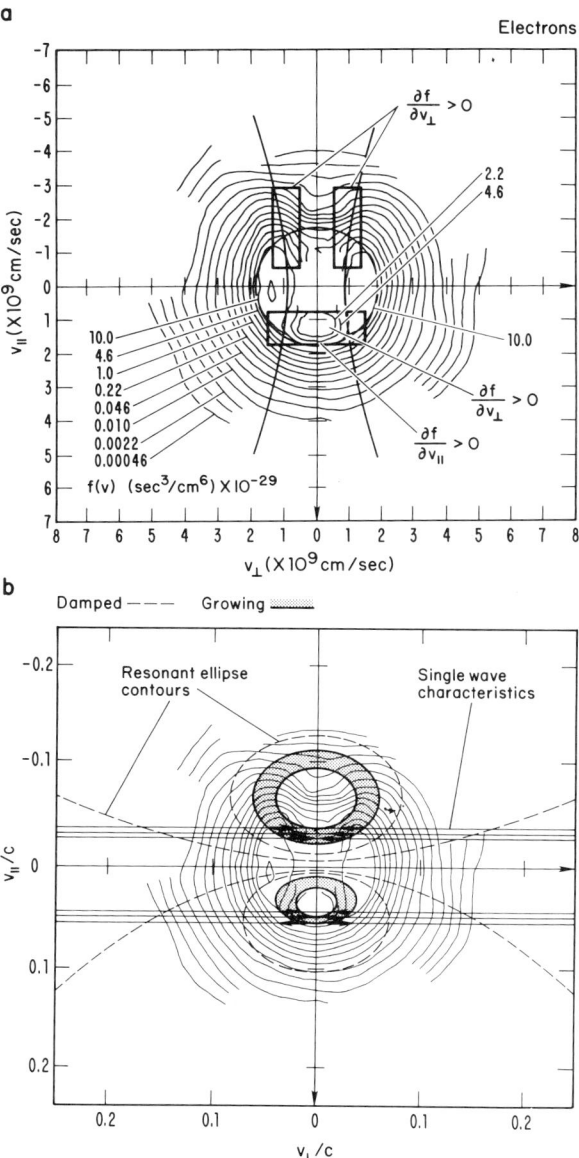

Fig. 5.5(a). An electron distribution function from the S3−3 satellite at 7330 km altitude (Croley *et al*. 1978) typical of that observed as a result of the parallel electric field acceleration along auroral field lines. The ellipse separates particles that are trapped below the top of the parallel electric field region from particles that can surmount the total parallel potential difference. The hyperbola demarcates the atmospheric loss cone. The rectangles illustrate potential sources of free energy for wave growth (from Dusenbery and Lyons, 1982). (b) Sample single-wave characteristics passing through the potentially unstable regions of the auroral electron distribution function. Arrows on the characteristics indicate the direction of net particle diffusion. Sample resonant ellipses are also schematically illustrated. Waves associated with ellipses outside the shaded regions are most likely damped, while waves associated with ellipses within the shaded region are most likely amplified (from Dusenbery and Lyons, 1982).
(© by American Geophysical Union)

where $h_\parallel = ck_\parallel/n\Omega$, and the ellipse is centered at $v_{\parallel 0}$ given by

$$\left|\frac{v_{\parallel 0}}{c}\right| = \frac{h_\parallel}{1+h_\parallel^2}\left|\frac{\omega}{n\Omega}\right|$$

and where the semi-minor axis 'a' is given by

$$\frac{a^2}{c^2} = \frac{1}{(1+h_\parallel^2)^2}\left[(1+h_\parallel^2) - \left(\frac{\omega}{n\Omega}\right)^2\right].$$

The semi-minor axis $a < |v_{\parallel 0}|$ for $\omega > |n\Omega|$.

The upper frequency branch of the $R-X$ mode has $\omega/k_\parallel > c$ so that for $\omega \approx |n\Omega|$, $|h_\parallel| < 1$. For $\omega/k_\parallel \gg c$, we have $|h_\parallel| \ll 1$. Under such conditions $|v_{\parallel 0}/c|$ can become small enough that the resonant ellipse defined by (5.24) can be centered within the region of velocity space of the auroral electrons. Such resonant ellipses are schematically illustrated in Figure 5.5b along with the single-wave characteristics, and it can be seen that these ellipses can be entirely within the region of velocity space where $\partial f/\partial v_\perp > 0$. (For the purposes of this discussion, the region in velocity space between the two rectangles in the upgoing direction where $\partial f/\partial v_\perp$ is small, but $\geqslant 0$, is included as part of the region where $\partial f/\partial v_\perp > 0$.) For such resonant ellipses, $\hat{G}$ is positive over the entire surface of resonance in velocity space. Thus the growth rate (Equation (5.3)) must be positive for any wave with an ω and $k_\parallel$ giving such a resonant ellipse.

The next step is to determine whether $R-X$ waves exist, and if so under what conditions, having ω and $k_\parallel$ such that the resonant ellipse becomes entirely within the region where the auroral electron distribution has $\partial f/\partial v_\perp > 0$. The $R-X$ mode has $\omega > \omega_x = (1/2)\,[(\Omega_e^2 + 4\omega_{pe}^2)^{1/2} + \Omega_e] > \Omega_e$. AKR observations showing that ω_x is only slightly above Ω_e are consistent with the observation that AKR generation requires $\omega_{pe}^2 \ll \Omega_e^2$. Making the assumption $\omega \gg \omega_{pe}$, we have $\mathscr{L} \approx \mathscr{P} \approx 1$, so that the cold plasma dispersion relation (5.14) yields

$$\tilde{\mu}^2 = \frac{2\mathscr{R}}{(\mathscr{R}+1)\sin^2\theta + 2\cos^2\theta}. \tag{5.25}$$

Figure 5.6 shows $k_\parallel c/\Omega_e$ versus ω/Ω_e at various θ's as obtained from (5.25) for $\omega_{pe}/\Omega_e = 0.05$ and 0.1. In this figure $k_\parallel$ is negative to denote upgoing waves. Figure 5.6 shows that waves exist at all frequencies $\omega > \omega_x$, ω_x/Ω_e increasing with ω_{pe}/Ω_e. At each ω for which waves exist, $k_\parallel$'s are possible between 0 and a maximum value that occurs for field-aligned propagation. This maximum value of $k_\parallel$ at each ω for which wave propagation is possible is given by

$$\frac{ck_{\parallel,\,\mathrm{max}}}{\Omega_e} = \frac{\omega}{\Omega_e}\sqrt{\mathscr{R}}. \tag{5.26}$$

From the relativistic resonance condition (5.24), resonance can occur only when the axes of the resonant ellipse are real, i.e., if $a^2 > 0$. Considering only the principal $n = -1$ resonance, we thus obtain that a minimum parallel wave number exists for which resonance can occur at each ω given by

$$\frac{ck_{\parallel,\,\mathrm{min}}}{\Omega_e} = \sqrt{\left(\frac{\omega}{\Omega_e}\right)^2 - 1}. \tag{5.27}$$

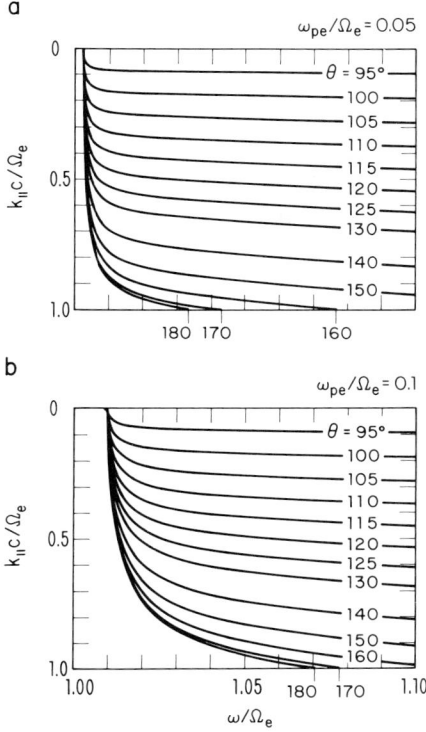

Fig. 5.6. Plot of $k_\| c/\Omega_e$ versus ω/Ω_e at various wave normal angles for ω_{pe}/Ω_e = 0.05 and 0.1. Note that at each ω for which waves exist, $k_\|$'s are possible between zero and a maximum value that occurs for field-aligned propagation (from Dusenbery and Lyons, 1982). (© by American Geophysical Union)

R–X waves for which resonance can occur with some particles exist whenever $k_{\|, \max} > k_{\|, \min}$. From (5.26), (5.27), and (5.13), we obtain that resonant R–X waves can exist at frequencies for which

$$\frac{\omega}{\Omega_e} > \frac{\omega_{\min}}{\Omega_e} = \left[1 - \left(\frac{\omega_{pe}}{\Omega_e}\right)^2\right]^{-1}.$$

For $\omega_{pe}/\Omega_e \ll 1$,

$$\frac{\omega_{\min}}{\Omega_e} - \frac{\omega_x}{\Omega_e} \approx \frac{5}{4}(\omega_{pe}/\Omega_e)^4,$$

so that the minimum possible resonant frequency is above, but very nearly equal to ω_x. For $\omega_x < \omega < \omega_{\min}$, R–X wave propagation is possible, but the waves cannot resonate with any electrons.

Figure 5.7 shows the minimum $k_\|$ for which resonance can occur as $ck_{\|, \min}/\Omega_e$ versus ω/Ω_e, as well as $ck_{\|, \max}/\Omega_e$ versus ω/Ω_e for various values of ω_{pe}/Ω_e. At each

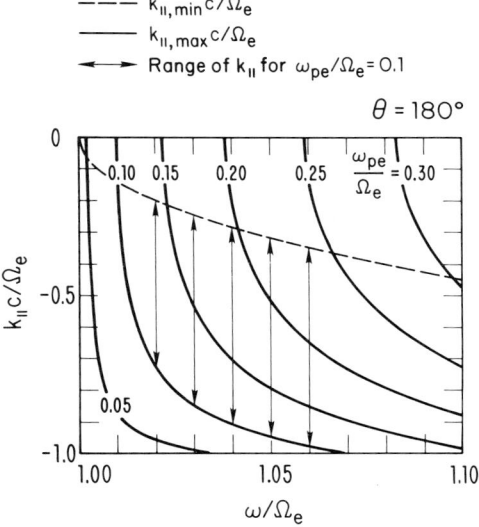

Fig. 5.7. Plot of $k_\parallel c/\Omega_e$ versus ω/Ω_e. Solid lines give the maximum $k_\parallel$ for which propagation is allowed for various values of ω_{pe}/Ω_e. The dashed line gives the minimum $k_\parallel$ for which resonance can occur, a quantity that does not depend upon ω_{pe}/Ω_e. At each ω waves exist that can resonate with some particles for $k_\parallel$ between $k_{\parallel,\,\text{min}}$ and $k_{\parallel,\,\text{max}}$ as indicated for $\omega_{pe}/\Omega_e = 0.1$ by the vertical arrows (from Dusenbery and Lyons, 1982). (© by American Geophysical Union)

ω, waves that can resonate with some particles exist between $k_{\parallel,\,\text{min}}$ and the values of $k_{\parallel,\,\text{max}}$ for the appropriate value of ω_{pe}/Ω_e as indicated for $\omega_{pe}/\Omega = 0.1$ by the vertical arrows in the figure.

Figure 5.8 shows computed resonant ellipses overlaid on the electron distribution function shown in Figure 5.5. In each panel, the ellipses are shown for values of $k_\parallel c/\Omega_e$ at the indicated values of ω/Ω_e. Also given on each panel is the maximum value of ω_{pe}/Ω_e for which $k_{\parallel,\,\text{max}} > k_{\parallel,\,\text{min}}$ so that wave propagation occurs at $k_\parallel$'s for which there is resonance with some particles at the indicated value ω/Ω_e. Contours are overlaid only on the upgoing portion of the distribution, as required to study wave generation by the velocity space regions where $\partial f/\partial v_\perp > 0$ associated with the upgoing loss cone. Corresponding contours for the downgoing portion of velocity space are simply symmetric reflections of the contours for the upgoing portion about the $v_\perp$-axis.

At each frequency, the resonant ellipses collapse to a point as $k_\parallel$ decreases to $k_{\parallel,\,\text{min}}$. In addition, the $|v_\parallel|$ at the center of ellipses, $v_{\parallel 0}$, increases with ω/Ω_e. As can be seen in Figure 5.8, there are resonance contours lying entirely within the region where $\partial f/\partial v_\perp > 0$ for $\omega/\Omega_e \lesssim 1.004$. Wave growth must occur at ω's and $k_\parallel$'s corresponding to these contours, provided the value of ω_{pe}/Ω_e allows wave propagation. The ellipses are entirely within the growth region in velocity space only for $k_\parallel$ close to $k_{\parallel,\,\text{min}}$. As $k_\parallel$ is increased, the length of the axes of the ellipses increases and a decreasing fraction of each contour lies within the region where $\partial f/\partial v_\perp > 0$. Contours that encounter regions in velocity space with $\partial f/\partial v_\perp > 0$ and regions with $\partial f/\partial v_\perp < 0$ may be associated with growing or damped waves. An evaluation of the growth rate integral (Equation (5.3)) would be required to

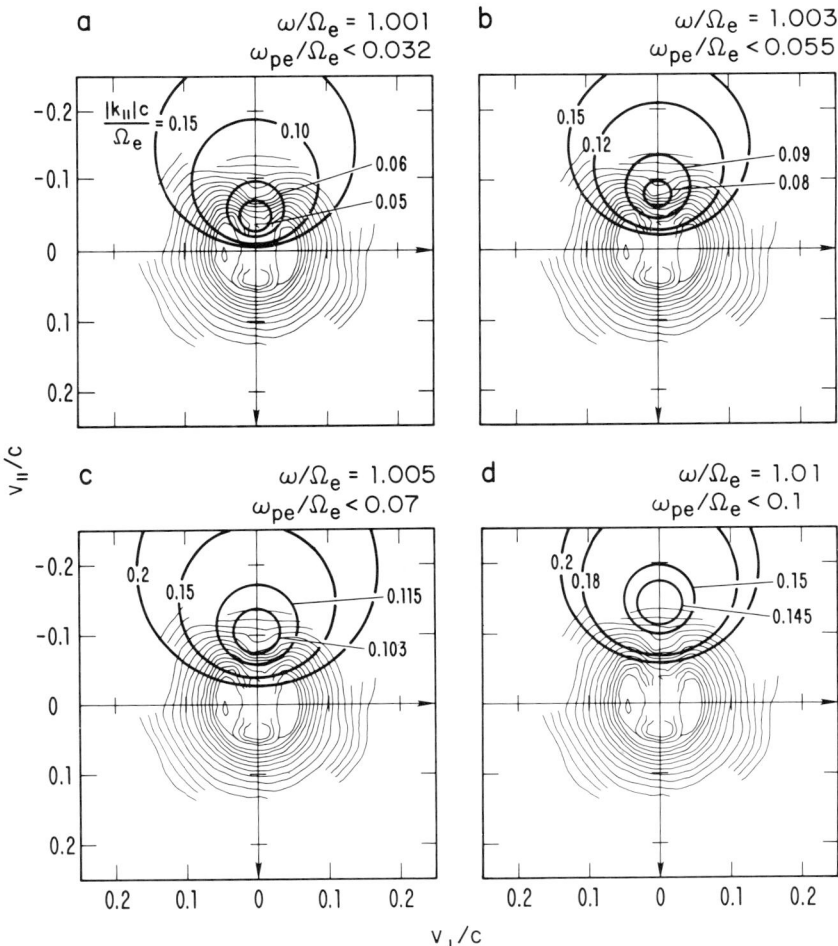

Fig. 5.8. Computed resonant ellipses overlaid on the electron distribution function shown in Figure 5.5. In each panel, ellipses are shown for values of $k_\parallel c/\Omega_e$ at the indicated values of ω/Ω_e. Also given on each panel is the maximum value of ω_{pe}/Ω_e for $k_{\parallel,\,max} > k_{\parallel,\,min}$ so that resonance with some particles is possible at the indicated value of ω/Ω_e (from Dusenbery and Lyons, 1982). (© by American Geophysical Union)

determine precisely which waves will be amplified and which will be damped. Nevertheless, all terms in the growth rate expression are positive definite for the waves with values of ω and $k_\parallel$ for which the entire resonant ellipse is within the region of $\partial f/\partial v_\perp > 0$. Thus such waves should have the largest growth rates and should therefore result in the most intense AKR.

Figure 5.9a is a plot of $ck_{\parallel,\,min}/\Omega_e$ and $ck_{\parallel,\,max}/\Omega_e$ for various ω_{pe}/Ω_e versus ω/Ω_e. This figure is similar to Figure 5.7, except that all regions where a positive growth rate is possible are shaded. For each point within the shaded region in Figure 5.9a, the

154   CHAPTER 5

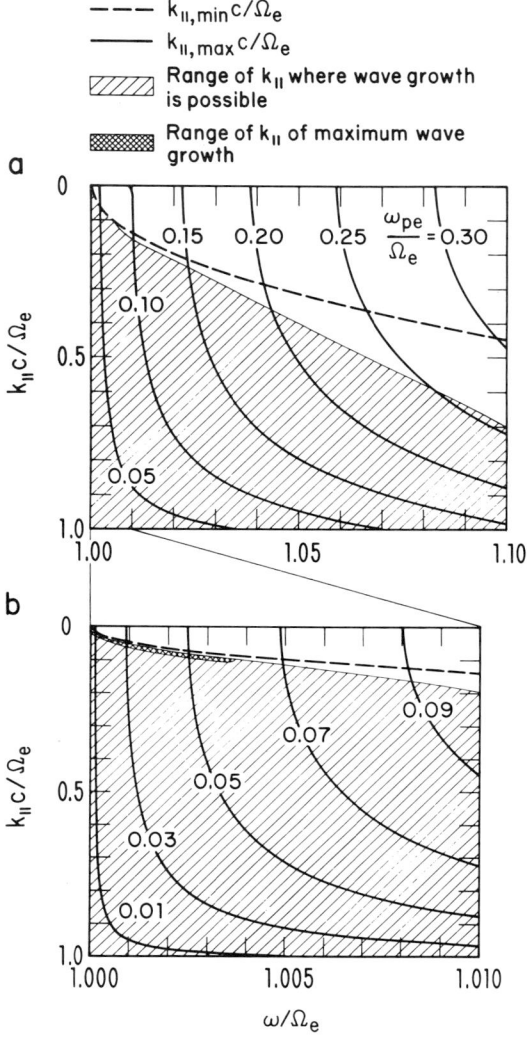

Fig. 5.9. Plot of $k_{\parallel,\min}c/\Omega_e$ and $k_{\parallel,\max}c/\Omega_e$ versus ω/Ω_e for various ω_{pe}/Ω_e similar to Figure 5.7. For each point within the shaded region in (a) the corresponding resonant ellipse has at least some portion within the region where the upgoing portion of velocity space has $\partial f/\partial v_\perp > 0$ so that wave growth is possible. The lightly shaded region in (b) is the same as in (a); however, the small, darkly shaded region shows those points in (ω, $k_\parallel$) space that correspond to resonant contours entirely within the region where $\partial f/\partial v_\perp > 0$. Thus the darkly shaded region should indicate the fastest growing waves (from Dusenbery and Lyons, 1982). (© by American Geophysical Union)

corresponding resonant ellipse has at least some portion within the region where the upgoing portion of velocity space has $\partial f/\partial v_\perp > 0$. Most of these contours also pass through regions where $\partial f/\partial v_\perp < 0$, so that most of the associated waves will be only weakly growing or damped. Resonant contours corresponding to the unshaded region in Figure

5.9 a are entirely above the rectangles in the upgoing portion of Figure 5.5a and thus do not encounter any portion of the electron distribution that can cause wave growth. Such waves must be stable.

Figure 5.9a accounts for the observation of Benson and Calvert (1979) that $\omega_{pe}/\Omega_e \lesssim 0.2$ in the density depletion regions where AKR is excited. As can be seen in the figure, no resonant waves exist for which wave growth is possible for $\omega_{pe}/\Omega_e \gtrsim 0.25$. Similar results to Figure 5.9a can be obtained for the region in the downgoing portion of velocity space where $\partial f/\partial v_\perp > 0$. However, the range of possible instability will be smaller than that shown in Figure 5.9a for the upgoing electrons, since the region where $\partial f/\partial v_\perp > 0$ is less extended in the $v_\parallel$ direction for the downgoing electrons.

Figure 5.9b is an enlargement of Figure 5.9a for frequencies very near Ω_e. The lightly shaded region is the same as in Figure 5.9a. However, the small darkly shaded region shows those ω's and $k_\parallel$'s that correspond to resonant contours lying entirely within the region where $\partial f/\partial v_\perp > 0$. These fastest growing waves are at frequency $\omega/\Omega_e \lesssim 1.004$ and can be generated only in regions where $\omega_{pe}/\Omega_e \lesssim 0.06$. This accounts for the observation by Calvert (1981b) that the peak AKR intensities are generated in the regions of smallest ($\lesssim 0.03$) values of ω_{pe}/Ω_e.

Figure 5.10 is similar to Figure 5.9b, except that only one value of ω_{pe}/Ω_e (= 0.05) is considered and $k_\parallel c/\Omega_e$ is shown for various wave normal angles from 95° to 180°. The range $\theta = 90°-180°$ was chosen when this study was performed, rather than the range $0°-90°$, to represent upgoing waves; however this distinction is not significant to the

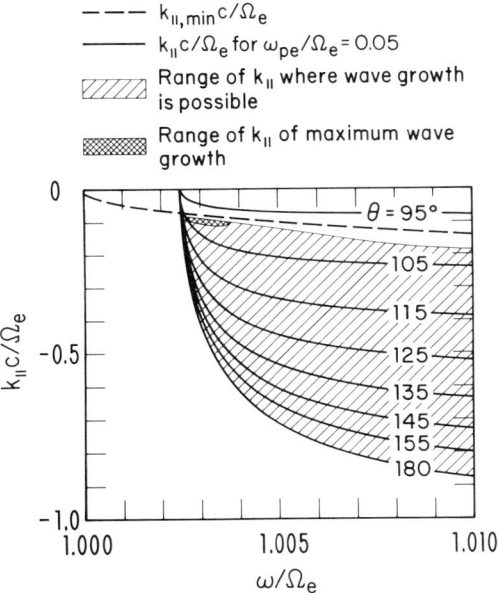

Fig. 5.10. Similar to Figure 5.9(b), but only one value of ω_{pe}/Ω_e (= 0.05) is considered and $k_\parallel c/\Omega_e$ is shown for various wave normal angles from 95° to 100°. Notice that $\theta \approx 100°$ over most of the frequency range of the fastest growing waves (from Dusenbery and Lyons, 1982). (© by American Geophysical Union)

present discussion. Notice that $\theta \approx 100°$ over most of the frequency range of the fastest growing waves. For other values of $\omega_{pe}/\Omega_e \lesssim 0.06$, the result is essentially the same, the fastest growing waves having wave normal angles between approximately 95° and 105°. Thus the observation that AKR is generated at wave normal angles nearly normal to the ambient magnetic field is easily explained by this analysis.

Discussion of the temporal wave growth rates is not complete without including the convective growth length defined as the group velocity v_g divided by the temporal growth rate. The group velocity of AKR has been calculated by Dusenbery and Lyons (1982). However, the basic result can be seen from Figure 5.11, where kc/Ω_e is shown versus

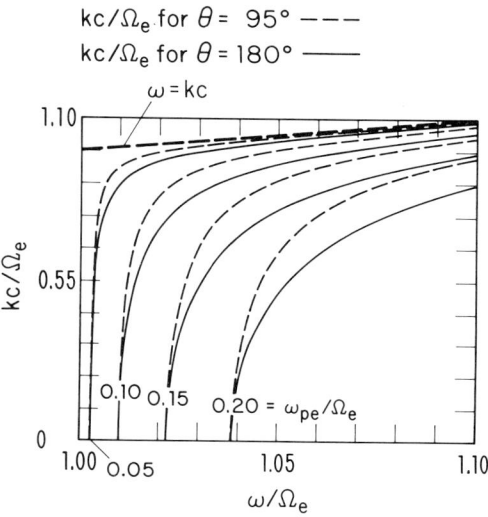

Fig. 5.11. Plot of kc/Ω_e versus ω/Ω_e for $\theta = 95°$ and 180° for three representative values of ω_{pe}/Ω_e. The magnitude of the group velocity v_g is roughly proportional to $\partial\omega/\partial k$, so that v_g decreases as ω decreases, becoming very small for ω close to ω_x. Since the temporal growth rate peaks for ω near ω_x, we find that v_g minimizes for waves with the largest growth rate (from Dusenbery and Lyons, 1982). (© by American Geophysical Union)

ω/Ω_e for $\theta = 95°$ and 180° and for three representative values of ω_{pe}/Ω_e. The magnitude of v_g is roughly proportional to $\partial\omega/\partial k$, so that v_g will be a minimum where $\partial\omega/\partial k$ minimizes. Figure 5.11 shows that for all θ, $\partial\omega/\partial k \to 0$ as $\omega \to \omega_x$, so that v_g decreases as $\omega \to \omega_x$, becoming very small for ω close to ω_x. Since the temporal growth rate peaks for ω near ω_x, we find that v_g minimizes for waves with the largest growth rate. This implies a very small convective growth length (large convective growth rate) for AKR, which should be important in understanding the large intensity of AKR.

5.5. Whistler-Mode Waves and the Radiation Belts

Pitch-angle diffusion driven by plasma wave turbulence can be an important process by which particles are lost from the earth's radiation belts, since such diffusion can drive trapped particles into the atmospheric loss cone. Dungey (1963) and Cornwall (1964) suggested that such diffusion could occur for radiation belt electrons as a result of cyclotron resonant interactions with lightning-generated whistlers. Later, Cornwall (1965) and Kennel and Petschek (1966) suggested that the trapped particle distributions in the radiation belts can themselves be unstable to the generation of whistler-mode and ion-cyclotron waves if their pitch-angle distribution is sufficiently anisotropic as a result of the loss cone along the direction of the ambient magnetic field. Satellite observations have since shown that lightning makes only a minor contribution to the total wave energy within the magnetosphere. Wave energy believed to be generated naturally within the magnetosphere, as suggested by Cornwall (1965) and Kennel and Petschek (1966), dominates the wave spectrum and thus should control the pitch angle diffusion of radiation belt particles.

5.5.1. GENERATION OF PLASMASPHERIC HISS

Kennel and Petschek (1966) investigated the interaction of radiation belt electrons with naturally generated whistler-mode waves by making the simplifying assumptions that $\Omega_p \ll \omega < \Omega_e$ and $\omega_{pe}^2/\Omega_e^2 \gg 1$. The assumption $\omega \gg \Omega_p$ can break down for whistler-mode waves close to the earth, but this assumption and the high density assumption $\omega_{pe}^2/\Omega_e^2 \gg 1$ are valid approximations throughout most of the radiation belts. All whistler-mode waves have $\omega < \Omega_e$. Under these assumptions, the cold plasma dispersion relation for wave propagation parallel to the ambient magnetic field, $\tilde{\mu}^2 = \mathcal{R}$, can be approximated as

$$\tilde{\mu}^2 = \frac{\omega_{pe}^2/\omega^2}{(\Omega_e/\omega - 1)}. \tag{5.28}$$

This allows us to write the parallel electron energy for cyclotron resonance with whistler-mode waves as

$$K_{\parallel,\mathrm{res}} = \frac{B^2}{8\pi N} \frac{\Omega_e}{\omega} \left(n + \frac{\omega}{\Omega_e}\right)^2 \left(1 - \frac{\omega}{\Omega_e}\right). \tag{5.29}$$

Equation (5.29) shows that $K_{\parallel,\mathrm{res}}$ is proportional to the magnetic energy per particle, $B^2/8\pi N$, which Kennel and Petschek (1966) referred to as the characteristic energy for cyclotron resonance. This characteristic energy is illustrated in Figure 5.12 as a function of L for equatorial value of B and N. Specifically, the dipole value of B was used and, on the basis of the observations of Chappell et al. (1970) shown in Figure 4.21, $N = 500 \, (r/L)^4 \, \mathrm{cm}^{-3}$ was taken within the plasmasphere and $N = 1 \, \mathrm{cm}^{-3}$ was taken outside the plasmasphere. The plasmapause is shown as a density discontinuity at $L = 5$ in the figure.

Kennel and Petschek wrote the growth rate for parallel propagating whistler-mode waves in terms of the pitch-angle gradient in the electron distribution function. From Section 5.1, we obtain for the wave growth rate

$$\frac{\gamma}{\omega} \propto \hat{G}f = -\frac{\partial f}{\partial \alpha} + \frac{\omega}{k_\parallel} \frac{\partial f}{\partial v_\perp},$$

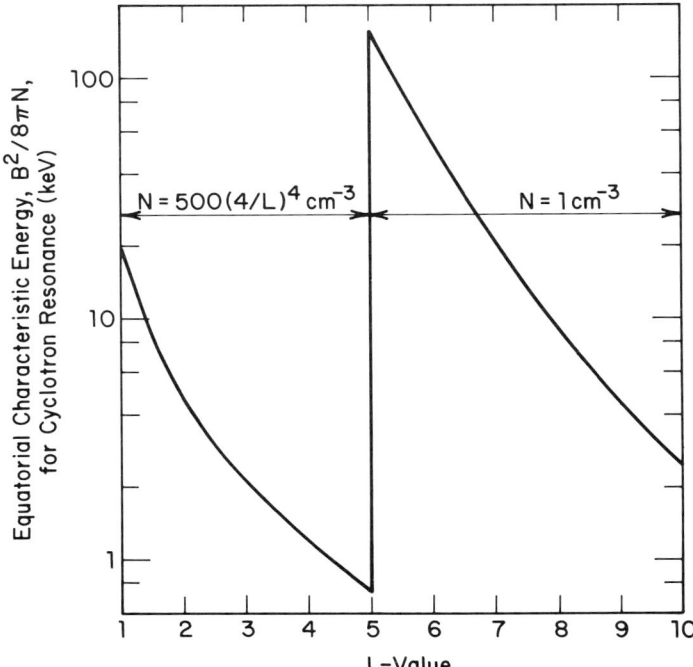

Fig. 5.12. Equatorial characteristic energy for cyclotron resonance, $B^2/8\pi N$, versus L. The dipole value of B was used. $N = 500(4/L)^4$ cm^{-3} was taken at $L < 5$, and $N = 1$ cm^{-3} was taken at $L \geq 5$.

where the $\partial f/\partial\alpha$ term dominates for $\omega \ll |n|\Omega$, and the $\partial f/\partial v_\perp$ term becomes increasingly important as $\omega/|n|\Omega$ is increased. We thus see that a pitch-angle distribution with a minimum along the field line direction can cause wave growth for the whistler mode at the negative harmonic resonances. However, the degree of pitch-angle anisotropy required for wave growth increases with $\omega/|n|\Omega$. We also see that for a given relative shape of the pitch-angle distribution, i.e., for a specified value of $(1/f)(\partial f/\partial\alpha)$, the growth is proportional to the magnitude of f.

Magnetospheric particle distribution functions generally monotonically decrease with increasing particle energy. Thus, without considering the effects of varying pitch-angle distributions, we expect wave growth to increase as $K_{\parallel,\text{res}}$ decreases. Equation (5.29) shows that $K_{\parallel,\text{res}}$ is lower at the principal $n = -1$ cyclotron resonance than at the other cyclotron resonances. In addition, whistler-mode waves are predominantly right-hand polarized so that Θ_{nk} (see Equation (5.4)) is maximum at the $n = -1$ cyclotron resonance for resonant interactions with electrons. Thus the wave growth should generally be dominated by the $n = -1$ cyclotron resonance, and the largest growth rates should occur in regions where the characteristic energy, $B^2/8\pi N$, is lowest. From Figure 5.12, we are then able to estimate that the whistler-mode wave growth will maximize just within the plasmapause, will minimize just outside the plasmapause, and may reach a secondary maximum well outside the plasmapause. As discussed shortly, the minimum in $B^2/8\pi N$

just within the plasmapause plays a crucial role in the generation of whistler-mode waves within the plasmasphere.

Kennel and Petschek (1966) noted that wave growth should maximize near the equator since $B^2/8\pi N$ increases with increasing latitude, and they suggested the possibility of an unspecified wave reflection process at higher latitudes that could return wave energy to the region of maximum growth near the equator. In this way, wave energy would bounce back and forth along the magnetic field lines and be amplified on each pass across the equator. Owing to the increase in the growth rate with increasing f, Kennel and Petschek suggested that, if the magnitude of the resonant electron distribution function exceeded a critical value f_c, the wave growth on each pass across the equator would exceed wave losses due to absorption and imperfect reflection. The resulting net growth of wave energy would then cause enhanced pitch-angle scattering of resonant electrons into the loss cone, thus reducing f toward f_c. If f is below f_c, net wave growth is not expected and no electron losses are predicted by this theory. Thus f_c represents a limit to the stably trapped electron distribution function that Kennel and Petschek found to be comparable with observations of 40-keV outer zone electrons. In addition, they suggested that a typical total wave amplitude near the equator should be of the order of 10^{-2} nT.

Although the concept of a well-defined limit to stably trapped electron fluxes throughout the radiation belts is not generally supported by observations, the analysis of Kennel and Petschek (1966) demonstrated that sufficient wave energy could be naturally generated by radiation belt particles for wave-particle interactions to be an important loss process for the particles. In fact, subsequent wave observations have found that 10^{-2} nT is a typical amplitude for whistler-mode waves in the vicinity of the equator. In addition, Kennel and Petschek's suggestion that wave reflection could substantially increase particle losses has proved to be crucial to understanding the structure of radiation belt electrons.

A significant advance in understanding the interaction of whistler-mode waves with radiation belt electrons resulted from studies of wave propagation. These studies found that waves could naturally reflect within the magnetosphere because of the variation of the refractive index as waves propagate away from the equator (Kimura, 1966; Thorne and Kennel, 1967). As a wave propagates away from the equator, its wave normal angle rapidly increases. Above the lower hybrid frequency ω_{LH} ($\omega_{LH} \approx (\Omega_i \Omega_e)^{1/2}$ for $\omega_{pe}^2 \gg \Omega_e^2$), wave propagation normal to **B** is prohibited so that wave reflection is impossible. However, as B increases with increasing latitude, waves with frequencies below Ω_e eventually reach a latitude where their frequency is below ω_{LH}. Then wave propagation normal to **B** is possible, and application of Snell's Law (Kimura, 1966; Thorne and Kennel, 1967; Lyons and Thorne, 1970) shows that the wave vector of whistler-mode waves will rotate through 90°, causing wave energy initially propagating away from the equator to return to the equator.

Lyons and Thorne (1970) found that whistler-mode waves below the equatorial lower-hybrid frequency readily propagate across magnetic field lines during the wave reflection process. Wave observations within the plasmasphere show that whistler-mode radiation is dominated by a band of turbulence, termed 'plasmaspheric hiss', centered near a few hundred Hz (Russell *et al.*, 1969; Thorne *et al.*, 1973) which is below the equatorial lower hybrid frequency throughout most of the plasmasphere. Thus earlier studies, (e.g., Kennel and Petschek, 1966), which assumed wave propagation essentially along the magnetic field lines, needed to be modified to include the effects of wave propagation across the field lines. Specifically, if waves are generated locally in one region of the plasmasphere, they can readily propagate across field lines and thus can fill the entire plasmasphere with waves (Lyons *et al.*, 1972).

Figure 5.13 shows examples of 500-Hz whistler-mode ray paths obtained using a Stanford ray-tracing computer program (Kimura, 1966), which illustrate how wave energy generated in the outer region of the plasmasphere can propagate so as to fill the entire plasmasphere with waves. Generation in the outer region of the plasmasphere was assumed because $B^2/8\pi N$ is minimum in this region. Wave observations within the plasmasphere support the suggestion that wave energy is generated within a rather narrow range of L-values, since the frequency spectrum is quite constant throughout the plasmasphere, compared with the large changes in the electron gyrofrequency.

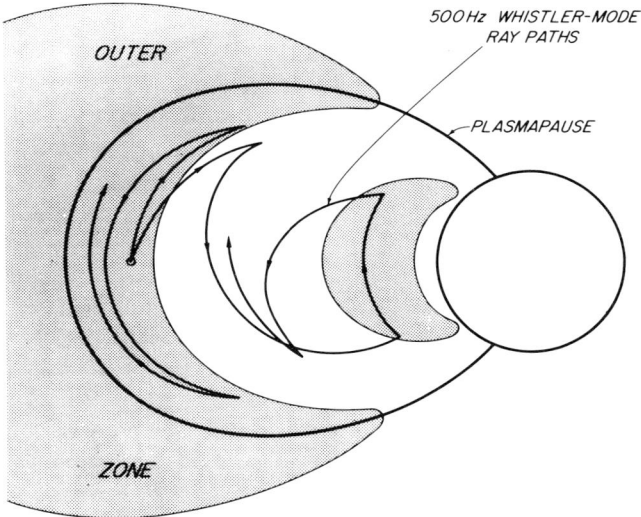

Fig. 5.13. Examples of 500-Hz whistler-mode ray paths obtained using a Stanford ray-tracing program (Kimura, 1966), which illustrate how wave energy generated in the outer region of the plasmasphere can propagate across field lines so as to fill the plasmasphere with waves (from Lyons *et al.*, 1972). (© by American Geophysical Union)

Thorne *et al.* (1979) investigated the propagation of whistler-mode waves within the plasmasphere in more detail, and in particular they included the effects of the plasmapause. A significant result of their study is the important role played by the plasmapause. Ray path calculations presented by Thorne *et al.* (1979) and obtained using the Stanford ray-tracing program are shown in Figure 5.14. A trajectory along which waves propagate away from the equator, magnetospherically reflect, and return directly to the equator is shown in Figure 5.14a. As indicated by the small arrows along the ray trajectory, the wave vector of such waves remains nearly normal to the ambient magnetic field after reflection. Thus further wave growth due to cyclotron resonance upon a wave's return to the equator would be minimal, since significant wave growth does not occur for wave vectors well away from the field line direction.

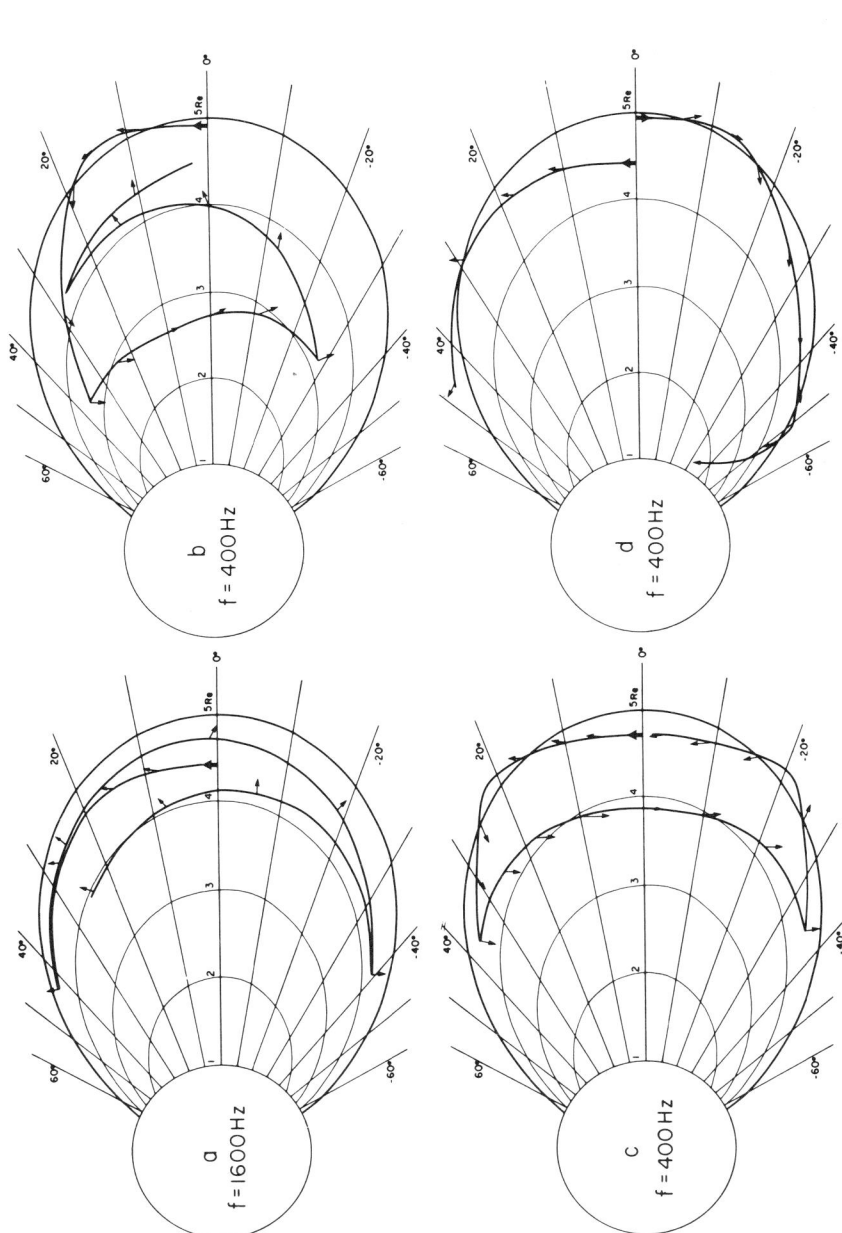

Fig. 5.14. Ray paths within the plasmasphere obtained using the Stanford ray-tracing program. Each wave originates (large arrow) in the outer plasmasphere propagating parallel to **B**. Subsequent wave normal directions are indicated by the small arrows. A trajectory along which waves propagate away from the equator, magnetospherically reflect, and return directly to the equator is shown in (a). The class of trajectories that reflect off the plasmapause and propagate to low L is illustrated by the trajectory in (b). The class of trajectories that reflect off the plasmapause and exhibit cyclic trajectories is illustrated in (c); (d) illustrates the small class of trajectories that penetrate the plasmapause at low altitudes or reach the ionosphere (from Thorne et al., 1979). (© by American Geophysical Union)

Thorne *et al.* (1979) found, however, that the majority of the waves generated in the outer region of the plasmasphere will reach the plasmapause. Such waves then reflect off the plasmapause because of the pronounced decrease in the refractive index associated with the large density gradient at the plasmapause. An important class of such waves exhibits cyclic trajectories like those shown in Figure 5.14c. When these waves return to the equatorial region of the outer plasmasphere, wave vectors are aligned with the magnetic field so that significant amplification by multiple transits of the growth region is possible in a manner somewhat analogous to that suggested by Kennel and Petschek (1966). A small class of wave trajectories penetrates the plasmapause at low altitudes or reaches the ionosphere (Figure 5.14d). However, an important group of trajectories reflects off the plasmapause and propagates to low L-values (Figure 5.14b). Such trajectories should be crucial to filling the entire plasmasphere with a rather uniform distribution of waves.

A summary of the different types of wave trajectories obtained by Thorne *et al.* (1979) as a function of wave frequency and starting L-value is shown in Figure 5.15 for a plasmapause density gradient lying between $L = 5$ and $L = 5.3$. The unshaded area in the figure includes all non-cyclic trajectories, such as those in Figure 5.14b, which reflect off the plasmapause and remain trapped within the plasmasphere. The class of cyclic trajectories (dark shading) is restricted to frequencies below 1 kHz. These trajectories occur in two zones in Figure 5.15, but the two zones represent parts of the same ray

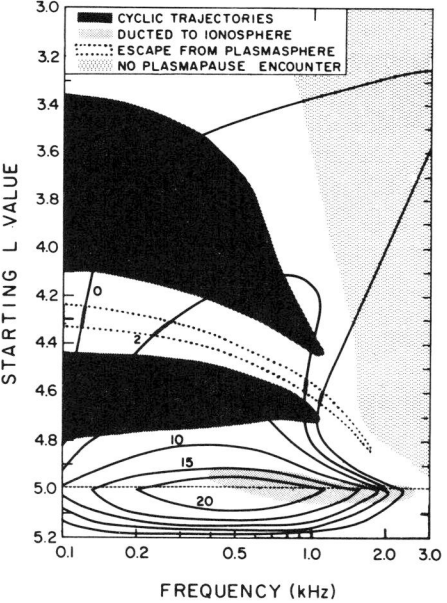

Fig. 5.15. A summary of the different types of whistler-mode wave trajectories as a function of wave frequency and starting L-value. The unshaded area in the figure indicates all non-cyclic trajectories, such as those in Figure 5.14(b), which reflect off the plasmapause and remain trapped within the plasmasphere. Shaded regions indicate other classes of trajectories. The solid contours give the net exponential gain in wave energy for a single wave transit of the equatorial region between $\pm 30°$ geomagnetic latitude (from Thorne *et al.*, 1979). (© by American Geophysical Union)

trajectories. Thorne et al. (1979) also calculated the net exponential gain in wave energy for a single transit of the equatorial region between ± 30° magnetic latitude. These gains are given by the solid contours in Figure 5.15, and they can be seen to peak at several hundred Hz in the outer region of the plasmasphere.

These results of Thorne et al. (1979) appear to account for the broadband electromagnetic emission, plasmaspheric hiss, that is almost always present within the plasmasphere (Russell et al. 1969; Thorne et al., 1973; Parady et al., 1975). The waves are observed to be effectively confined within the plasmasphere, as can be explained by the reflection of the waves off the plasmapause. The waves are observed throughout most of the plasmasphere, and the power and the frequency spectra of the waves show little variation with L, as expected from a localized source region. The waves are generally weakly polarized, as expected from a mixing of waves of a variety of different wave normal angles. Finally the peak power of the waves is generally at a frequency below 1000 Hz, which is consistent with the growth rate and wave reflection results shown in Figure 5.15.

5.5.2. GENERAL CONCEPTS ON PITCH-ANGLE DIFFUSION IN THE RADIATION BELTS

In order to discuss the effects of the plasmaspheric hiss on radiation belt electrons, we first present some general results concerning pitch-angle diffusion in the radiation belts.

Since wave-particle interactions are generally not confined to a specific latitude, we first consider pure pitch-angle diffusion and time-average the pitch-angle diffusion equation over a particle's bounce trajectory between mirror points. Assuming pure pitch-angle diffusion is equivalent to assuming that $D_{\alpha\alpha}$ is much greater than the other diffusion coefficients in Equation (5.7), an assumption which is valid for wave frequencies well below the gyrofrequency of the resonant particles. Although we consider only the limit of pure pitch-angle diffusion here, the other terms in Equation (5.7) can similarly be averaged over a bounce trajectory. Averaging the terms in (5.7) over a bounce trajectory is valid as long as the wave-particle interactions at each point in space depend upon the local plasma and field parameters and are independent of the gradient in these parameters.

We start with the pure pitch-angle diffusion equation,

$$\frac{\partial f}{\partial t} = \frac{1}{\sin \alpha} \frac{\partial}{\partial \alpha} \left[D_{\alpha\alpha} \sin \alpha \frac{\partial f}{\partial \alpha} \right], \tag{5.30}$$

and use Liouville's theorem to obtain that $f(\alpha) = f_0(\alpha_0)$ along a particle trajectory. Here f_0 is the equatorial particle distribution function, α_0 is the particle pitch angle at the equator, and the pitch-angle variation along a particle's bounce trajectory is given by

$$\sin^2 \alpha = \sin^2 \alpha_0 \frac{B}{B_0}, \tag{5.31}$$

where B_0 is the equatorial magnetic field strength. Differentiating (5.31) gives

$$\frac{\partial \alpha}{\partial \alpha_0} = \frac{\tan \alpha}{\tan \alpha_0}.$$

The particle bounce period (the time for a particle to travel from one mirror point to the other and back again) $\tau_B = \tau_0 s(\alpha_0)$ for a dipole magnetic field, where $\tau_0 = 4LR_e/v$ and $s(\alpha_0)$ gives the small variation of τ_B with pitch angle (Equation (2.26)).

Taking the time average of (5.30) gives

$$\tau_0 s(\alpha_0) \frac{\partial f_0}{\partial t} = \int_0^{\tau_B} \frac{1}{\sin \alpha} \frac{\partial}{\partial \alpha} \left[D_{\alpha\alpha} \sin \alpha \frac{\partial f}{\partial \alpha} \right] dt.$$

We next transform the integral over time in the above equation to an integral over geomagnetic latitude λ, using $dt/d\lambda = (d\ell/d\lambda)/v_\parallel$, where $d\ell/d\lambda$ along dipole field lines is given by Equation (2.8) and ℓ is distance along field lines. We also transform all local pitch angles α to α_0, noting that $D_{\alpha\alpha} \sim (v\Delta\alpha)^2/(2\Delta t)$ can be written as a diffusion coefficient in equatorial pitch angle by using

$$(v\Delta\alpha_0)^2/2\Delta t = [(v\Delta\alpha)^2/(2\Delta t)] \, (\partial \alpha_0/\partial \alpha)^2.$$

This yields

$$\frac{\partial f_0}{\partial t} = \frac{1}{s(\alpha_0) \sin \alpha_0 \cos \alpha_0} \frac{\partial}{\partial \alpha_0} \left[s(\alpha_0) \sin \alpha_0 \cos \alpha_0 \langle D_{\alpha\alpha} \rangle \frac{\partial f_0}{\partial \alpha_0} \right], \quad (5.32)$$

where $\langle D_{\alpha\alpha} \rangle$ is the time average of $D_{\alpha\alpha}(\partial \alpha_0/\partial \alpha)^2$ over a bounce orbit that may be written

$$\langle D_{\alpha\alpha} \rangle = \frac{1}{2s(\alpha_0)} \int_{\lambda_{M1}}^{\lambda_{M2}} D_{\alpha\alpha} \frac{\cos \alpha}{\cos^2 \alpha} \cos^7 \lambda \, d\lambda, \quad (5.33)$$

and where λ_{M1} and λ_{M2} are the geomagnetic latitudes of the two mirror points. Equation (5.32) is the pitch-angle diffusion equation for particles in a dipole field (Schulz and Lanzerotti, 1974).

To allow for the effects of the atmospheric loss cone, Equation (5.32) must include a term representing the loss of particles to the atmosphere. This can be accomplished by adding an atmospheric loss term to (5.32) which gives

$$\frac{\partial f_0}{\partial t} = \frac{1}{s(\alpha_0) \sin \alpha_0 \cos \alpha_0} \frac{\partial}{\partial \alpha_0} \left[s(\alpha_0) \sin \alpha_0 \cos \alpha_0 \langle D_{\alpha\alpha} \rangle \frac{\partial f_0}{\partial \alpha_0} \right] - \frac{f_0}{\tau_{\text{atm}}}, \quad (5.34)$$

where τ_{atm} is the time scale for losses to the atmosphere. Assuming that losses occur only from within the loss cone and that the loss cone is emptied twice per bounce period, and neglecting spatial variations of f within the loss cone due to the emptying of the loss cones at low altitudes,

$$\tau_{\text{atm}} = \begin{cases} \tau_B/2, & \text{for } \alpha_0 \text{ within the loss cone} \\ \infty, & \text{for } \alpha_0 \text{ outside the loss cone} \end{cases}.$$

Given $\langle D_{\alpha\alpha} \rangle$ and an initial pitch-angle distribution for particles of a given energy, (5.34) can be solved numerically for the evolution of the pitch-angle distribution with time. However, more physical insight into the pitch-angle diffusion process can be obtained by assuming that the shape of the pitch-angle distribution is independent of time, so that the distribution function is separable into two components:

$$f_0(t, \alpha_0) = F(t)g(\alpha_0),$$

where $F(t)$ represents the equatorial density of particles for a given energy and $g(\alpha_0)$ specifies the shape of the pitch-angle distribution and is independent of time. The assumption of a steady state shape for the pitch-angle distribution is not unreasonable. Roberts (1966, 1969) has in fact demonstrated that any arbitrary pitch-angle distribution that might arise from an injection event will soon decay to a lowest normal mode.

Substituting $F(t)g(\alpha_0)$ for f_0 in (5.34) yields

$$\left(\frac{1}{\tau_{atm}} + \frac{1}{F}\frac{\partial F}{\partial t}\right) g(\alpha_0) s(\alpha_0) \sin\alpha_0 \cos\alpha_0$$

$$= \frac{\partial}{\partial \alpha_0}(\alpha_0) \sin\alpha_0 \cos\alpha_0 \langle D_{\alpha\alpha}\rangle \frac{\partial g}{\partial \alpha_0}. \tag{5.35}$$

Integrating from $\alpha_0 = 0$ to $\pi/2$, assuming $\partial g/\partial \alpha_0 = 0$ at $\alpha_0 = 0$ and $\alpha_0 = \pi/2$, relates the lifetime from precipitation into the atmosphere,

$$\tau_p = -\left(\frac{1}{F}\frac{\partial F}{\partial t}\right)^{-1},$$

to $g(\alpha_0)$ by

$$\tau_p = \tau_0/2\eta \tag{5.36}$$

where

$$\eta = \left(\int_0^{\alpha_{Lc}} g(\alpha_0) \sin\alpha_0 \cos\alpha_0 \, d\alpha_0\right) \Big/ \left(\int_0^{\pi/2} g(\alpha_0) s(\alpha_0) \sin\alpha_0 \cos\alpha_0 \, d\alpha_0\right).$$

The factor of $\cos\alpha_0$ that appears in the integrands for η reflects the magnetic mirror geometry of the magnetic field. One expects the precipitation lifetime to depend upon the fraction of the total number of particles that are within the loss cone. Since the number of particles per unit volume within a pitch-angle interval $d\alpha_0$ is proportional to $g(\alpha_0) \sin\alpha_0 \, d\alpha_0$, the factor $\cos\alpha_0$ does not appear when a uniform magnetic field is assumed. However, in a magnetic mirror geometry, the total number of particles within a flux tube must be considered. Following the derivation of the mirror point density distribution of Hess (1968), we consider an element of volume $dA \, d\ell$ at the geomagnetic equator, where dA is the area of a flux tube and $d\ell$ is an element of length along the field lines. Each particle spends a time $dt = d\ell/(v \cos\alpha_0)$ within the volume. Thus the number of particles within the volume is a fraction $2dt/\tau_B$ of the total number of particles within the flux tube, so that the total number of particles within $d\alpha_0$ is proportional to $g(\alpha_0) s(\alpha_0) \sin\alpha_0 \cos\alpha_0 \, d\alpha_0$. The expression for τ_p (5.36) therefore states that the lifetime is inversely proportional to the fraction of the total number of particles within a flux tube that is within the loss cone.

distribution consistent with observations of Thorne et al. (1973). The frequency of maximum wave intensity was taken to be 600 Hz, the half-width of the frequency distribution was taken to be 300 Hz, and waves were assumed to propagate over a range of wave normal angles from 0° to near 90°. Weak pitch-angle diffusion was assumed so that $f(\alpha_{LC}) \to 0$. Under these conditions, $\tau_p \propto B_{\text{wave}}^2$ and the total wave amplitude B_{wave} can be left as a scaling factor for the precipitation lifetimes. B_{wave} thus need not be specified to obtain the shape of the electron pitch-angle distribution nor the variation of the precipitation lifetime τ_p with energy at a given L. The wave amplitude was assumed to be independent of L to obtain the variation of τ_p with L. Finally, the cold plasma density was taken to be 1000 cm^{-3} $(4/L)^4$, a factor of 2 larger than that used in Figure 5.12. The density was taken to be constant along magnetic field lines, consistent with diffusive equilibrium well away from the ionosphere.

The bounce-orbit averaged, pitch-angle diffusion coefficients due to the cyclotron resonances and the Landau resonance are shown as a function of equatorial pitch angle in the left-hand panel of Figure 5.21 for 20-, 200-, and 2000-keV electrons. Notice that significant diffusion into the loss cone occurs over the entire energy range, though the diffusion coefficients generally decrease with increasing electron energy at energies high

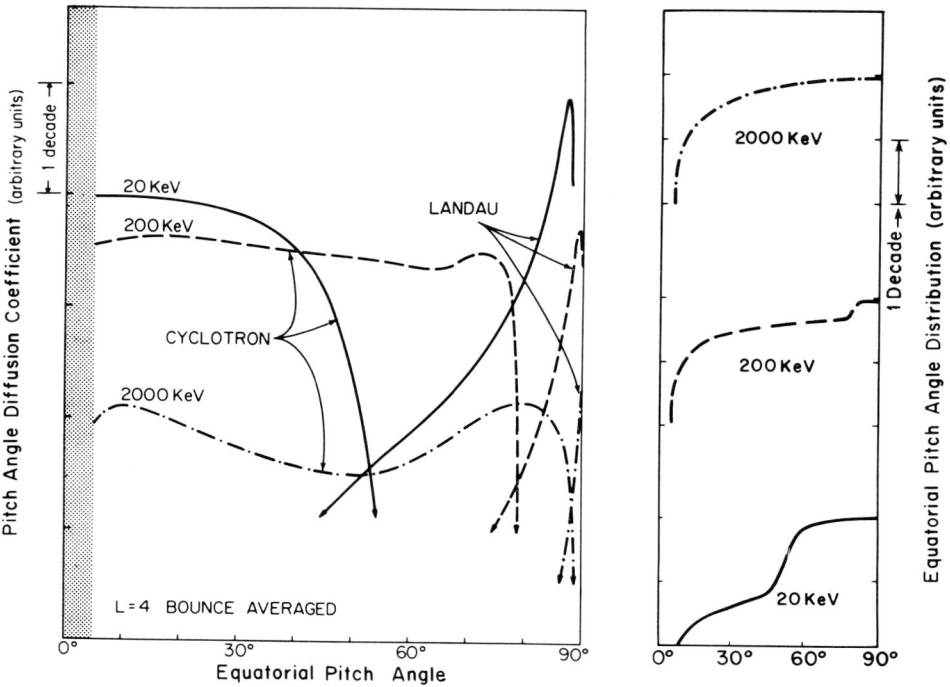

Fig. 5.21. Bounce-orbit averaged cyclotron and Landau resonant pitch-angle diffusion coefficients as a function of equatorial pitch angle at $L = 4$ for 20, 200, and 2000 keV electrons. For each energy, the line coding used for the Landau resonant diffusion coefficients corresponds to that for the sum of the cyclotron-harmonic resonances, where the particle energy is indicated. Equatorial pitch-angle distributions, calculated using the diffusion coefficients shown in the figure, are displayed on the right-hand side (from Lyons et al., 1972). (© by American Geophysical Union)

enough for cyclotron resonance. Landau resonance occurs only at local pitch angles near 90°, but the off-equatorial interactions cause the range of equatorial pitch-angle range affected by the Landau resonance to overlap the pitch-angle range of cyclotron resonance. Strictly, there is no Landau resonance at exactly 90° pitch angle, but the region devoid of diffusion is much less than 1° wide.

Although the combined bounce-averaged cyclotron and Landau resonances provide diffusion over essentially all pitch angles, there is a significant reduction in the total diffusion rate at pitch angles near the overlap of the two regions. For example, in Figure 5.21 the 20-keV electron diffusion coefficient is considerably smaller between $\alpha_0 = 50°$ and 70° than at other pitch angles. At higher energies, cyclotron resonance extends to higher pitch angles, and the region of relatively slow diffusion moves to higher pitch angles and covers a smaller pitch-angle range.

The existence of a region of slow diffusion produces a characteristic feature in the shape of the equatorial pitch-angle distributions, as can be seen in the calculated distributions shown in the right-hand panel of Figure 5.21. Qualitatively, the variation of the slope of the pitch-angle distribution as a function of pitch angle is inversely proportional to the variation of the magnitude of the pitch-angle diffusion as a function of pitch angle. Thus the slope of the pitch-angle distributions will be large over the pitch-angle range of small diffusion coefficients. This causes the equatorial pitch-angle distributions to develop bumps surrounding 90° pitch angle, and these bumps decrease in size and pitch-angle extent with increasing electron energy. From Equation (5.29), the minimum parallel energy for cyclotron resonance with waves, when $\omega \ll \Omega_e$, varies as $B^3/N \propto L^{-5}$ with $N \propto L^{-4}$. Thus the bumps should also decrease with increasing L-value for a fixed electron energy.

After calculating these pitch-angle distributions, Lyons et al. obtained unpublished pitch-angle distribution data from H. West. These equatorial pitch-angle distributions of 79-, 158-, 882-, and 1530-keV electrons measured at four L-values during the decay period following a strong injection of electrons into the slot region are shown in Figure 5.22, where they are compared with the theoretically calculated pitch-angle distributions (solid lines) for the appropriate L-values and energies.

The data in Figure 5.22 exhibit significant flux enhancements surrounding 90° pitch angle as predicted by the calculations. Although the calculations tend to exaggerate the magnitude of the enhancements, there is very good agreement with the pitch-angle width of the observed enhancements. Furthermore, the calculations predict the correct trend that the enhancements decrease in magnitude and pitch-angle extent with increasing energy and with increasing L. The overestimate of the magnitude of the flux enhancements surrounding 90° pitch angle could be corrected by reasonable alterations of the distribution of wave energy. In particular, the width of the frequency spectrum assumed by Lyons et al. (1972) was based on individual wave observations. Over a time period of several days, corresponding to the time scale of the pitch-angle diffusion, the frequency spectrum varies somewhat. Thus a wave distribution with a frequency width a factor of 2 wider than that used by Lyons et al. (1972) would be more representative of the average wave distribution over a several-day period. Lyons et al. (1972) showed that a broader frequency distribution would decrease the magnitude of the 90° flux enhancement without significantly affecting the pitch angle extent of the enhancement. However, no reasonable alteration of the wave distribution could change how the flux enhancements

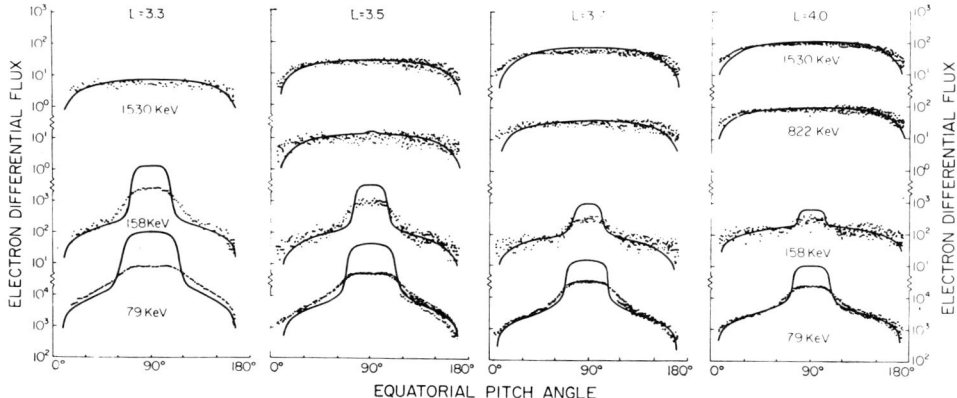

Fig. 5.22. Comparison of calculated equatorial pitch-angle distributions with equatorial distributions observed within the electron slot. The data (electrons/cm²-s-ster-keV) are plotted as a function of pitch angle at the L values and energies indicated on the figure. The vertical placements of the corresponding calculated distributions, shown by solid lines, are arbitrary on a logarithmic scale and have therefore been adjusted for best fit with the data (from Lyons *et al*., 1972). (© by American Geophysical Union)

vary with L and energy. Additional OGO–5 data showing pitch-angle distributions like those in Figure 5.22 were published by West *et al*. (1973).

Explorer 45 observations have now shown that the pitch-angle distributions with enhancements surrounding $\alpha_0 = 90°$ as first observed by West *et al*. (1973) are a general feature of the quiet-time plasmasphere (Lyons and Williams, 1975a). Electron pitch-angle distributions observed on Explorer 45 during a quiet period are shown in Figure 5.23 every 0.2 in L from $L = 2$ to $L = 5$ for four energy channels. Dashes are the observations and solid lines are the corresponding theoretically calculated pitch-angle distributions. The slopes of the pitch-angle distributions in Figure 5.23 are less steep than in Figure 5.22, since the angular resolution of the Explorer 45 measurements ($\sim 22°$) was significantly larger than that on OGO–5 ($\sim 5°$–$10°$). Nevertheless, the enhancements surrounding 90° pitch angle show in the observations, and the size and pitch-angle extent of the enhancements decrease with increasing electron energy and L, in agreement with the calculations.

Since the minimum energy for cyclotron resonance increases significantly with decreasing L, cyclotron resonance does not occur within the inner zone for electrons of the energies observed by Explorer 45. Thus electron scattering by Coulomb collisions with the cold plasmaspheric particles has been included in the calculations shown in Figure 5.23, and the rounding of the calculated pitch-angle distributions within the inner zone occurs when Coulomb collisions dominate the pitch-angle diffusion. The observed pitch-angle distributions in the figure become rounded within the inner zones, the enhancements surrounding $\alpha_0 = 90°$ disappearing at the same L-values as predicted by the calculations ($L \lesssim 2$ for 240–500 keV electrons, and $L \lesssim 3$ for 35–70 keV electrons). This comparison supports the concept that pitch-angle diffusion is dominated by resonant interactions with plasmaspheric hiss within the slot region and by Coulomb collisions within the inner zone. The Coulomb collision rates, however, are very much less than the

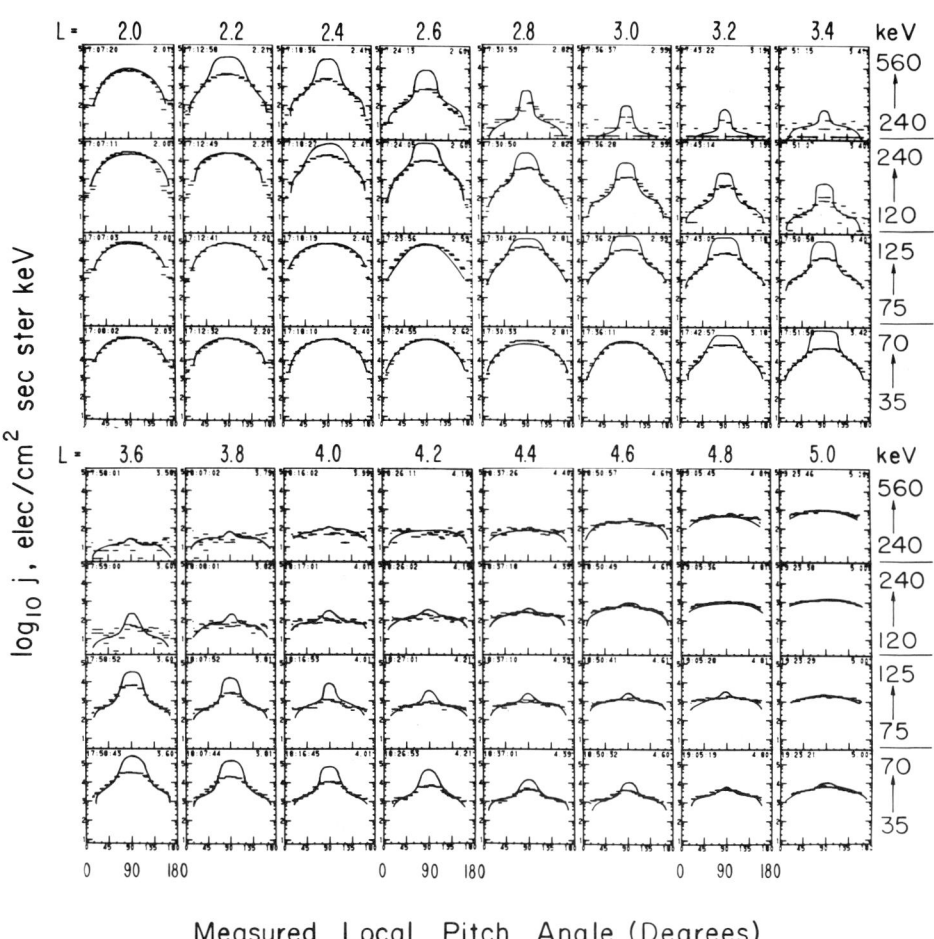

Fig. 5.23. Equatorial pitch-angle distributions observed on December 15, 1971 (orbit 95) and those predicted by Lyons et al. (1972) to result from resonant interactions with the plasmaspheric whistler-mode wave band. Distributions are shown every 0.2 in L from $L = 2$ to $L = 5$, and the four energies are stacked vertically at each L with the lowest energy at the bottom. The dashes give the measured electron flux with their horizontal extent indicating the pitch-angle scan for each measurement. The solid curves show the theoretically predicted pitch-angle distribution for the geometric mean energy of each interval. Diffusion from Coulomb collisions has been added to the theoretical calculations, which accounts for the rounding of the pitch-angle distributions within the inner zone. The vertical positionings of the theoretical distributions are arbitrary on a logarithmic scale and have thus been adjusted to best illustrate the comparison with the observations (from Lyons and Williams, 1975a).
(© by American Geophysical Union)

slot-region loss rates from the wave-particle interactions. This accounts for the much faster loss of slot-region electrons than of inner zone electrons.

Examples of Explorer 45 observations of electron pitch-angle distributions obtained just preceding, during, and following the large geomagnetic storm of December 17, 1971 (Lyons and Williams, 1975b) are shown in Figure 5.24. The observations show that the storm injection results in pitch-angle distributions greatly distorted from their quiet-time shape with no evidence of the bumps surrounding 90°. Following the storm, the distributions return to their pre-storm shape over a period of several days and maintain that shape as the electron fluxes decay back to their quiet-time values. These observations imply that resonant interactions with plasmaspheric hiss become the dominant pitch-angle diffusion mechanism for slot-region electrons almost immediately following storm-time injections.

Calculated precipitation lifetimes for radiation belt electrons in resonance with the plasmaspheric hiss are shown in Figure 5.25. The values shown in the figure are for a total wave amplitude $B_{wave} = 3.5 \times 10^{-2}$ nT ($35m\gamma$) throughout the entire plasmasphere. This amplitude is somewhat larger than average and was chosen because it gives the best agreement with measured decay rates of slot-region electrons following a large magnetic storm. However, the lifetimes in Figure 5.25 can easily be adjusted for other values of B_{wave}, since the lifetimes are proportional to $(B_{wave})^{-2}$, assuming weak pitch-angle diffusion. On the basis of observations of Thorne et al. (1973), a reasonable average value for B_{wave} is 10^{-2} nT ($10m\gamma$), which gives lifetimes approximately a factor of 10 longer than those in Figure 5.25. The variation of the lifetimes with L and energy, however, is independent of B_{wave} for B_{wave} constant throughout the plasmasphere.

For all electron energies, Figure 5.25 shows a relatively sharp L-boundary below which the lifetimes become extremely long. This boundary separates the region of cyclotron resonant loss in the outer region of the plasmasphere from the inner zone region where cyclotron resonance does not occur. Notice that this boundary moves to lower L-values with increasing electron energy. This accounts for the observed fact that the slot extends to lower L-values with increasing electron energy (Russell and Thorne, 1970). One additional feature of the lifetimes is that they increase towards the outer part of the plasmasphere at the higher energies.

Before comparing the calculated precipitation lifetimes with observations, it should be noted that any direct comparison neglects the effects of particle sources. As will be discussed in Section 5.8, electrons continually diffuse in their radial position, and such diffusion will modify the actual particle decay rates. Despite this important limitation, we compare observed decay rates (A. L. Vampola, Personal Communication, 1971) of electron fluxes measured on the OV3–3 satellite with the calculated lifetimes in Figure 5.26. The data used were taken during the period of steady decay of the electron fluxes following the storm-time electron injection on September 4, 1966, shown in Figure 5.17. Because of the increasing importance of radial diffusion with increasing L-value, a detailed comparison of the variation of the decay rates with L is inappropriate. However, the calculated large change in the decay rates, which occurs at the boundary between the inner zone and the slot region, and its variation in position with electron energy compare favorably with the observations.

In addition to the comparisons with equatorial electron measurements, low altitude measurements of electron precipitation from the slot region have been found to be

WAVE-PARTICLE INTERACTIONS 179

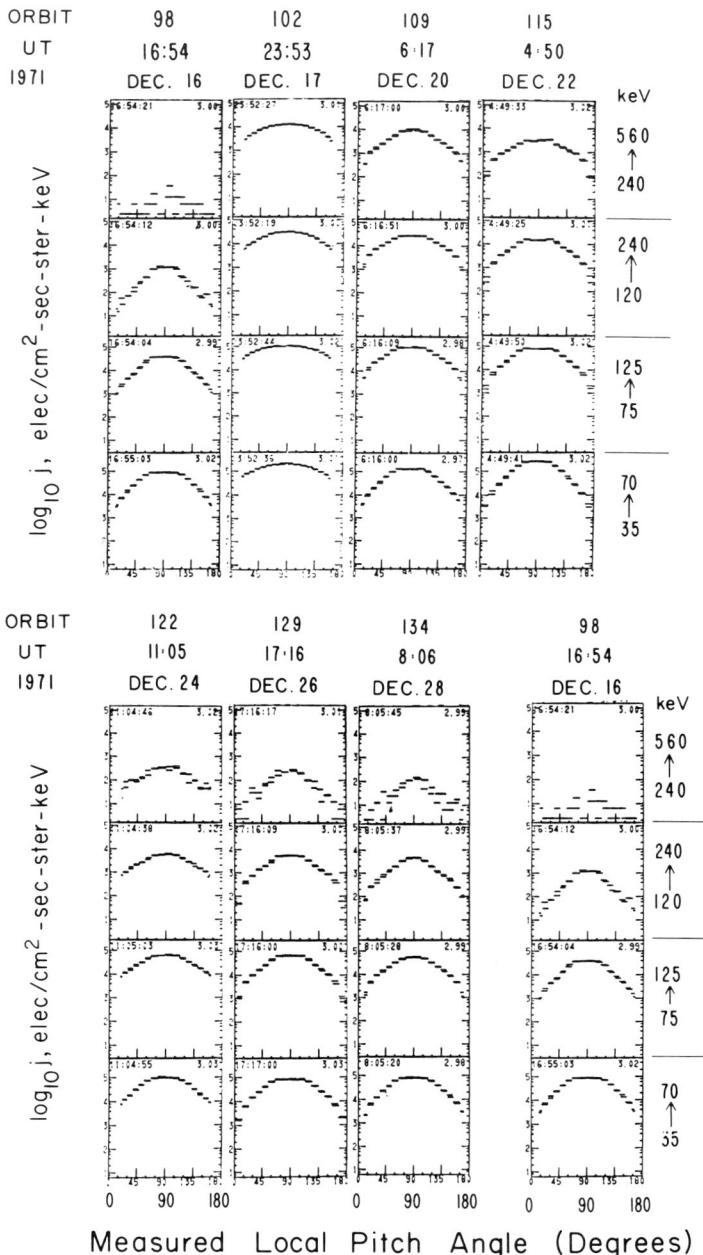

Fig. 5.24. Observed pitch-angle distributions at $L = 3$ for the periods preceding, during, and following the storm of December 17, 1971. Pre-storm distributions are repeated at end of figure to indicate return to pre-storm shapes and intensities (from Lyons and Williams, 1975b). (© by American Geophysical Union)

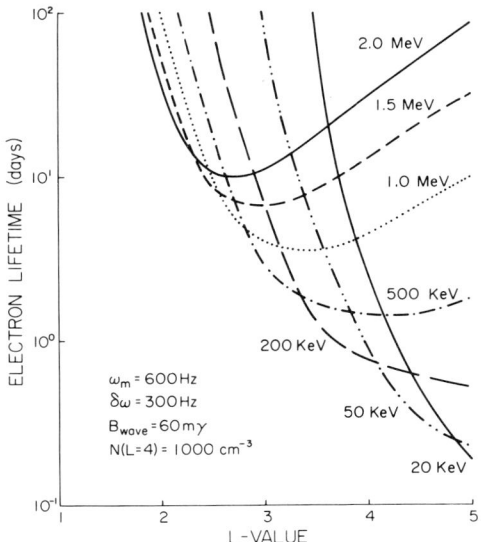

Fig. 5.25. Theoretical precipitation lifetimes versus L-curves for several electron energies. The results are not valid at L-values beyond the plasmapause (from Lyons *et al.*, 1972). (© by American Geophysical Union)

consistent with precipitation expected from resonant interactions with plasmaspheric hiss (Imhof *et al.*, 1974, 1982). The latter study used measurements of the wave spectra and plasma density near the equator obtained simultaneously with the low altitude measurements of electron precipitation.

The comparisons presented here demonstrate a quantitative understanding of the formation of the electron slot by resonant electron interactions with plasmaspheric hiss. The results are consistent with the long-term stability of the inner zone, the outward extension of the inner zone as electron energy decreases, and the removal of electrons to low flux levels in the slot. These results are combined with radial diffusion in Section 5.8 to provide a quantitative explanation for the radial profiles of the electron fluxes throughout the quiet-time plasmasphere.

5.6. Loss of Ring Current Ions by Ion-Cyclotron Waves

Radiation belt ions can cause the growth of ion-cyclotron waves in much the same manner that electrons cause the growth of whistler-mode waves. A major difference, however, is that the group velocity of ion-cyclotron waves is far more collimated along the ambient magnetic field direction at all wave normal angles and frequencies than is the group velocity of whistler-mode waves. Thus wave propagation across magnetic field lines should be far less important for ion-cyclotron waves than for whistler-mode waves, so that interactions at a given L-value should result only from waves generated near that L-value.

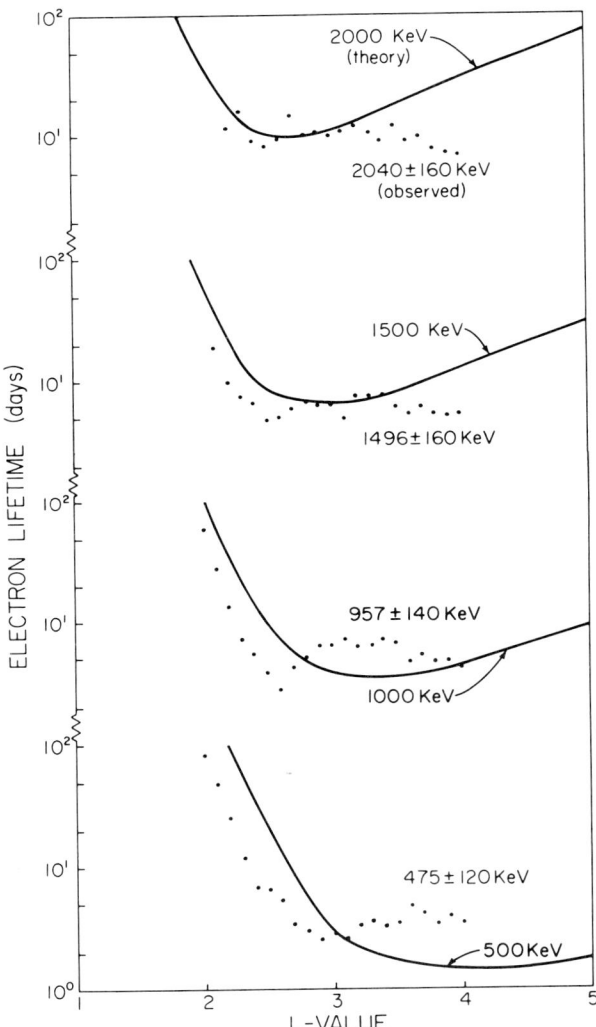

Fig. 5.26. Comparison of slot-region electron decay rates at four energy channels observed following a large injection event (Vampola, personal communication, 1971) with the theoretical precipitation lifetimes versus L-curves (from Lyons *et al.*, 1972). (© by American Geophysical Union)

The cold plasma dispersion relation for parallel propagating ion-cyclotron waves $\widetilde{\mu}^2 = L$ becomes, for $(\omega_{pe}^2/\Omega_e^2) \gg 1$,

$$\widetilde{\mu}^2 = \frac{\omega_{pi}^2}{\Omega_i^2}\left(1 - \frac{\omega}{\Omega_i}\right)^{-1},$$

where $\omega < \Omega_i$ for ion-cyclotron waves. This gives the parallel ion energy for cyclotron resonant interactions with parallel propagating ion cyclotron waves as

$$K_{\parallel,\,\text{res}} = \frac{B^2}{8\pi N} \left(\frac{\Omega_i}{\omega}\right)^2 \left(n + \frac{\omega}{\Omega_i}\right)^2 \left(1 - \frac{\omega}{\Omega_i}\right). \tag{5.40}$$

Thus the resonant energies for ion interactions with ion-cyclotron waves depend upon the same characteristic energy $B^2/8\pi N$ as do the resonant energies for electron interactions with whistler-mode waves (Kennell and Petschek, 1966). Also, as in the case of electron-whistler interactions, cyclotron resonance with ion-cyclotron waves when wave energy is distributed over some band of frequencies, occurs for all parallel ion velocities greater than some minimum value, provided interactions over a wide range of latitudes and $|n| > 1$ harmonic resonances are included.

Cornwall *et al.* (1970) noted that $B^2/8\pi N$ minimized just within the plasmapause as shown in Figure 5.12, so that maximum instability of ion-cyclotron waves should occur just within the plasmapause. This instability was predicted to be most intense following the storm-time injection of ions into the inner regions of the radiation belts which gives rise to the storm-time ring current (see Chapter 3). The injection of ring current ions occurs together with, and over approximately the same L-value range as, the storm-time injection of electrons into the slot-region discussed in the previous section. The ions are often referred to as 'ring current' ions rather than simply 'radiation belt' ions because they form most of the current surrounding the earth responsible for the storm-time magnetic field decrease on the ground.

Cornwall *et al.* (1970) suggested that the pitch-angle diffusion of ions resonant with the ion cyclotron waves generated within the plasmapause should be an important loss process for the ring current. Lifetimes of the resonant ions were predicted to be less than a day, which is less than the time scale over which the plasmasphere refills following a storm. Thus losses were predicted to occur first at the lower L-values within the location of the depressed, storm-time plasmapause. The ion losses were predicted to then extend outwards to higher L with the refilling of the plasmasphere during the recovery phase of storms.

Williams and Lyons (1974a, b) investigated the theoretical predictions of Cornwall *et al.* (1970) by performing a detailed analysis of Explorer 45 measurements of the pitch-angle distributions of ring current ions obtained during the recovery phase of the large geomagnetic storm on December 17, 1971. Examples of the Explorer 45 measurements of the equatorial pitch-angle distributions of ring current ions obtained near local midnight during the recovery phase of the storm are shown in Figures 5.27 and 5.28. The observations in Figures 5.27 and 5.28 were obtained ~ 16 hr (orbit 103) and ~ 24 hr (orbit 104) after the minimum D_{st} of the storm main phase. In each figure, pitch-angle distributions for selected ion energy channels are stacked vertically every 0.4 in L from $L = 3.0$ to $L = 5.0$.

Note in both figures the transition from nearly isotropic pitch-angle distributions at the lower energies to rounded pitch-angle distributions, peaked at 90° pitch angle, at the higher energies. This transition occurs at higher ion energies with increasing L-value. A careful examination of the nearly isotropic distributions shows that when the pitch-angle scans reach the loss cone, significant intensity decreases are seen; these decreases imply a nearly empty loss cone (Williams and Lyons, 1974a). Such isotropic distributions with

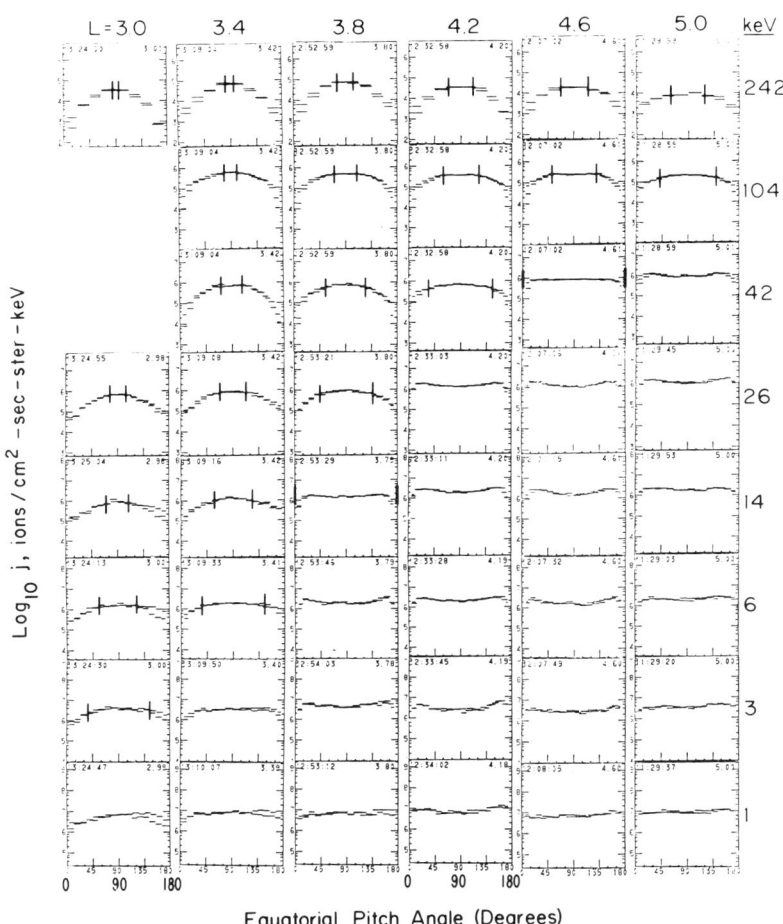

Fig. 5.27. Equatorial proton pitch-angle distributions observed on Explorer 45 orbit 103 inbound ~ 16 hr after the minimum D_{st} of the December 17, 1971, storm main phase. Distributions are shown every 0.4 in $L = 3$ to $L = 5$, and selected proton energy channels are stacked vertically at each L. No data suffering from the saturation problem with the solid-state detector (Williams et al., 1973) are shown. Elevated fluxes at pitch angles $90°-180°$ for energies ⩽ 14 keV are due to reflected sunlight. Ticks are at constant values of $K_{\parallel}$ for each L, the chosen value of $K_{\parallel}$ at each L being equal to the energy of one of the Explorer 45 channels (from Joselyn and Lyons, 1976). (© by American Geophysical Union)

an empty loss cone are indicative of a stably trapped particle population undergoing negligible losses from pitch-angle diffusion. The isotropy must have originated from either an isotropic injection or strong pitch-angle diffusion during the storm main phase. However, the empty loss cones demonstrate that significant pitch-angle diffusion was not occurring for these ions at the time of their observation during the storm recovery phase.

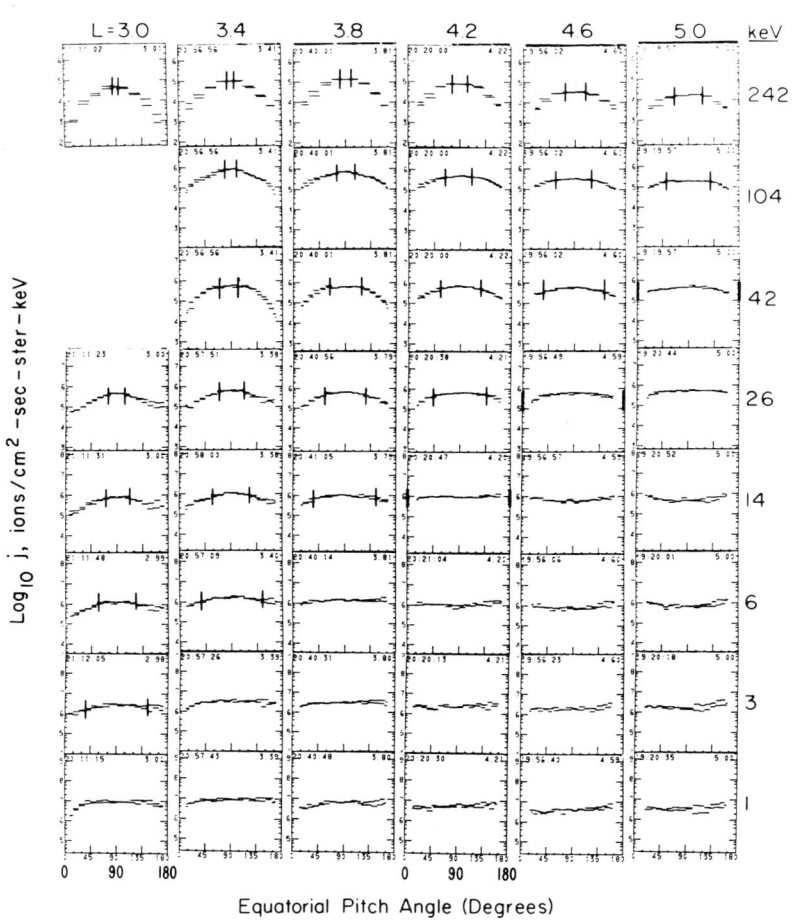

Fig. 5.28. Same as Figure 5.27, but for Explorer 45 orbit 104 inbound, ~ 24 hr after the minimum D_{st} of the storm main phase (from Joselyn and Lyons, 1976). (© by American Geophysical Union)

Pitch-angle distributions with fluxes increasing monotonically towards 90° pitch angle are expected under conditions of pitch-angle diffusion into the loss cone, except in the limit of strong pitch-angle diffusion. Williams and Lyons (1974a) noted that isotropic distributions evolved to rounded distributions during the storm recovery phase in such a manner that the rounded distributions extend to decreasing energies with time. Such a rounding of initially isotropic distributions can be seen to have occurred between the times of the observations in Figures 5.27 and 5.28 for 14 keV ions at $L = 3.8$, for 26 keV ions at $L = 4.2$, and for 42 keV ions at $L = 4.6$.

The evolution from isotropic distributions to rounded distributions can also be seen in Figure 5.29. In this figure, contours of constant distribution function are plotted in the $(v_\perp, v_\parallel)$-plane for $L = 3.2, 4.0$, and 5.0. Observations are shown for orbit 101, just after

WAVE-PARTICLE INTERACTIONS 185

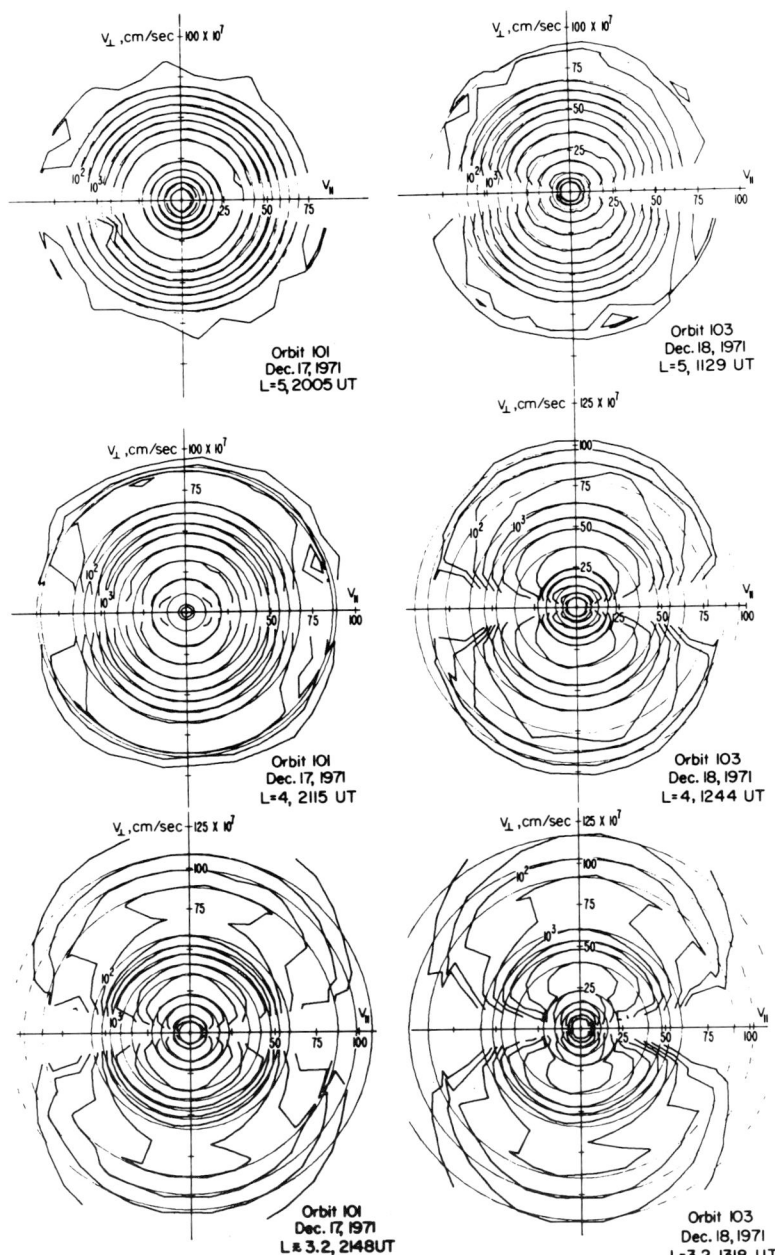

Fig. 5.29. Contours of constant j/E, which is proportional to the phase space density f, plotted in $(v_\perp, v_\parallel)$-plane. Circles centered at $v_\perp = v_\parallel = 0$ (thin lines) are added for reference. Contours are generated every 0.5 unit in $\log_{10} j/E$ by computer; no smoothing is applied to the data. Contours are generated for $v_\perp > 0$ and $v_\perp < 0$ separately, and region of no available data separates the two sets of contours. These plots show the spatial evolution of f just following the main phase particle injection (orbit 101) and during the recovery phase (orbit 103) (from Williams and Lyons, 1974a). (© by American Geophysical Union)

the particle injection associated with the storm main phase, and for orbit 103, approximately 16 hr into the storm recovery phase. Notice that just following the main phase particle injection, the particle distributions are isotropic at the lower energies and rounded at the higher energies, and the transition energy decreases with decreasing L. Comparison of the distributions for orbits 101 and 103 shows the decrease in this transition energy with time.

A logical process for causing isotropic distributions to evolve to rounded distributions above a given ion energy is pitch-angle diffusion driven by resonant wave-particle interactions. If this is the case, then cyclotron resonant interactions should occur for all parallel particle energies $K_\parallel$ greater than a minimum value $K_{\parallel,\,min} = 1/2 m v_{\parallel,\,min}^2$.

Joselyn and Lyons (1976) investigated whether the ion distributions were consistent with pitch-angle diffusion occurring for $K_\parallel > K_{\parallel,\,min}$. For each altitude, the highest Explorer 45 energy channel showing a nearly isotropic pitch-angle distribution was chosen to be $K_{\parallel,\,min}$, and the pitch angles corresponding to this chosen value of $K_\parallel$ were calculated for all higher energy channels, using $\cos^2 \alpha_0 = K_{\parallel,\,min}/K_\parallel$. These pitch angles are indicated by the vertical ticks on the pitch-angle distributions in Figures 5.27 and 5.28. In addition, the ion pitch-angle distributions from orbit 103 for two sample L-values, 3.4 and 4.6, are shown in Figure 5.30 for every available energy channel over the range of energies for which the transition to rounded distributions occurs. The chosen value of $K_{\parallel,\,min}$ is equal to the energy of the Explorer 45 channel with ticks shown at 0° and 180°. Notice that within the accuracy of the pitch-angle measurements (22° for $K \leq 104$ keV, 33° for $K > 104$ keV), the pitch-angle distributions appear to be nearly isotropic between the ticks, i.e., for parallel energies less than $K_{\parallel,\,min}$. At larger parallel energies (pitch angles approaching 0° and 180°) the pitch-angle distributions are rounded. Thus the rounding of the pitch-angle distributions appears to occur for $K_\parallel > K_{\parallel,\,min}$, as expected from wave-particle interactions.

From Equation (5.40), we can use the measured value of $K_{\parallel,\,min}$ to estimate the plasma density as a function of radial distance. At the $n = -1$ resonance we have

$$\frac{N}{F(\omega/\Omega_i)} = \frac{B^2}{8\pi K_{\parallel,\,min}},$$

where

$$F\left(\frac{\omega}{\Omega_i}\right) = \left(\frac{\Omega_i}{\omega}\right)^2 \left(1 - \frac{\omega}{\Omega_i}\right)^3.$$

The highest value of ω/Ω_i for which there is significant wave energy should be used when evaluating $F(\omega/\Omega_i)$, since $K_{\parallel,\,res}$ decreases with increasing ω/Ω_i. Unfortunately, measurements of ω/Ω_i in the plasmasphere are not available because of technical difficulties in measuring low amplitude waves at these frequencies. However, growth rate calculations using the observed ion distributions (Joselyn and Lyons, 1976) indicate a peak value of $\omega/\Omega_i \approx 0.3$. This can be used to obtain an order-of-magnitude estimate for $F(\omega/\Omega_i)$ given by $F(0.3) = 4$. Of course, it must be remembered that the real value of ω/Ω_i may vary as a function of L and time, and $F(\omega/\Omega_i)$ is a somewhat sensitive function of ω/Ω_i.

Figure 5.31 shows $N/F(\omega/\Omega_i)$ as obtained by Williams and Lyons (1974b) for orbit 101, just following the storm main phase, and for orbits 103 and 104 during the recovery

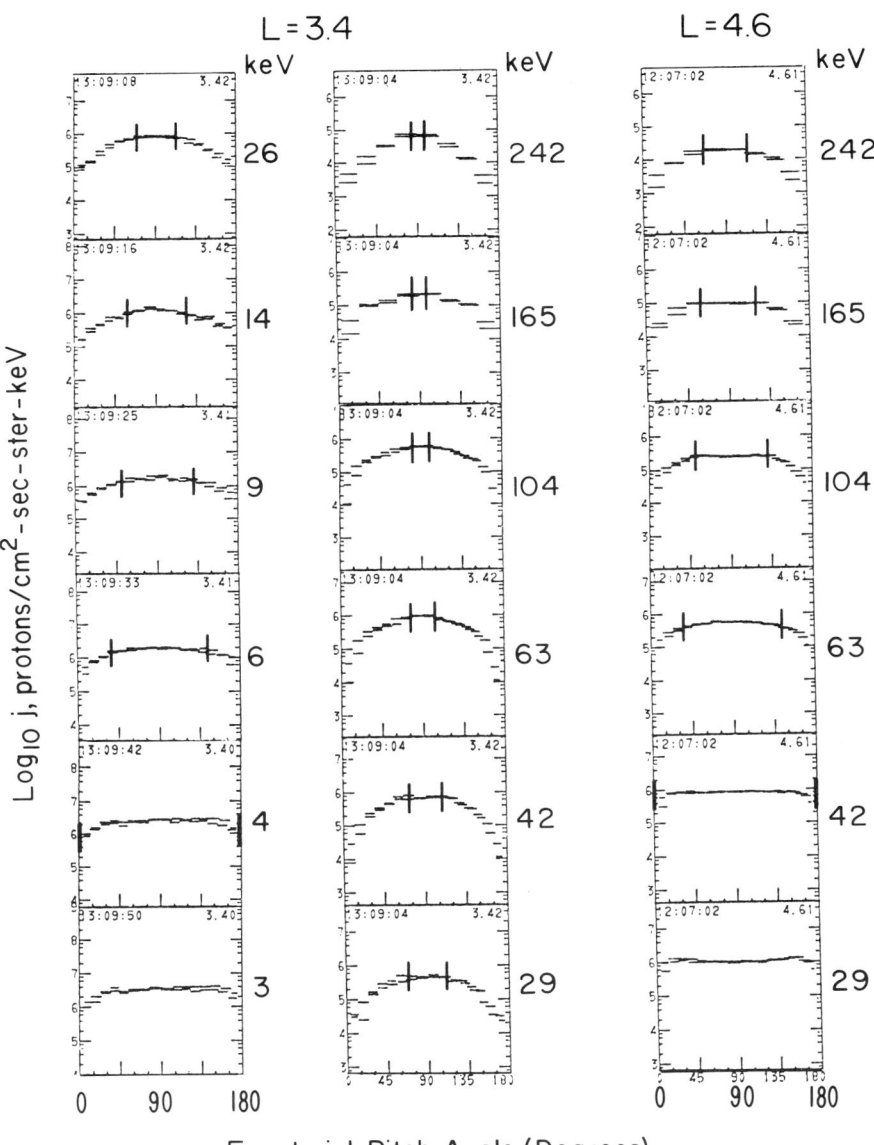

Fig. 5.30. Pitch-angle distributions from orbit 103 inbound for $L = 3.4$ and 4.6. Every available energy channel is shown over the range of proton energies for which the transition from flat (with an empty loss cone) to rounded (peaked at 90° pitch angle) distributions occurs. Vertical ticks shown on each pitch-angle distribution are at constant values of $K_\parallel$ for each L. Note that to within the accuracy of the measurements, the pitch-angle distributions are nearly isotropic between the ticks and rounded outside the ticks (from Joselyn and Lyons, 1976). (© by American Geophysical Union)

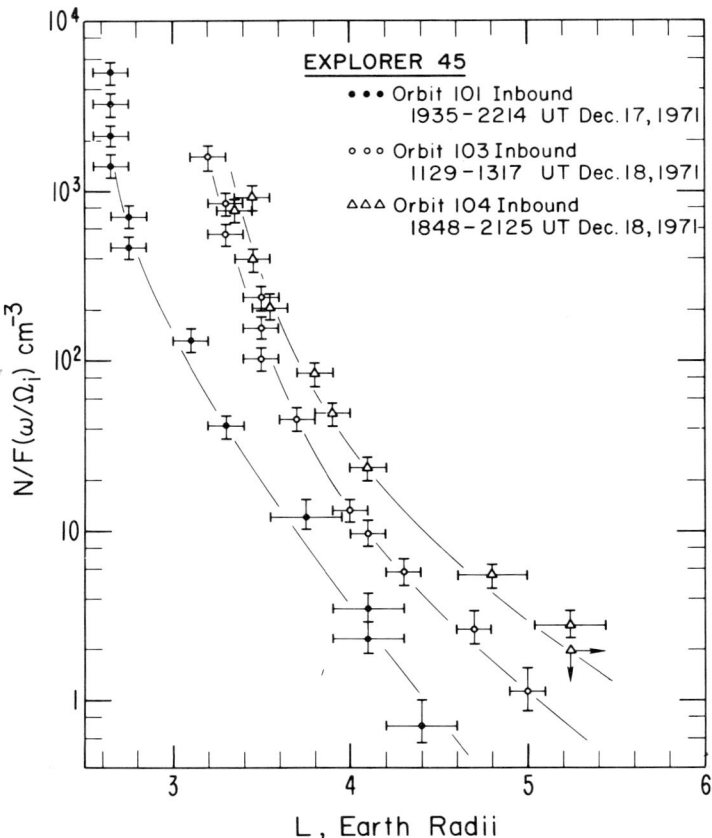

Fig. 5.31. $N/F(\omega/\Omega_i)$ for orbit 101, just following the storm main phase, and for orbits 103 and 104 during the storm recovery phase. The measured value of B and the value of $K_{\|,\,min}$ obtained from the observed pitch-angle distributions have been used (from Williams and Lyons, 1974b). (© by American Geophysical Union)

phase of the storm. The satellite was significantly farther off the equator during orbit 102 than during the other three orbits, making observations from orbit 102 much more difficult to interpret. The measured value of B and the value of $K_{\|,\,min}$ obtained from the observed pitch-angle distributions were used in obtaining the results shown in Figure 5.31. In spite of the uncertainties in $F(\omega/\Omega_i)$, the general shape of the density profiles is reasonable on the basis of the depression of the plasmapause during the storm main phase and the subsequent refilling of the plasmasphere (Chappell *et al.*, 1970; Park, 1970).

Using $F(\omega/\Omega_i) = 4$, estimates of the plasmaspheric refilling rates obtained by Williams and Lyons (1974b) from the density profiles shown in Figure 5.31 are 4.8×10^7 to 8.8×10^8 ions/cm²-s. These estimates compare favorably with the estimate of plasmaspheric refilling rates of 4.5 to 6.6×10^7 ions/cm²-s obtained by Chappell *et al.* (1970) from in situ cold plasma observations, of 1 to 3×10^8 ions/cm²-s obtained by Park

(1970, 1974) from ground recordings of whistlers, and with the maximum rates of 7×10^7 to 2×10^8 ions/cm²-s given by Banks (1972).

The spatial, parallel energy, and temporal dependences of the pitch-angle distributions, and the inferred values of $N/F(\omega/\Omega_i)$, strongly indicate that the rounding of the pitch-angle distribution is due to resonant interactions with ion-cyclotron waves generated by the interaction of the energetic ring current ions with the cold plasmaspheric plasma as predicted by Cornwall et al. (1970). In order to be easily observable, this process must be an important loss process for ring current ions.

For any ring current ion energy at a given time, rounded distributions resulting from pitch-angle diffusion lie at lower L than do the isotropic distributions whose empty loss cones are indicative of stably trapped cones. This led Williams and Lyons (1974a) to predict that a region of anisotropic ion precipitation into the ionosphere should exist during a storm recovery phase. This anisotropic precipitation was predicted to be at lower latitudes and distinctly separated from the isotropic ion precipitation observed at auroral latitudes well outside the plasmapause. This isotropic, auroral precipitation was expected to originate from altitudes above those of the stable ion distributions observed on Explorer 45.

Such a two-zone precipitation structure is schematically illustrated in Figure 5.32. Williams and Lyons presumed the isotropic, auroral ion precipitation was due to strong pitch-angle diffusion beyond the apogee of the Explorer 45 satellite. However, the results of Lyons and Speiser (1982) imply that the isotropic precipitation is at least partially a result of current sheet energization of ions in the geomagnetic tail. This two-zone precipitation structure might also be expected during quiet periods, though the ion precipitation within the plasmapause should be much weaker than during a storm recovery. Such a two-zone precipitation structure of precipitating ions has since been observed at 1–6 keV by Hultqvist et al. (1976) and at $>$ 100 keV by Lundblad et al. (1979), thus providing additional support to the idea that ion-cyclotron wave interactions within the plasmasphere are an important loss process for ring current ions.

The arguments of Williams and Lyons (1974a, b) imply that the evolution of initially isotropic distributions with an empty loss cone to rounded distributions peaked at $\alpha_0 = 90°$ occurs as the density increases during a storm recovery, thus causing $K_{\|, \min}$ to decrease. However, their argument does not explain how the isotropic distributions, which apparently were not initially undergoing pitch-angle diffusion, could become resonant with growing waves. Although $K_{\|, \min}$ lowers with increasing density, the isotropic distributions, even with an empty loss cone, have too small a pitch-angle anisotropy to grow waves. Joselyn and Lyons (1976) suggested a resolution to the problem by noting that $K_{\|, \text{res}} \sim B^4/N$ along field lines for a wave of a given frequency. Thus $K_{\|, \text{res}}$ must increase markedly with increasing latitude. Thus waves at the highest frequencies, which must cause the observed rounding of the ion pitch-angle distributions to extend to decreasing values of $K_\|$ with increasing plasma density, can be generated off the equator. Under this suggestion, the highest frequency waves are generated off the equator at values of $K_\| > K_{\|, \min}$ where the distributions are rounded. These waves then propagate towards the equator and resonate with decreasing value of $K_\|$, causing the $K_\|$ of rounded distributions to decrease.

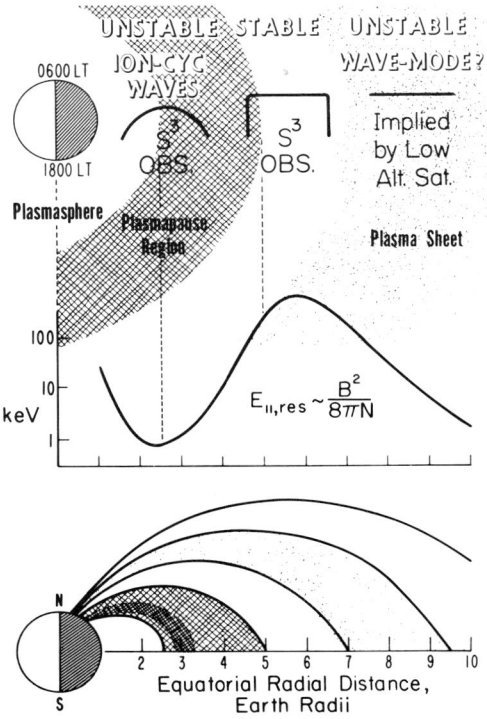

Fig. 5.32. Schematic of geomagnetic storm recovery phase showing hot ring-current plasma behavior in the geomagnetic field and its interaction with the cold plasmaspheric plasma. The top panel shows an equatorial projection, and the bottom panel shows a midnight-meridian projection. Midnight-meridian features are projected to low altitudes to indicate expected low-altitude latitudinal structure. A semiquantitative resonant energy-versus-altitude plot that used Explorer 45 results is shown for reference. Moderate pitch-angle diffusion occurs for the hot ring-current plasma in the plasmapause due to the amplification of ion-cyclotron waves. Each energy of the ring-current plasma begins its interaction in the plasmapause region at the appropriate value of $B^2/8\pi N$. Heavier shading in the plasmapause region is used to indicate that energy flow to the ionosphere from the hot ring-current plasma may be peaked at preferential locations and will depend on the plasmapause shape and hot plasma energy spectrum. More than one peak in the energy flow is possible. Above the plasmapause region the hot ring-current plasma is stably trapped with negligible or no losses due to pitch-angle scattering. Above this region, and above Explorer 45 apogee during recovery phase, there is the region of proton auroral precipitation exhibiting a full loss cone as implied by low-altitude measurements. This isotropic precipitation region was presumed to result from strong pitch-angle diffusion by Williams and Lyons (1974a, b), but probably results at least partially from current sheet energization of ions in the tail (Lyons and Speiser, 1982) (From Williams et al., 1976). (© by American Geophysical Union)

5.7. Electrostatic Waves Outside the Plasmapause

The examples of magnetospheric wave turbulence discussed in the previous section all had dispersion relations adequately described by cold plasma theory. However cold plasma theory does not describe the propagation characteristics of all wave modes in the magnetosphere. An important class of such waves consists of the intense electrostatic waves detected outside the plasmapause, which have peak intensities at frequencies near $p + 1/2$ times the local electron gyrofrequencies, where $p = 1, 2, 3, \ldots$ (Kennel et al., 1970; Fredricks and Scarf, 1973; Scarf et al., 1973; Shaw and Gurnett, 1975; Christiansen et al., 1978; Gurnett et al., 1979). Typical amplitudes for these waves are 1–10 mV m^{-1}, though amplitudes up to 100 mV m^{-1} have been observed. Kennel et al. (1970) suggested that these waves may be an important cause of pitch-angle diffusion for 0.1–10 keV electrons on auroral magnetic field lines, and Scarf et al. (1973) presented evidence that the waves can cause significant pitch-angle diffusion of electrons up to 80 keV in energy during geomagnetic disturbances.

Theoretical studies of the generation of these waves (e.g., Fredricks, 1971; Young et al., 1973, Ashour-Abdalla et al., 1975) have made use of the general dispersion relation for electrostatic waves (Harris, 1959):

$$D(\omega^*, \mathbf{k}) = 1 + \sum_j \frac{\omega_{pj}^2}{N_j k^2} \sum_{n=-\infty}^{\infty} \int d^3v \, \frac{J_n^2 [(k_\perp v_\perp)/\Omega_j]}{\omega^* - k_\parallel v_\parallel + n\Omega_j} \cdot$$
$$\cdot \left[-\frac{n\Omega_j}{v_\perp} \frac{\partial f_j}{\partial v_\perp} + k_\parallel \frac{\partial f_j}{\partial v_\parallel} \right] = 0, \tag{5.41}$$

where $\omega^* = \omega + i\gamma$ is the complex wave frequency.

For small growth rates $\gamma \ll \omega$, this dispersion relation can be expanded about $\gamma = 0$ to obtain

$$D(\omega^*, \mathbf{k}) \approx D(\omega, \gamma = 0, \mathbf{k}) + \gamma \frac{\partial D}{\partial \gamma} \approx 0.$$

Assuming D is an analytic function and applying the Cauchy-Rieman conditions, the above equation can be separated into real (Re) and imaginary (Im) parts, yielding

$$\text{Re } D(\omega, \gamma = 0, \mathbf{k}) - \gamma \frac{\partial}{\partial \omega} \text{Im } D(\omega, \gamma = 0, \mathbf{k}) \approx 0 \tag{5.42}$$

$$\text{Im } D(\omega, \gamma = 0, \mathbf{k}) + \gamma \frac{\partial}{\partial \omega} \text{Re } D(\omega, \gamma = 0, \mathbf{k}) \approx 0. \tag{5.43}$$

From Equation (5.43), we obtain

$$\gamma \approx -\frac{\text{Im } D(\omega, \gamma = 0, \mathbf{k})}{(\partial/\partial \omega) \text{Re } D(\omega, \gamma = 0, \mathbf{k})}, \tag{5.44}$$

and for $\gamma \to 0$, Equation (5.42) gives the relation between ω and $\mathbf{k}$ as

$$\text{Re } D(\omega, \gamma = 0, \mathbf{k}) \approx 0. \tag{5.45}$$

The real and imaginary parts of D necessary to evaluate (5.44) and (5.45) are given by

$$\operatorname{Re} D(\omega, \gamma = 0, \mathbf{k}) = 1 + \sum_j \frac{\omega_{pj}^2}{N_j k^2} \sum_{n=-\infty}^{\infty} P\int d^3v \frac{J_n^2[(k_\perp v_\perp)/\Omega_j]}{\omega - k_\| v_\| + n\Omega_j} \cdot$$

$$\cdot \left[-\frac{n\Omega_j}{v_\perp} \frac{\partial f}{\partial v_\perp} + k_\| \frac{\partial f}{\partial v_\|} \right], \quad (5.46)$$

$$\operatorname{Im} D(\omega, \gamma = 0, \mathbf{k}) = -2\pi^2 \sum_j \frac{\omega_{pj}^2}{N_j k^2} \sum_{n=-\infty}^{\infty} \int dv_\perp\, J_n^2 \left(\frac{k_\perp v_\perp}{\Omega_j}\right) \cdot$$

$$\cdot \hat{G} f \bigg|_{v_\| = \frac{\omega + n\Omega_j}{k_\|}}, \quad (5.47)$$

where $\operatorname{Im} D$ is written in terms of $\hat{G}$ to emphasize the correspondence of the electrostatic wave growth rate expression (5.44) with the more general electromagnetic growth rate expression (5.3). As in Equation (5.3), $\omega/k_\|$ is taken to be positive, and $v_{\|,\text{res}}$ is allowed to have either sign. From (5.44) and (5.47), $\hat{G}f > 0$ is necessary for wave growth whenever $(\partial/\partial\omega)\operatorname{Re} D(\omega, \gamma = 0, k) > 0$. Solutions of (5.45) exist for $(\partial/\partial\omega)\operatorname{Re} D(\omega, \gamma = 0, \mathbf{k}) < 0$; however, Dusenbery and Kaufmann (1980) have shown that, for a Maxwellian particle distribution function, such solutions are spurious since the expansion, (5.41) and (5.43), of $D(\omega^*, \mathbf{k})$ does not converge sufficiently for the resulting values of ω and γ to satisfy the dispersion relation (5.41) for $\omega^* = \omega + i\gamma$. Although we do not know of a study showing that all solutions of (5.45) with $(\partial/\partial\omega)\operatorname{Re} D < 0$ are spurious for an arbitrary particle distribution function, the energy analysis of Section 5.1 suggests that this is the case.

Analyses of the $(p + 1/2)\Omega_e$ electrostatic waves have generally considered the effects of regions of the electron distribution function where $\partial f/\partial v_\perp > 0$, which can occur as the result of loss-cone type of pitch-angle distributions, and the analyses have generally assumed $\partial f/\partial |v_\||| < 0$ at all $\mathbf{v}$. Since $\hat{G}f = [(\omega/k_\| - v_\|)(\partial f/\partial v_\perp) + v_\perp(\partial f/\partial v_\|)]$, a necessary, but not sufficient, condition for wave growth when $\partial f/\partial |v_\||| < 0$ is that $\omega/k_\| - v_{\|,\text{res}} > 0$ in some resonant region of the electron distribution where $\partial f/\partial v_\perp > 0$. The resonance condition gives us that

$$\frac{\omega}{k_\|} - v_{\|,\text{res}} = -\frac{v_{\|,\text{res}}}{(\omega/n\Omega) + 1}.$$

This can be greater than zero only for $v_{\|,\text{res}} > 0$ and $\omega/n\Omega < -1$. (For $v_{\|,\text{res}} < 0$, we must have $\omega/n\Omega < -1$, which implies $n \geq 0$ for $\omega > \Omega$. This gives $v_{\|,\text{res}} = (\omega/k_\|)(1 + n\Omega/\omega) > 0$, so that $\omega/k_\| - v_{\|,\text{res}}$ cannot be less than zero for $v_{\|,\text{res}} < 0$). For $p\Omega < \omega < (p+1)\Omega$, we thus find that wave growth is possible only at cyclotron harmonic resonances for which $-1 \geq n \geq -p$. For a given value of ω/Ω, $v_{\|,\text{res}}$ minimizes for $n = -p$. Thus for a given value of $(1/f)(\partial f/\partial v_\perp) > 0$, $\hat{G}f$ will maximize at $n = -p$ if f monotonically decreases with increasing $|v_\||$. However, the peak values of J_n^2 decrease with increasing $|n|$, so that an evaluation of γ is necessary to determine which values of n will dominate wave growth for a particular particle distribution function.

All cyclotron harmonic resonances $n \geq 1$ and $n \leq -p-1$ contribute damping for $\partial f/\partial v_\perp > 0$ or growth for $\partial f/\partial v_\perp < 0$. However, for any n, $v_{\|,\text{res}}$ is greater at the $+n$ resonance than at the $-n$ resonance, so that the effects of the negative resonance will generally dominate the effects of the corresponding positive resonances. Additionally, although for n between -1 and $-p$ there is a value of $n \leq -p-1$ with approximately the same $|v_{\|,\text{res}}|$, the decrease in the magnitude of J_n^2 with $|n|$ decreases the effects of the $n \leq -p-1$ resonances relative to the effects of the $-1 \geq n \geq p$ resonances.

The above considerations have not included the real part of the wave dispersion relation which must be considered to determine whether or not wave propagation can occur at values of ω and $k_\|$ that give resonance at $v_\|$'s with $\partial f/\partial v_\perp > 0$. However, relations for $\omega(\mathbf{k})$ as general as those from cold plasma theory are not possible for the $(p+1/2)\Omega_e$ waves since $\omega(\mathbf{k})$ depends upon the electron distribution function. Thus most studies of the $(p+1/2)\Omega_e$ waves have assumed a form for the electron distribution function and have solved Equations (5.44)–(5.47) for $\omega(\mathbf{k})$ and $\gamma(\mathbf{k})$ simultaneously.

A common assumption for the electron distribution function is that it consists of a cold Maxwellian portion with density N_C and temperature T_C and a hot portion with a specified density and temperature, and $\partial f/\partial v_\perp > 0$ for $v_\perp$ between 0 and some maximum value (Young et al., 1973; Ashour-Abdalla et al., 1975; Ashour-Abdalla and Kennel, 1978; Hubbard and Birmingham, 1978). A general result of these studies is that instability is possible for frequencies in harmonic bands between $p\Omega_e$ and $(p+1)\Omega_e$ in agreeent with the wave observations. The highest unstable frequency is expected to be near (Ashour-Abdalla and Kennel, 1978) or somewhat above (Hubbard and Birmingham, 1978; Rönmark et al., 1978) the upper hybrid frequency $\omega_{UH}^c = (\Omega_e^2 + \omega_{pc}^2)^{1/2}$ of the cold electrons. Here $\omega_{pc}^2 = 4\pi N_c e^2/m_e$ is the plasma frequency of the cold electrons. An important result of Ashour-Abdalla and Kennel (1976, 1978) is the need to consider the spatial wave growth rate γ/v_g, since v_g can approach zero and thus strongly affect the wave amplification.

Observations have now indicated that electron distribution functions outside the plasmapause having $\partial f/\partial v_\perp > 0$ at resonant $v_\|$'s are associated with the electrostatic waves at $(p+1/2)\Omega_e$ (Sentman et al., 1979; Kurth et al., 1979a, 1980). In addition, wave growth rate calculations based on satellite observatons of the electron distribution function have successfully explained the frequency peaks in simultaneously observed wave spectra (Ronmark, 1978; Kurth et al., 1979b).

Kennel et al. (1970) estimated that the $(p+1/2)\Omega_e$ waves observed in the magnetosphere were sufficiently intense to drive electrons of $\sim 0.1-10$ keV onto strong pitch-angle diffusion, thus causing precipitation of these electrons into the auroral zone with isotropic pitch-angle distributions. Pitch-angle diffusion rates at higher energies, due to higher harmonic cyclotron resonances and off-equatorial interactions, will generally decrease with increasing electron energy (Lyons, 1974c). Thus auroral electron precipitation driven by the $(p+1/2)\Omega_e$ waves should be isotropic at the lowest energies for which precipitation occurs. At some higher energy, there should be a transition to anisotropic pitch-angle distributions, and the anisotropy should increase with increasing electron energy above the transition energy. The energy spectra of the precipitation will monotonically decrease with increasing electron energy, assuming the source magnetospheric electron distribution is not peaked in energy.

Diffuse auroral electron precipitation, including that associated with pulsating aurora, generally has pitch-angle distributions and energy spectra as expected from resonant interactions with the $(p + 1/2)\Omega_e$ waves. An example of such precipitation is shown in Figure 5.33. The triangles and circles in this figure give sounding rocket observations of the electron fluxes obtained by Whalen and McDiarmid (1973) during a post-breakup aurora at two local pitch angles, 71° and 42°, within the loss cone. The energy flux

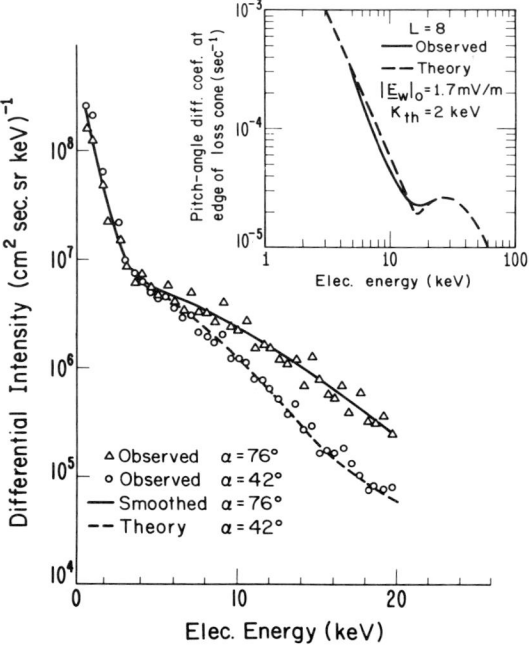

Fig. 5.33. Sounding-rocket observations of electron energy spectra obtained within the loss cone at two local pitch angles ($\alpha = 76°$ and $\alpha = 42°$) during a postbreakup aurora (from Whalen and McDiarmid, 1973). In the upper right, the pitch-angle diffusion coefficient at the edge of the loss cone is shown versus electron energy as obtained from the anisotropy of the observations (solid line) and as obtained from the theoretical calculations for $E_{\text{wave}} = 1.7$ mV m^{-1} and $K_{th} = 2$ keV (dashed line). The main part of the figure shows the smooth line through the $\alpha = 76°$ energy spectrum (solid line) and the theoretical energy spectrum at $\alpha = 42°$ (dashed line) obtained using the theoretical diffusion coefficients and the smoothed $\alpha = 76°$ spectrum (from Lyons, 1974c). (© by American Geophysical Union)

carried by the precipitating electrons associated with diffuse aurora is generally less than that associated with discrete aurora. In addition, the electron distribution responsible for discrete aurora is significantly different from that responsible for the diffuse aurora, since the electrons responsible for discrete aurora are accelerated by an electric field parallel to the magnetic field lines (see Section 4.4). However, the electron distributions incident upon the parallel electric field may well be basically the same as those responsible for diffuse aurora.

On the basis of the suggestion of Kennel et al. (1970), Lyons (1974c) calculated the pitch-angle diffusion coefficient $\langle D_{\alpha\alpha} \rangle$, time-averaged over the electron bounce trajectory, for electrons resonant with electrostatic waves at $\omega = (3/2)\Omega_e$. To determine whether the data in Figure 5.33 are consistent with the calculated pitch-angle diffusion coefficients, Lyons (1974c) applied Equation (5.39), which relates the pitch-angle distribution within the loss cone to $\langle D_{\alpha\alpha} \rangle$ within the loss cone. Smooth curves were first drawn through the energy spectra for $\alpha = 76°$ (solid curve in Figure 5.33) and $\alpha = 42°$ (not shown). Then at energies $\geqslant 5$ keV, where the observed fluxes are anisotropic in pitch angle, the ratio $j(42°)/j(76°)$ was obtained and used with Equation (5.39) to calculate $\langle D_{\alpha\alpha} \rangle$ as a function of electron energy (solid curve in upper panel of figure). The theoretically calculated diffusion coefficients are proportional to E_{wave}^2/K_{th} and are functions of K/K_{th}, where K_{th} is the thermal energy of the resonant, magnetospheric electron distribution. It was found that for $E_{wave} = 1.7$ mV m^{-1} and $K_{th} = 2$ keV, the theoretically calculated diffusion coefficients at the edge of the loss cone (dashed line in upper panel of figure) agree well with the diffusion coefficients obtained from the observations. Finally, using the theoretical diffusion coefficients, the ratio $j(42°)/j(76°)$ was calculated from Equation (5.39) for energies between 1 and 20 keV. This yielded the theoretical $\alpha = 42°$ energy spectrum shown as the dashed line in the main part of the figure.

Notice that the theoretical $\alpha = 42°$ energy spectrum lies well within the scatter of the observations. That such good agreement could be obtained using values of E_{wave} and K_{th} that are within the range of observations shows that resonant interactions with the magnetospheric electrostatic waves could have caused the electron precipitation. It is not possible to rule out the possibility that some other wave mode gives diffusion coefficients versus energy sufficiently similar to those from the observed electrostatic waves to account for the observed precipitation. However, resonant wave-particle interactions are by far the most likely candidate for causing precipitation with an energy and angular distribution of the form associated with diffuse aurora as shown in Figure 5.33.

The preceding discussion has been restricted to electron interactions with electrostatic waves at frequencies of the order of the electron gyrofrequency. Similar resonant interactions are expected between ions and electrostatic waves with frequencies of the order of the ion gyrofrequency (Ashour-Abdalla and Thorne, 1977, 1978), and these interactions may be an important source of auroral ion precipitation. Such ion precipitation should be distinguishable from the auroral ion precipitation resulting from energization in the current sheet (see Section 4.3), since ions precipitating directly from the current sheet should be isotropic in pitch angle at all energies for which precipitation occurs. The precipitation from resonant wave-particle interactions, on the other hand, should be isotropic only at the lowest energies for which precipitation occurs. Above some energy, there should be a transition to anisotropic distributions, the anisotropy increasing with increasing particle energy above the transition energy. Of course, the energy to which isotropy extends may vary with space and with time in association with variations of E_{wave}, since isotropy extends to increasing energies with increasing E_{wave}.

5.8. Balance Between Radial Diffusion and Radiation Belt Particle Losses

The previous sections of this chapter have discussed the interactions of particles with frequencies on the order of the particles' gyrofrequency. Such interactions violate the first adiabatic invariant, and the resulting pitch-angle diffusion is an important loss process for radiation belt particles. In addition, particles can diffuse across L-shells within the radiation belts as a result of fluctuations in the large scale magnetospheric magnetic (Parker, 1960) and electric (Dungey, 1965; Fälthammar, 1965) fields. Fluctuations in these fields with frequencies of the order of the frequency for particle drift around the earth cause violation of the third adiabatic invariant. This results in radial diffusion of the particles, with the first and second adiabatic invariants being conserved.

Radial diffusion is generally described using an equation for diffusion in L at constant first and second adiabatic invariant:

$$\frac{\partial f}{\partial t} = L^2 \frac{\partial}{\partial L}\left[D_{LL} L^{-2} \frac{\partial f}{\partial L}\right] \tag{5.48}$$

where D_{LL} is the radial diffusion coefficient (see Dungey, 1965; Fälthammar, 1965, 1968; Haerendel, 1968; Schulz and Lanzerotti, 1974). The effects of fluctuations in the geomagnetic field and of fluctuations in the magnetospheric electric potential field are normally considered separately in evaluations of D_{LL}.

Fluctuations in the geomagnetic field give rise to an induced electric field giving radial diffusion with a diffusion coefficient (Fälthammar, 1968; Cornwall, 1972):

$$D_{LL}^M = L^{10} \sum_{m=1}^{\infty} A_m \omega_D^2 p^M(m\omega_0).$$

Here $p^M(m\omega_D)$ is the power spectrum of the geomagnetic field fluctuations evaluated at harmonics of the drift period ω_D. A_m are weighting functions for each m, which Fälthammar (1968) evaluated for $m = 1$ as

$$A_1 = 26 \, \Gamma_w(\alpha_0) \quad (R_e^2/s),$$

where $\Gamma_w(\alpha_0)$ gives the pitch angle variation of D_{LL}^M (Figure 5.34). Note that $\Gamma_w(\alpha_0)$ is strongly peaked at $\alpha_0 = 90°$. The radial diffusion coefficient for fluctuations of the magnetospheric electric potential field has been written as (Fälthammar, 1968; Cornwall, 1972)

$$D_{LL}^E = 2.9 \times 10^3 L^6 \sum_{m=1}^{\infty} P^E(m\omega_D) \quad (R_e^2/s),$$

independent of equatorial pitch angle. Here $P^E(m\omega_D)$ is the power spectrum of the fluctuations of the azimuthal magnetospheric electric field.

The diffusion coefficient for magnetic fluctuations had been considered by Parker (1960), Davis and Chang (1962), and Tverskoy (1964, 1965). Later the diffusion coefficient was explicitly evaluated for equatorially mirroring particles by Nakada and Mead (1965) using observations of magnetic field changes on the ground that yielded

$$D_{LL}^M = 3 \times 10^{-10} L^{10} \quad (R_e^2/\text{day}).$$

Lanzerotti et al. (1978) evaluated D_{LL}^M using power spectra of magnetic field variations observed within the radiation belts at synchronous orbit. They found values of D_{LL}^M

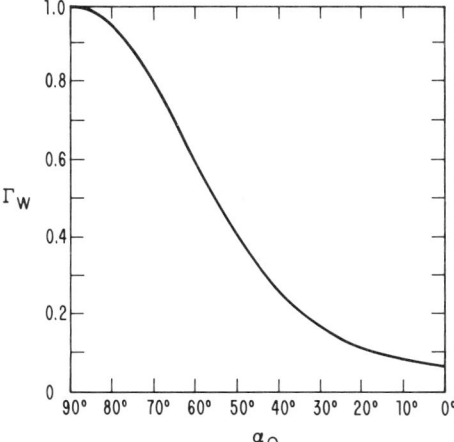

Fig. 5.34. The factor $\Gamma_W(\alpha_0)$ expressing the variation with equatorial pitch angle of the radial diffusion coefficient from magnetic field fluctuations.

varying from approximately that obtained by Nakada and Mead for geomagnetically quiet periods to a value a factor of ~ 20 higher for geomagnetically disturbed periods. This gives

$$D_{LL}^M = (3 - 60) \times 10^{-10} L^{10} \quad (R_e^2 \text{ day}^{-1}) \tag{5.49}$$

for equatorially mirroring particles.

Cornwall (1968, 1972) considered convection electric field fluctuations with a rapid rise and exponential decay with lifetime T as might arise as the result of substorm activity. He obtained

$$D_{LL}^E = \frac{c^2 T L^6 \langle E^2 \rangle}{4 B_e^2 [1 + (2\pi \Omega_{DB} T/2)^2]} \quad (\text{cm}^2/\text{s}), \tag{5.50}$$

where $\langle E^2 \rangle$ is the mean square fluctuating field strength and the equatorial magnetic field strength at the earth's surface $B_e = 0.31$ G. The drift frequency Ω_{DB} for equatorially mirroring particles can be written

$$2\pi \Omega_{DB} = \frac{3\mu c}{\gamma_r q L^2 R_e^2} = 7.3 \times 10^{-4} \frac{\mu(\text{MeV G}^{-1})}{\gamma_r Z L^2} \quad (\text{rad s}^{-1})$$

where μ is the first adiabatic invariant. (The magnetic moment is $\mu/\gamma_r = K/B$.) Evaluating the constants in (5.50), letting T be the value 2.5×10^3 s (42 min) chosen by Cornwall (1972), and expressing E in mV m^{-1}, we obtain

$$D_{LL}^E = 1.4 \times 10^{-3} \langle E^2 (\text{mV m}^{-1}) \rangle \frac{L^6}{1 + (0.91/L^4)[\mu(\text{MeV G}^{-1})/(\gamma_r Z)]^2}. \tag{5.51}$$

Note that for a dipole magnetic field, $0.91\mu(\text{MeV G}^{-1})/\gamma_r$ equals the particle kinetic energy at $L = 7$ expressed in keV for nonrelativistic particles. Cornwall estimated $\langle E^2 \rangle$ to be on the order of $0.01-0.1$ $(\text{mV m}^{-1})^2$ on the basis of limited available data. This gives an estimate for the constant in (5.51):

$$D_{oe} = 1.4 \times 10^{-3} \langle E^2(\text{mV m}^{-1}) \rangle = 1.4 \times 10^{-5} - 1.4 \times 10^{-4} \quad (R_e^2 \text{ day}^{-1}).$$

Electric field measurements obtained from balloons (Mozer, 1971) and from the radial motion of whistler ducts (Carpenter *et al.*, 1972; Carpenter and Seely, 1976) indicate that these values of $\langle E^2 \rangle$ are reasonable.

The radial diffusion coefficients D_{LL}^M and D_{LL}^E are shown in Figure 5.35 as a function of $(\mu/\gamma_r)B$, which nonrelativistically equals the particle kinetic energy, for $L = 3$ and $L = 6$ as presented by Spjeldvik (1979). Values of D_{LL}^M are shown for the constant in Equation (5.49), D_{om}, varying from slightly below to slightly above the values given in

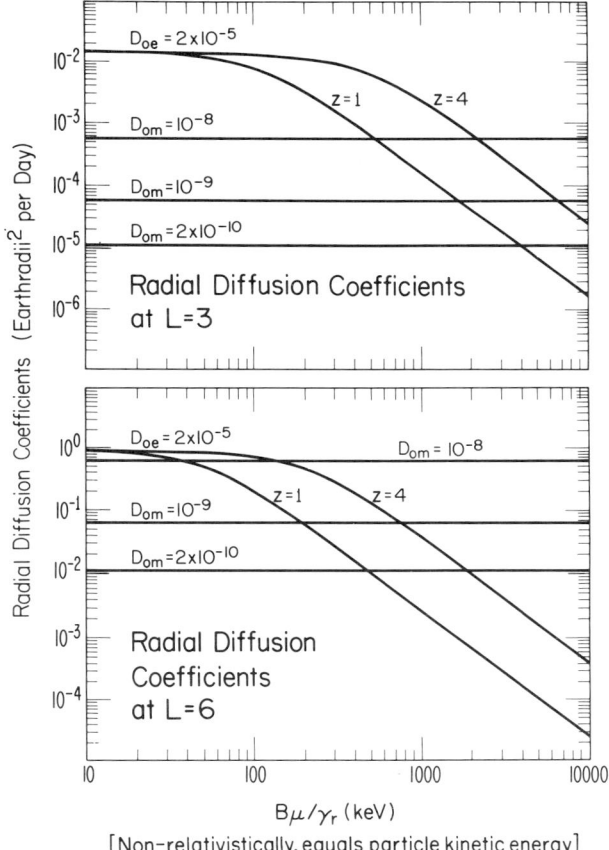

Fig. 5.35. Radial diffusion coefficients as a function of particle energy at $L = 3$ and $L = 6$. The coefficients for magnetic field fluctuations are shown for $D_{om} = 10^{-8}$, 10^{-9}, and $2 \times 10^{-10} R_e^2$ day^{-1}, and the coefficients for electric field fluctuations are shown for $D_{oe} = 2 \times 10^{-5} R_e^2$ day^{-1} and charged states $Z = 1$ and $Z = 4$ (from Spjeldvik, 1979).

Equation (5.49). The curves for $D_{om} = 10^{-9} R_e^2$ day^{-1} are perhaps reasonable values of D_{LL}^M. Values of D_{LL}^E are shown for $D_{oe} = 2 \times 10^{-5} R_e^2$ day^{-1}, a value near the lower limit obtained above and probably reasonable for quiet times. Curves of D_{LL}^E are shown for two charge states, $Z = 1$ and $Z = 4$.

From Figure 5.35, we see that D_{LL}^E is generally significantly greater than D_{LL}^M at the lower L-values and the lower particle energies. However D_{LL}^E decreases with increasing particle energy above ~ 100 keV, while D_{LL}^M is independent of energy. Thus there is a transition energy above which D_{LL}^M is greater than D_{LL}^E. In addition D_{LL}^M increases with L more rapidly than does D_{LL}^E, so that diffusion by magnetic field fluctuations becomes increasingly important with increasing L relative to that from electric potential field fluctuations.

In order to combine the effects of particle losses from pitch-angle diffusion with radial diffusion, it is generally necessary to solve an equation that includes both diffusion processes. This yields a two-dimensional diffusion equation that is significantly more complicated to solve than the one-dimensional pitch angle or radial diffusion equation. However, Walt (1970) has shown that if pitch-angle distributions are in their lowest normal mode, the pitch-angle diffusion coefficients are independent of L, and the radial diffusion coefficients are independent of pitch angle, then particle losses from pitch-angle diffusion can be included in the radial diffusion equation (5.48) as simply a loss term, $-f/\tau_p$:

$$\frac{\partial f}{\partial t} = L^2 \frac{\partial}{\partial L} \left[D_{LL} L^{-2} \frac{\partial f}{\partial L} \right] - \frac{f}{\tau_p}. \tag{5.52}$$

Here τ_p is the trapped particle lifetime given by Equation (5.36) for loss of particles to the atmosphere. The assumption concerning the radial diffusion coefficient is valid for electric potential field fluctuations, but becomes questionable for magnetic field fluctuations. The assumption of L-independence for the pitch-angle diffusion coefficients is not strictly valid for any pitch-angle diffusion processes discussed in this chapter. Nevertheless, Equation (5.52) has been successfully applied to the radiation belts. However, it must be remembered that the effects of violations of the above assumptions have not been quantitatively evaluated.

5.8.1. EQUILIBRIUM STRUCTURE OF RADIATION BELT ELECTRONS WITHIN THE PLASMASPHERE

Initial studies of the balance between electron loss and radial diffusion (e.g., Haerendel, 1968; Newkirk and Walt, 1968) required assumptions concerning the electron lifetimes. The use of such assumptions prevented an explanation of the two-zone structure of radiation belt electrons. Lyons and Thorne (1973) used the results of the electron lifetime calculations shown in Figure 5.25 for electron interactions with plasmaspheric hiss. They were thus able to include realistic variations of the electron loss rate with energy and L.

Since cyclotron resonant interactions with plasmaspheric hiss do not occur within the inner zone because of the increase in the minimum resonant electron energy with decreasing L, Lyons and Thorne (1973) included the effects of electron loss from Coulomb collisions with cold plasmaspheric particles. Such collisions cause radiation belt particles to be scattered in both energy and pitch angle. Equation (5.52) includes only particle loss from pitch-angle scattering into the loss cone and does not allow for the effects of energy scattering; thus this one-dimensional diffusion equation can be employed only if the energy scattering can be neglected. Wentworth *et al.* (1959a) calculated electron lifetimes from Coulomb collisions, including energy scattering and neglecting energy scattering. They found that the energy scattering did not significantly affect the electron lifetimes, and they obtained lifetimes for < 2 MeV electrons approximately given by

$$\tau_c = 3 \times 10^8 \, [K(\text{keV})]^{3/2}/N \quad (\text{s}),$$

where K is expressed in keV and N in particles cm^{-3}.

On the other hand, Wentworth *et al.* (1959a) found that ion lifetimes from Coulomb collisions are significantly affected by energy scattering, so that diffusion in energy as well as diffusion in L must be included when balancing radial diffusion with Coulomb collisions for radiation belt ions. In addition, for electrons, Coulomb collisions cause energy loss at a rate comparable with the rate of loss from scattering into the loss cone (Walt and MacDonald, 1962), though the energy loss does not appear to affect the electron loss rate significantly. However, the effects of this energy loss have not been included quantitatively in a study of the balance between radial diffusion and loss for radiation belt electrons.

Electron lifetimes due to resonant interactions with plasmaspheric hiss and due to Coulomb scattering are shown in Figure 5.36 for three representative values of the first

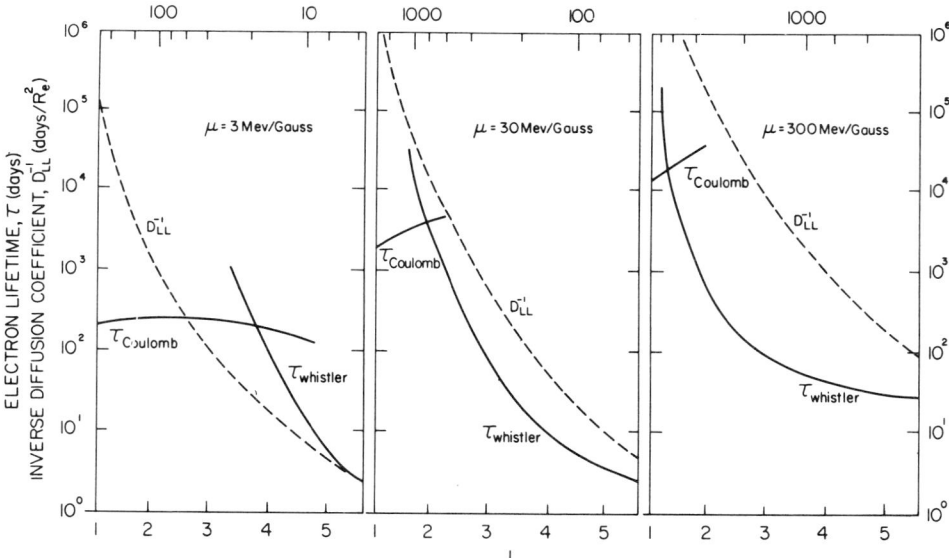

Fig. 5.36. Electron lifetimes versus L due to Coulomb scattering and due to the whistler-mode hiss within the plasmasphere ($B_{\text{wave}} = 10^{-2}$ nT) for three representative values of the first adiabatic invariant. Also shown is D_{LL}^{-1} for $\langle E^2 \rangle^{1/2} = 0.1$ mV m^{-1} for comparison with the lifetimes (from Lyons and Thorne, 1973). (© by American Geophysical Union)

adiabatic invariant. The lifetimes are for a total wave amplitude 10^{-2} nT, a reasonable average value, and a plasmaspheric density $N = 1000\,(4/L)^4$. Notice that collisional losses dominate within the inner zone, and losses from wave interactions dominate throughout the slot region. The dashed lines in Figure 5.36 give $(D_{LL}^E)^{-1}$ versus L for $\langle E^2 \rangle = 0.01$ (mV m^{-1})^2, which corresponds to $D_{oe} = 1.3 \times 10^{-5} R_e^2$ day^{-1}. This gives an estimate of the time scale for radial diffusion which can be compared with the time scale for electron loss as given by the lifetimes. It can be seen in the figure that radial diffusion dominates electron loss at the lower μ values in the outer plasmasphere, and electron loss dominates at low L and high μ.

An equilibrium structure for equatorially mirroring electrons can be obtained by setting $\partial f/\partial t = 0$ in Equation (5.52) and solving

$$L^2 \frac{\partial}{\partial L}\left[D_{LL} L^{-2} \frac{\partial f_p}{\partial L}\right] = \frac{f_p}{\tau_p} \qquad (5.53)$$

for f_p as a function of L at constant μ. Here f_p is the particle distribution in momentum space, which must be used for radiation belt electrons since electrons of relativistic energies are an important component of the radiation belts. The results of this calculation, using the physical boundary condition $f_p = 0$ at $L = 1$, are shown in Figure 5.37a for representative values of the first adiabatic invariant μ. The results have been converted to differential flux j per unit energy versus L at constant μ using the relation (2.84):

$$j = p^2 f_p. \qquad (5.54)$$

Since $p = (2mB\mu)^{1/2}$ for equatorially mirroring particles, we have that $p \propto L^{-3/2}$ at constant μ for a dipole magnetic field. Thus the loss-free solution of Equation (5.53), $f_p = $ const, becomes $j \propto L^{-3}$ for equatorially mirroring particles by virtue of (5.54).

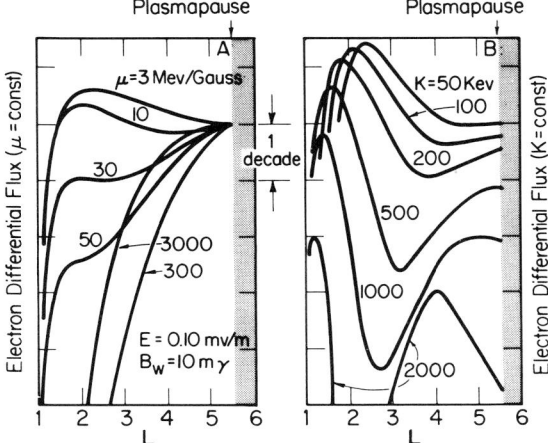

Fig. 5.37. Equilibrium electron differential flux profiles obtained by using large-scale electric field fluctuations with amplitude $\langle E^2 \rangle^{1/2} = 0.10$ mV m^{-1} and wide-band whistler-mode turbulence with amplitude $B_{\text{wave}} = 10^{-2}$ nT. Curves on the left are for electrons of constant μ. Curves on the right are converted to electron flux profiles at constant energy by prescribing the source energy spectrum at $L = 5.5$ (from Lyons and Thorne, 1973). (© by American Geophysical Union)

All curves in Figure 5.37a have been normalized to $j = 1$ at $L = 5.5$. These results are valid only within the plasmasphere, since resonant interactions with plasmaspheric hiss were considered, and the results were obtained under the assumption that electron sources other than radial diffusion can be neglected. Thus the entirety of the curves is valid for a plasmapause at $L \gtrsim 5.5$, and for geomagnetically quiet periods having no direct injections of electrons within $L = 5.5$. During average, non-storm-time conditions, however, geomagnetic activity results in injections of electrons to $L = 4$ to 5 (see Figure 5.18 and the quiet-period electron fluxes versus time shown in Figure 5.38). Under such circumstances, the flux profiles in Figure 5.37a are valid only within $L = 4$ to 5.

The shapes of the flux versus L profiles at each μ in Figure 5.37a are controlled by the relative rates of inward diffusion and loss. At low μ, and consequently low electron energy throughout the outer plasmasphere, the rate of radial diffusion exceeds that of electron loss in the outer plasmasphere as shown in Figure 5.36. Thus j approaches its loss-free diffusion solution $j \propto L^{-3}$. The pronounced reduction in j below $L = 2$ for low μ results from the dominance of Coulomb scattering over radial diffusion. At higher μ (e.g., $\mu = 300$ MeV G^{-1}), slot region losses from interactions with the hiss predominate over radial diffusion in the outer plasmasphere, thus causing a marked decrease in flux with decreasing L. At still higher μ (e.g., $\mu = 3000$ MeV G^{-1}), the rate of loss diminishes relative to the rate of radial diffusion, resulting in a slightly increased penetration of these electrons within the plasmasphere.

Outside the plasmapause, energetic electrons are strongly affected by injections associated with geomagnetic activity as can be seen in Figures 5.18 and 5.38 (see also Owens and Frank, 1968; Williams et al., 1968). These injections are an important, and perhaps the major, source of radiation belt particles; however, they cannot be treated using the equilibrium radial diffusion equation (5.53). To avoid treating this source explicitly, Lyons and Thorne simply adopted a typical outer zone electron energy spectrum near an assumed average location of the quiet time plasmapause as an external boundary condition. The results in Figure 5.37a thus represent inward diffusion from a plasmapause 'source' balanced with precipitation losses. The quiet-time plasmapause was taken to be at $L = 5.5$, and the average source energy spectrum at the location was taken from quiet-time observations of Pfitzer et al. (1966) for energies exceeding 50 keV. At lower energies, observed energy spectra vary considerably and a compromise with $j \sim K^{-3/4}$ was adopted between quiet-time observations of Schield and Frank (1970), DeForest and McIlwain (1971) and Sharp et al. (1971).

The profiles of Figure 5.37a have been converted in Figure 5.37b to profiles of equatorially mirroring j versus L at constant electron energy by using the prescribed electron energy spectrum at $L = 5.5$. The converted profiles exhibit a pronounced, two-zone structure, and the inner zone fluxes are considerably larger than those in the outer zone at the same energy. There is also a systematic inward displacement of the inner zone flux maxima and slot-region minima with increasing electron energy. These features are all in agreement with observations. In addition, for geomagnetically quiet periods, the flux maxima of low energies, outer zone electrons are observed to occur near the plasmapause (Rothwell and Lynam, 1969). Above 1 MeV, on the other hand, outer zone flux maxima are observed within the plasmapause, and a systematic inward displacement of the maxima occurs with increasing electron energy (Vernov et al., 1969). Figure 5.37b shows that the theoretical results reproduce these observed trends of the outer zone flux maxima.

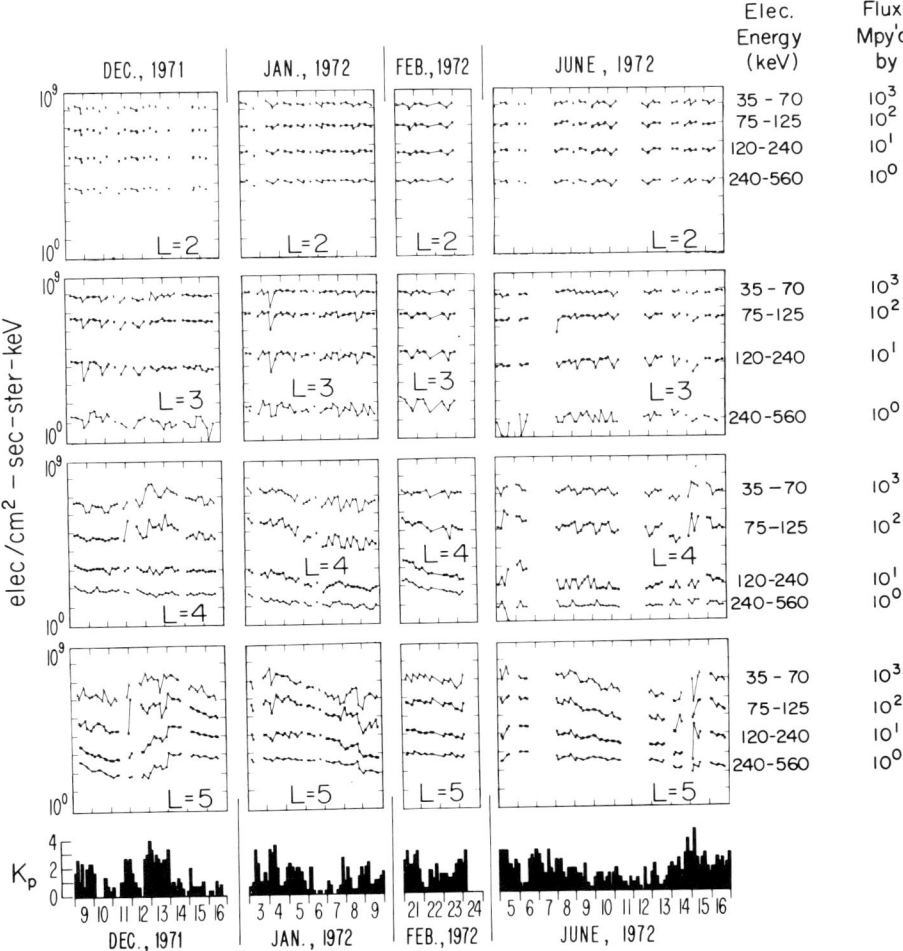

Fig. 5.38. Perpendicular (90° local pitch angle) electron fluxes obtained near the geomagnetic equator versus universal time for the quiet periods of December 9–16, 1971, and January 3–9, February 21–24, and June 5–16, 1972. All available data points from both the inbound and the outbound portions of the Explorer 45 orbit are shown. Each panel shows the observations at the indicated L-value for the four energy channels during one of the time periods. The 120 to 240 keV, 75 to 125 keV, and 35 to 70 keV fluxes have been multiplied by 10^1, 10^2, and 10^3, respectively. Observations at $L = 2, 3, 4$, and 5 are stacked vertically for each time period, and the 3-hr-range K_p index (Lincoln, 1973) is shown versus time at the bottom of each column (from Lyons and Williams, 1975a). (© by American Geophysical Union)

The marked decrease in the energetic electron fluxes at medium L-values in Figure 5.37b results from the pronounced electron losses within the slot region. That inner zone fluxes become enhanced over those in the outer zone at the same energy, despite the slot-region losses, can be understood by considering the source electron energy spectrum and the increase in kinetic energy with decreasing L. As an example, 500-keV electrons at the inner zone peak near $L = 1.5$ have $\mu = 8$ MeV G^{-1}. As shown in Figures 5.36 and 5.37a, 8 MeV G^{-1} electrons undergo very little loss during their inward radial diffusion from the plasmapause. On the other hand, 500 keV electrons at $L = 5.5$ have $\mu = 400$ MeV G^{-1}, and these electrons are significantly depleted as they diffuse inward through the slot region. Furthermore, electrons with 8 MeV G^{-1} have an energy of 15 keV at the plasmapause. Thus the plasmapause flux of 15-keV electrons controls the 500-keV inner zone flux at $L = 1.5$, and this flux is well above the 500-keV electron flux at the plasmapause.

The theoretical profiles are compared in Figure 5.39 with quiet-time observations of equatorially mirroring electrons obtained from the OGO–1 satellite during solar minimum conditions by Pfitzer *et al.* (1966). It can be seen from this figure that the theory quantitatively accounts for the two-zone structure of radiation belt electrons. The observed 2-MeV electron fluxes in the inner zone are approximately an order of magnitude greater than the theoretical predictions. However, artificial nuclear blast contaminants were present at the time (1964) of the measurements (Bostrom *et al.*, 1970) and might account for the enhanced inner zone electron fluxes above 1 MeV.

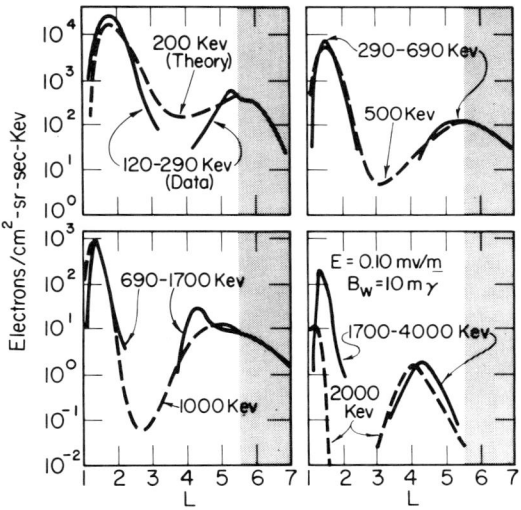

Fig. 5.39. Theoretical flux profiles for $\langle E^2 \rangle^{1/2} = 0.10$ mV m^{-1} and $B_{\text{wave}} = 10^{-2}$ nT are compared with quiet-time solar minimum observations of Pfitzer *et al.* (1966) (from Lyons and Thorne, 1973).
(© by American Geophysical Union)

Explorer 45 observations of equatorially mirroring electrons within the plasmasphere were compared with the theoretical equilibrium structure by Lyons and Williams (1975a) as shown in Figure 5.40. The outer boundary energy spectrum was taken at $L = 5.2$ (the Explorer 45 apogee) from the average of observations for one week. The values of B_{wave} and $\langle E^2 \rangle$ were adjusted to maximize agreement with the observations, and the theoretical equilibrium profiles in Figure 5.40 (dashed lines) are for $B_{\text{wave}} = 10^{-2}$ nT and $\langle E^2 \rangle^{1/2} = 0.05$ in V m^{-1}. This value of B_{wave} is the same as that used by Lyons and Thorne (1973),

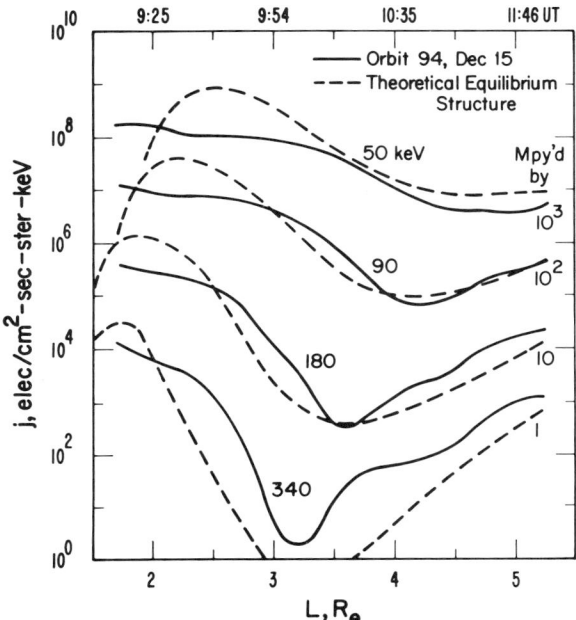

Fig. 5.40. Comparison between the radial profiles observed on the outbound portion of orbit 94 on December 15, 1971 (solid lines), and the theoretical equilibrium radial profiles (dashed lines) as obtained from the analysis of Lyons and Thorne (1973) using $B_{\text{wave}} = 10^{-2}$ nT, $\langle E^2 \rangle^{1/2} = 0.05$ mV m^{-1}. The theoretical profiles are for the indicated energies, which are the geometric mean of the four Explorer 45 energy channels. The 180, 90, and 50 keV fluxes have been multiplied by 10^1, 10^2, and 10^3, respectively (from Lyons and Williams, 1975a). (© by American Geophysical Union)

and the value of $\langle E^2 \rangle^{1/2}$ is a factor of 2 smaller. The simple theory, which neglects variations of B_{wave} and $\langle E^2 \rangle$ with L, can again be seen to reproduce the overall structure of radiation belt electrons within the plasmasphere reasonably well, the values of the fluxes being correct to within about an order of magnitude at all L-values and energies. Such accuracy is consistent with our knowledge of B_{wave} and $\langle E^2 \rangle$ and their spatial variations.

The constancy of the equilibrium structure of the electrons is illustrated in Figure 5.41, which shows Explorer 45 data from a 6-month period. Here radial profiles of the equatorially mirroring electron fluxes from three quiet periods (solid lines) are compared with the observations from Figure 5.40 for December 15, 1971 (dashed lines). The comparisons in Figure 5.41, together with those in Figures 5.39 and 5.40 suggest that an equilibrium structure for radiation belt electrons is a general feature within the plasmasphere. The agreement between theory and observations gives strong evidence that this equilibrium structure is maintained by a balance between the average rates of radial diffusion and scattering losses due to whistler-mode hiss in the slot region and Coulomb collisions in the inner zone. In addition, the quantitative agreement indicates that the theoretical rates of radial diffusion and loss are quantitatively reasonable.

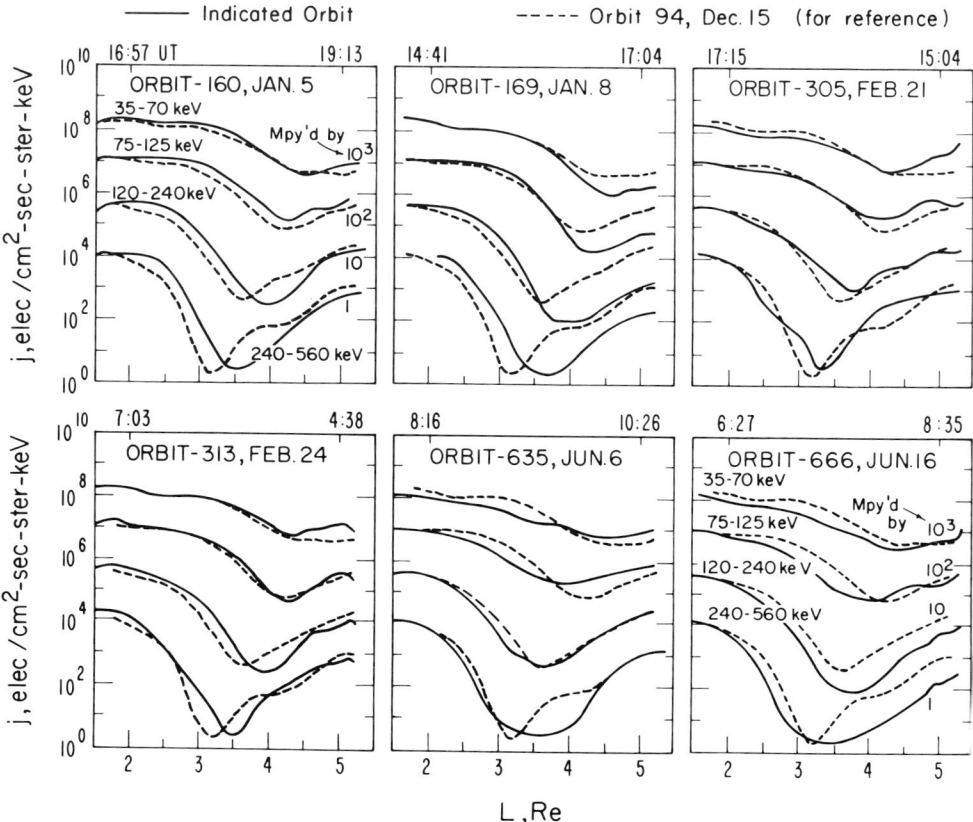

Fig. 5.41. Examples of radial profiles of the perpendicular flux obtained near the geomagnetic equator during the quiet periods of January 3–9, February 21–24, and June 5–12, 1972. Solid lines give the profiles from the orbits indicated in each panel, and dashed lines give the reference profiles from December 15, 1971 (orbit 94). The 120 to 240 keV, 75 to 125 keV, and 35 to 70 keV fluxes have been multiplied by 10^1, 10^2, and 10^3, respectively (from Lyons and Williams, 1975a). (© by American Geophysical Union)

5.8.2. EQUILIBRIUM STRUCTURE OF RADIATION BELT IONS WITHIN THE PLASMASPHERE

Evaluating the equilibrium structure for ions in the radiation belts is more complex than doing so for electrons in two respects. First, energetic ions can collide with neutral exospheric hydrogen of the geocorona. Such collisions lead to a charge exchange interaction between the cold hydrogen atoms and the energetic ions. For singly charged ions, such charge exchange leads to an energetic neutral atom and a cold proton. The addition of cold protons by means of charge exchange is of little consequence to the cold plasma population. However, the energetic neutral is no longer confined by the geomagnetic field and rapidly leaves the radiation belts. Thus charge exchange can be an important loss process for radiation belt ions (Dessler and Parker, 1959; Stuart, 1959; Wentworth *et al.*, 1959a, b), causing ion loss with a lifetime

$$\tau_{CE} = \langle \sigma v [H] \rangle. \tag{5.55}$$

Here σ is the charge exchange cross section, v is the ion speed, and $[H]$ is the neutral hydrogen density. The angular brackets denote an average over the ion bounce trajectory.

The second complexity for ions is that energy losses from Coulomb collisions cannot be ignored, since they dominate the angular scattering. Following Nakada and Mead (1965) and Cornwall (1972), we may write the evolution of the ion distribution function with time associated with the energy loss from Coulomb collisions as

$$\frac{\partial f}{\partial t} = G(L)\mu^{-1/2} \frac{\partial f}{\partial \mu}, \tag{5.56}$$

where the weighting factor $G(L)$ is given by

$$G(L) = \frac{50(2)^{1/2} \pi e^4 L^{9/2} Z m^{1/2}}{m_e B_e^{3/2}} (N + 0.08 N_b) \tag{5.57}$$

and m_e is the electron mass. N is the density of free electrons and N_b is the density of bound electrons. Cornwall (1972) included only the electrons bound in neutral hydrogen, an assumption giving $N_b = [H]$. Within the plasmasphere, bound electrons make a negligible contribution to $G(L)$, but Cornwall concluded that they can have a significant effect beyond the plasmapause where the plasma density is low.

Incorporating charge exchange as given by Equation (5.55) and Coulomb collisional energy loss as given by (5.56) into the radial diffusion equation (5.52) gives the following equation for the time evolution of radiation belt ions:

$$\frac{\partial f}{\partial t} = L^2 \frac{\partial}{\partial L} \left[D_{LL} L^{-2} \frac{\partial f}{\partial L} \right] + G(L) \mu^{-1/2} \frac{\partial f}{\partial \mu} - \frac{f}{\tau_{CE}} - \frac{f}{\tau_p}. \tag{5.58}$$

Additional terms may be included in (5.58) to describe the exchange of charge between ions in various charge states (Spjeldvik, 1979). This equation is a two-dimensional, differential equation, which requires solution as a function of L and μ.

Spjeldvik (1977) applied (5.58) to obtain an equilibrium structure of radiation belt protons. He took $\partial f/\partial t = 0$ and neglected the effects of precipitation losses from wave-particle interactions (i.e., he took $f/\tau_p = 0$ in Equation (5.58)). Wave-particle interactions were neglected because of the lack of a reliable estimate of τ_p as a function of L and energy. Spjeldvik argued that such interactions might be negligible at energies $\gtrsim 100$ keV but recognized that their neglect was most likely invalid at lower energies. Spjeldvik used charge exchange cross sections obtained from Fite *et al.* (1960), Claflin (1970), and Brinkmann and Kramers (1930). He used a neutral hydrogen density model from Tinsley (1976) for a mean exospheric temperature of 950 K, and the cold plasma density was taken to be

$$N = \begin{cases} 250(4.1/L)^{4.64} \text{ cm}^{-3} \text{ for } L \leq 4.1 \\ 13(4.1/L)^{4.64} \text{ cm}^{-3} \text{ for } L \geq 4.1 \end{cases}.$$

This density profile is within a factor of ~ 4 of that used by Lyons and Thorne (1973) with the plasmasphere; however it includes an abrupt reduction at an assumed plasmapause located at $L = 4.1$. Spjeldvik extended his calculations to high energies requiring that radial diffusion from both magnetic and electric field fluctuations be included. He took $D_{om} = 2 \times 10^{-10} R_e^2$ day^{-1}, near the lower limit in Equation (5.49), and $D_{oe} = 2 \times 10^{-5} R_e^2$ day^{-1}, which is ~ 1.5 times as large as the D_{oe} used by Lyons and Thorne (1973).

Boundary conditions employed by Spjeldvik were $f = 0$ at $L = 1$ and $f(\mu_{\max}) = 0$ at all L. The upper limit of μ, $\mu_{\max}$ was taken to correspond to an energy of 1000 MeV at $L = 7$. This upper energy is well outside the range of interest for the study, but was chosen to avoid unphysical effects from the high energy boundary condition. The most important boundary condition was the specification of an outer boundary proton energy spectrum. This was obtained from quiet-time ion observations from the geostationary ATS–6 satellite at $L = 6.6$ compiled by Fritz *et al.* (1977).

Energy spectra calculated by Spjeldvik (1977) from the numerical solution of (5.58) are shown as the solid lines in Figure 5.42. The data points shown at $L = 6.6$ are the ATS–6 observations used as the outer boundary condition, though the solution was extended to $L = 7$. Data at other L-values are quiet-time observations of equatorially mirroring ions from the Explorer 45 satellite. These observations extend from ~ 25 keV to ~ 1 MeV at $L \geq 4$, but the lower energy cutoff increases to more than 100 keV at lower L because of an instrumentation problem. It can be seen from Figure 5.42 that the theory reproduces the observed energy spectra reasonably well over the energy range covered by the Explorer 45 observations.

Figure 5.43 shows the calculated proton fluxes displayed as radial profiles at energies from 10 keV to 1000 MeV. Note, however, that only the fluxes between ~ 100 keV and 1 MeV have been verified by direct comparison with observations.

Ion fluxes at $L = 4.5$ as a function of energy from a number of satellites are shown in Figure 5.44 and compared with the theoretically predicted energy spectrum. The > 100-keV observations from Explorer 12 and Mariner 4 can be seen to agree well with the Explorer 45 observations and the theoretical spectrum. Below 40 keV, the Explorer 45 data are questionable; however OGO–3 observations from Pizzela and Frank (1971) are also shown. Note that the $\lesssim 50$-keV flux observations are well above the theoretical predictions.

The excellent agreement between theory and observations from 100 keV to 1 MeV indicates that the dominant source and loss processes for quiet-time protons within this energy range have been correctly identified and successfully evaluated quantitatively for $2 \lesssim L \lesssim 6.6$. Wave-particle interactions were not included in the calculations, indicating that such interactions are not important for equatorially mirroring protons at energies > 100 keV during quiet times. In addition, since the ion measurements did not distinguish between protons and other ions, the agreements in Figure 5.42 imply that the measured ion fluxes were indeed dominated by protons. The loss rates used in calculating the theoretical profiles were for protons. These rates and the resulting equilibrium structures are significantly different from those for other ion species.

On the other hand, the calculations show a marked reduction in flux at $L < 4$ and energies < 100 keV. This reduction shows up somewhat at $L = 4.5$, and is in disagreement with the observed fluxes in Figure 5.44. More recent ISEE–1 satellite observations of

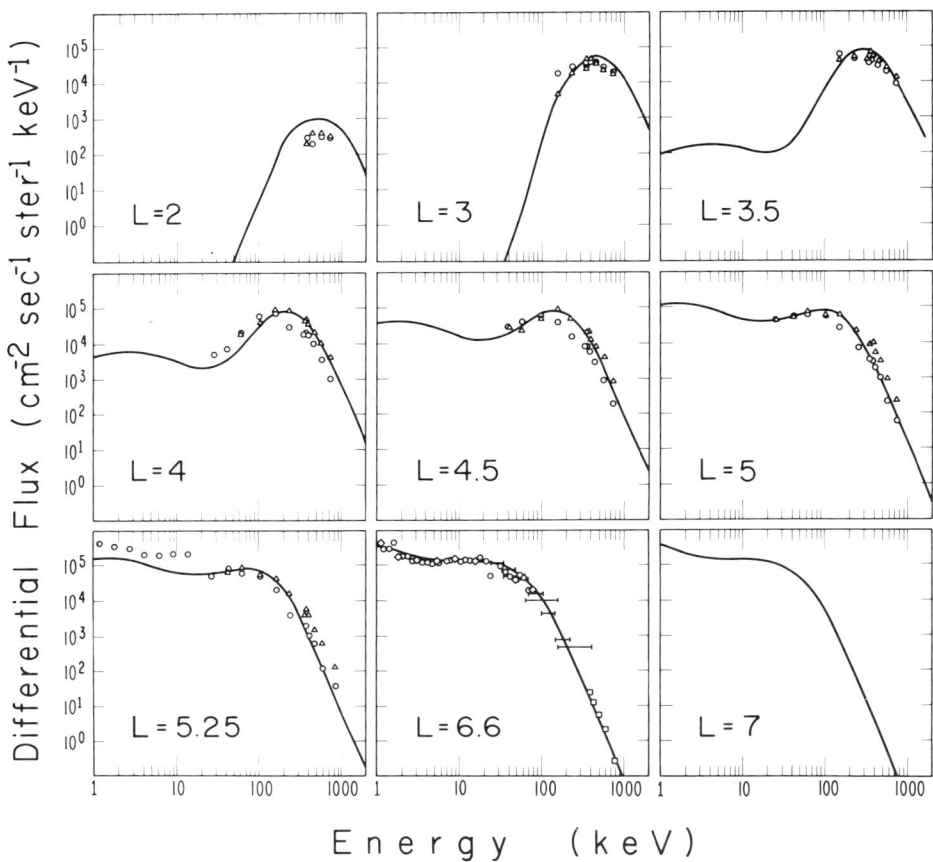

Fig. 5.42. Comparison of theoretical and observed energy spectra at L values of 2, 3, 3.5, 4, 4.5, 5, 5.25, 6.6, and 7. The data at $L = 6.6$ are taken from a number of experiments on the satellite ATS–6 in geostationary orbit (Fritz et al., 1977). The data at the lower L values are taken from quiet-time observations on board the satellite Explorer 45. Data from orbit 97 are denoted by circles, and data from orbit 667 are denoted by triangles (from Spjeldvik, 1977). (© by American Geophysical Union)

ion energy spectra are shown in Figure 5.45 (from Williams, 1981). Each panel in this figure shows the distribution function f (s^3/cm^6) of equatorially mirroring ions versus the first adiabatic invariant μ (MeV/G), and distributions are shown every 0.1 in L from $L = 8.0$ to $L = 3.0$. The x's are post-storm observations, and the solid lines are quiet-time observations that can be compared with Spjeldvik's calculations. Although the observations are shown versus μ, the lowest energy shown in each panel is 24 keV. Note in particular that the marked reduction in the calculated ion fluxes in Figure 5.42 at low energies and low L is in significant disagreement with the observations. The earlier discussion (Section 5.6) of ion interactions with ion-cyclotron waves indicates that ion loss from such interactions can be important for these ions. However, such interactions represent an additional loss process which, if included, would reduce the theoretically

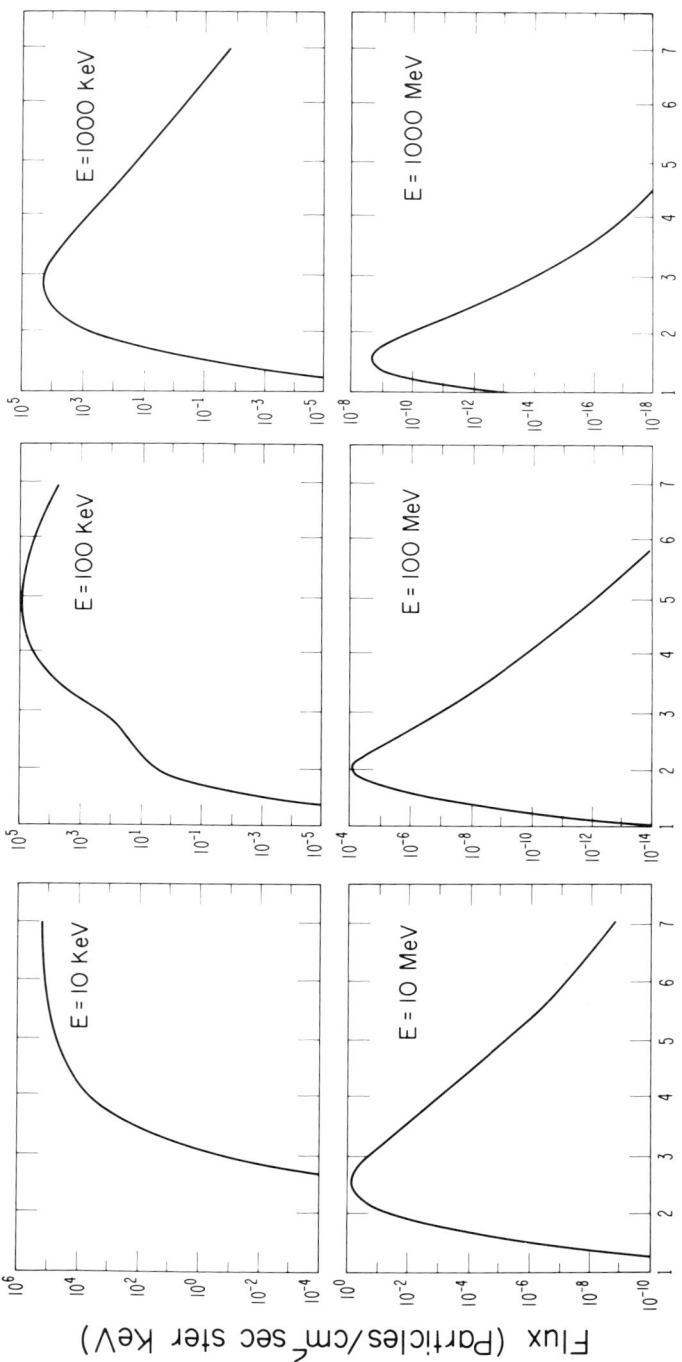

Fig. 5.43. Radial profiles of equatorially mirroring radiation belt protons for energies of 10 keV, 100 keV, 1000 keV, 10 MeV, 100 MeV, and 1000 MeV computed by assuming an equilibrium balance between radial diffusive transport driven by fluctuations in the large-scale electric and magnetic fields and losses due to charge exchange and Coulomb interactions. The results above 1.1×10^3 MeV G^{-1} (~ 1 MeV at $L = 7$, ~ 5 MeV at $L = 4$, and ~ 40 MeV at $l_* = 2$) stem from an extrapolated part of the outer boundary energy spectrum at $L = 6.6$ (from Spjeldvik, 1977). (© by American Geophysical Union)

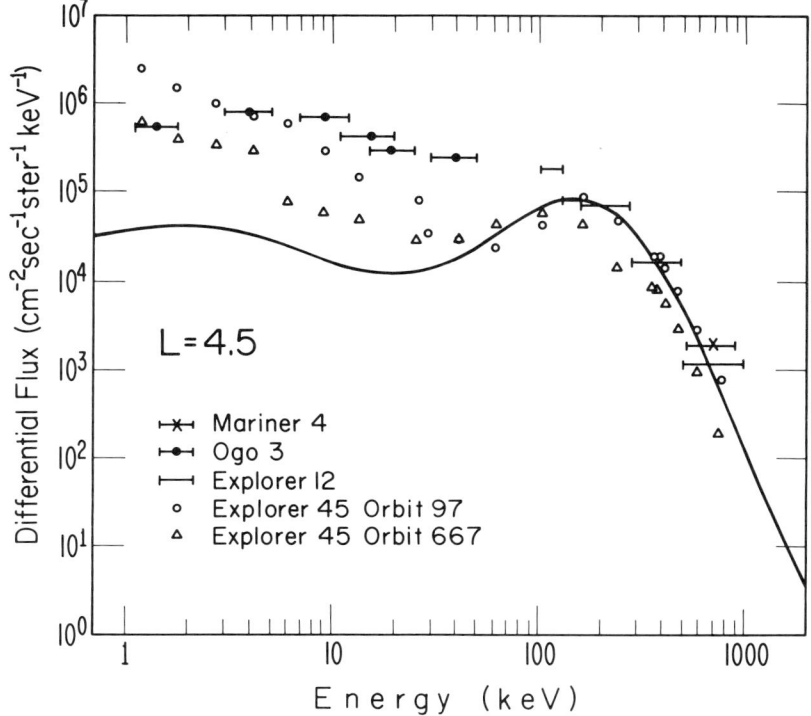

Fig. 5.44. Comparison of measured ion fluxes at $L = 4.5$ as a function of energy from a number of satellites and the theoretically predicted energy spectrum of equatorially mirroring protons at $L = 4.5$. The Explorer 45 observations below 40 keV are questionable (from Spjeldvik, 1977). (© by American Geophysical Union)

calculated fluxes from their values in Figure 5.42 and would increase the disagreement with the observations.

This difficulty had been addressed earlier by Tinsley (1976) and Lyons and Evans (1976), who considered the loss of < 50 keV ions from the storm-time ring current. The loss of < 50-keV protons at $L \lesssim 4$ is dominated by charge exchange in the equilibrium calculations, the charge exchange lifetime being on the order of hours for equatorially mirroring protons. This loss rate should increase markedly with increasing mirror latitude for particles of a given energy on a given L-shell because of the increase in neutral hydrogen density with decreasing altitude. Lyons and Evans investigated the question of how the nearly isotropic ion pitch-angle distributions at the lower energies following a storm-time injection of ions can remain isotropic in the presence of such charge exchange (see Figures 5.27 and 5.28).

Figures 5.46 and 5.47 compare the observed equatorial pitch-angle distributions of 2- and 10-keV ions at $L = 3.0$ and $L = 3.5$ following a storm-time injection with those expected to evolve from proton charge exchange with neutral hydrogen. At both L-values, the charge exchange calculations predict that the pitch-angle distributions should become

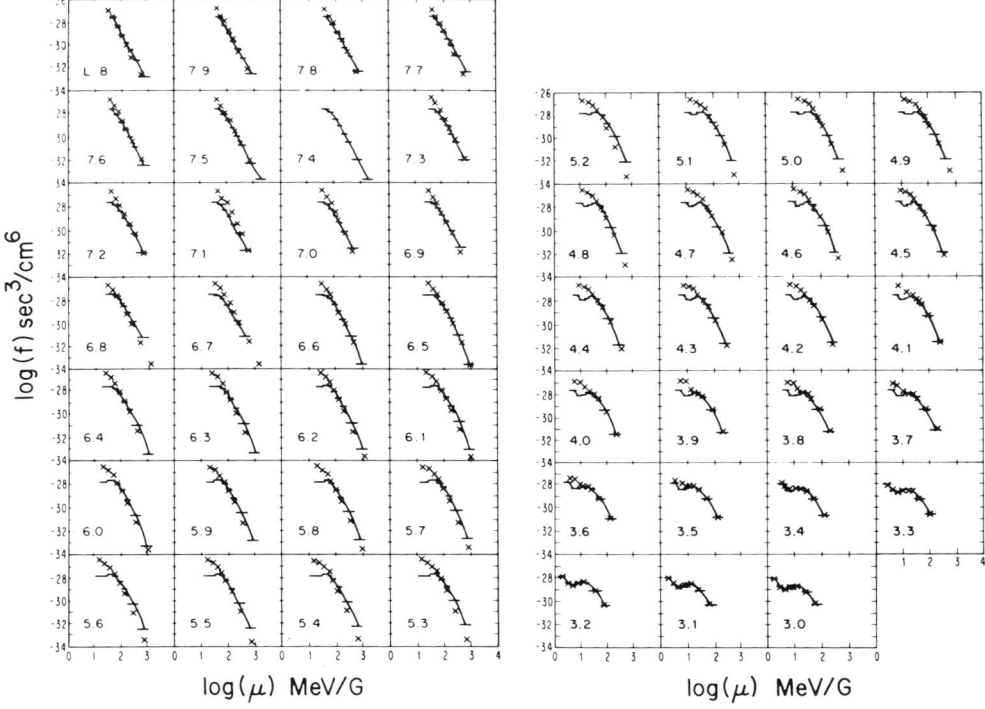

Fig. 5.45. $f(s^3 \text{ cm}^{-6})$ versus μ (MeV G^{-1}) for equatorially mirroring ions as measured on the ISEE-1 satellite. Plots are shown for each 0.1 in L from $L = 8.0$ to $L = 3.0$. Line plots in each panel are from the pre-storm pass of November 24-25, 1977. The superimposed crosses are from the post-storm pass of November 27, 1977. All μ values were obtained from the simultaneously measured magnetic field magnitude. The lowest energy channel is 24 keV in each panel (from Williams, 1981).

greatly anisotropic in < 8 hr. This rounding of the distributions is not observed! At $L = 3.5$, the observed distributions remain essentially isotropic, but at $L = 3.0$ some rounding of the distribution occurs. This rounding is apparently the result of the resonant interactions with ion-cyclotron waves as indicated in Figures 5.27 and 5.28. In addition to the disagreement between the shapes of the calculated and observed pitch-angle distributions, charge exchange predicts that even the 90° pitch-angle fluxes should decay much more rapidly than is observed. The discrepancy is particularly dramatic at $L = 3$.

Lyons and Evans concluded that neither a strong continual proton source at $L < 4$ nor a large error in the neutral hydrogen density was a reasonable explanation for the discrepancy between the observations and the charge exchange predictions. They concluded that the most likely explanation was that the ring current ions at energies $\lesssim 50$ keV and $L \lesssim 4$ were dominated by some ion species other than protons during the storm recovery phase. Such ions were predicted to have much longer lifetimes from charge exchange with hydrogen than do protons. They and Tinsley (1976) suggested He$^+$ as an attractive alternative, its charge exchange lifetime being a factor of ~ 50 longer than that for protons at energies between ~ 1 and ~ 30 keV.

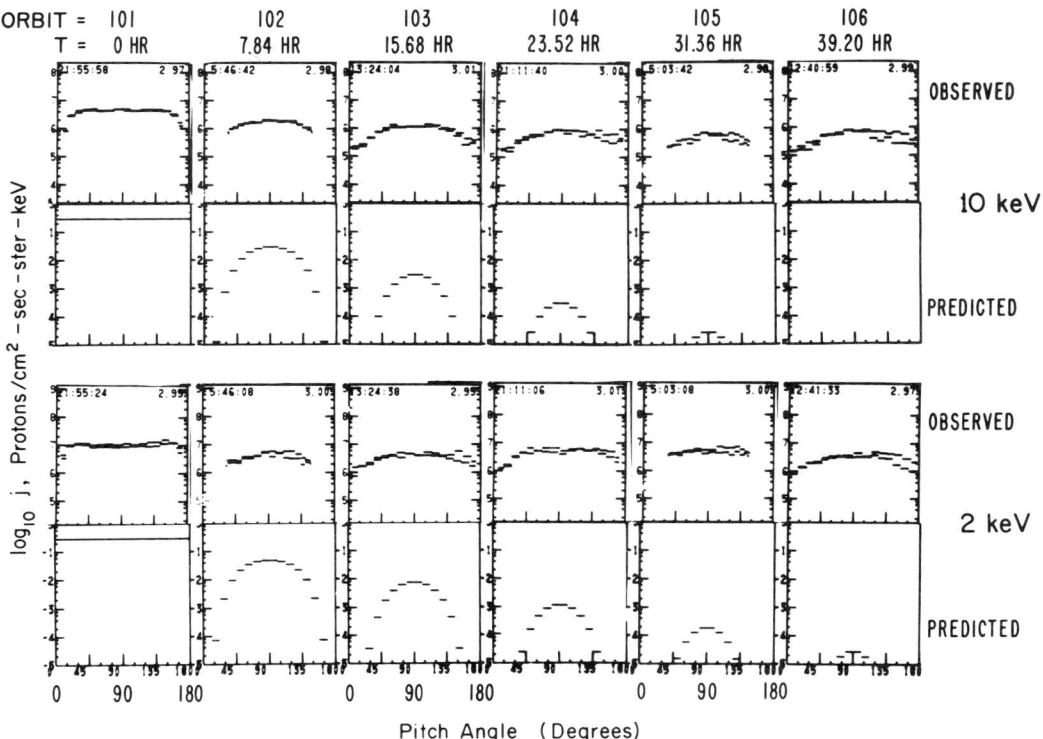

Fig. 5.46. Equatorial pitch-angle distributions at $L = 3$ observed during the recovery phase of the December 17, 1971, storm, compared with those expected to evolve from proton charge exchange with neutral hydrogen. Observations are shown from the inbound portion of Explorer 45 orbits 101, 102, ... ; the observations from orbit 101 were made less than 1 hr after the minimum of D_{st}. The appearance of elevated fluxes over the pitch-angle range of 90° to 180° is an instrumental error resulting from reflected sunlight. The pitch-angle distributions expected to evolve from charge exchange were obtained by neglecting possible sources, assuming an isotropic proton pitch-angle distribution at the time of the orbit 101 observations ($t = 0$), and using the charge exchange lifetimes given by Tinsley (1976). The initial fluxes for the calculations were arbitrarily normalized to approximately 3×10^{-1} (from Lyons and Evans, 1976). (© by American Geophysical Union)

It had generally been assumed that protons were the dominant component of the ion population throughout most of the magnetosphere, an assumption that had not been tested by direct measurements prior to 1972. Low altitude satellite observations of precipitating ions from Shelley *et al.* (1972) showed large fluxes of O^+ ions in the 0.7–12 keV energy range, which sometimes exceeded the H^+ fluxes. This led to the suggestion that H^+ is not always the dominant ion at all energies throughout the magnetosphere (Johnson *et al.*, 1975). Sharp *et al.* (1976a, b) presented mass spectrometer observations from a low altitude satellite that were obtained following the same magnetic storm as were the data analyzed by Lyons and Evans (1976). They found significant O^+ and H^+ precipitating from the outer ring current, but no detectable He^+. Sharp *et al.* (1977)

214 CHAPTER 5

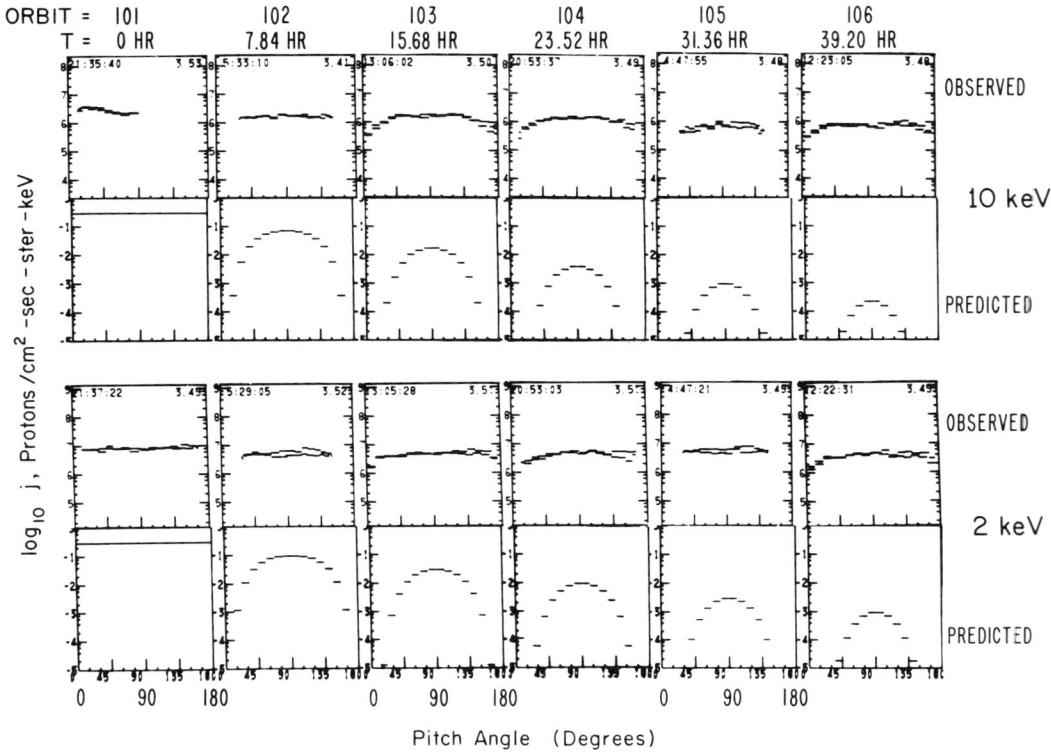

Fig. 5.47. Same as Figure 5.46, except that $L = 3.5$ (from Lyons and Evans, 1976). (© by American Geophysical Union)

examined the precipitating ion observations in more detail. They found O^+ fluxes exceeding H^+ fluxes, consistent with Lyons and Evans's conclusion that the ions could not have been dominated by protons but must have been dominated by an ion with a significantly longer charge exchange lifetime. (The charge exchange lifetime of ~ 7–30 keV O^+ is ~ 7 times longer than that of H^+, but ~ 7 times shorter than that of He^+). However He^+ was not found.

Johnson et al. (1977) presented the first observations of ⩽ 50-keV trapped ring current ions at $L \lesssim 4$. These observations were obtained on the polar orbiting S3–3 spacecraft at altitudes below 8000 km. Although these were the highest altitude composition observations at energies below 50 keV yet reported, the majority of the ring current ions mirror at higher altitudes and could not be measured. Limited measurements above background levels were obtained during the main phase of three geomagnetic storms as shown in Figure 5.48. O^+ and H^+ can be seen that have approximately equal fluxes, within a factor of 2, and He^+ fluxes are approximately an order of magnitude smaller. On the basis of the charge exchange arguments, it is expected that the relative abundance of H^+ at these energies should decrease markedly following the storm main phase. In

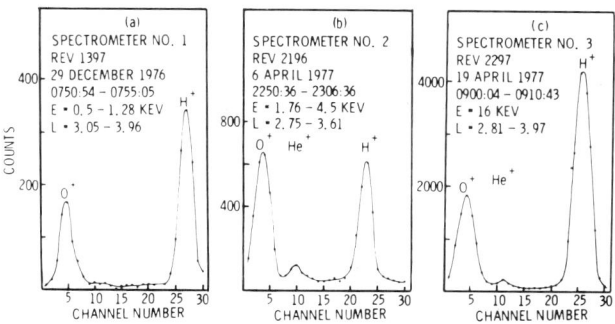

Fig. 5.48. Examples of mass spectra from the S3−3 satellite obtained during the main phases of three geomagnetic storms. Observations are of trapped ions but are at altitudes below 8000 km (from Johnson et al., 1977). (© by American Geophysical Union)

addition, the relative abundance of He^+ should increase with respect to H^+ and O^+ because of its long charge exchange lifetime. An example of measurements obtained during the recovery phase of one of the storms was presented by Johnson et al. (Figure 5.49). Figures 5.48 and 5.49 show the expected evolution of the relative species' abundances. O^+ became the dominant ion, He^+ was second, and H^+ had the smallest flux of the three.

In addition to these composition measurements, Smith et al. (1981) studied the decay of storm-time ring current ions, assuming a three-component (H^+, He^+, and O^+) ion population. They found that the observed decay could be accounted for with such an ion mixture.

Observations of the equatorial composition of low energy ring current ions have now become available which consistently show that protons are not the dominant ring current ion at energies ≤ 17 keV and $L \leq 4$ (Geiss et al., 1978; Lundin et al., 1980; Balsiger et al., 1980; Lennartsson et al., 1981). Observations of the energy density versus L of 0.2−17 keV ions obtained near the equator from a Swedish instrument on board the PROGNOZ−7 satellite are shown in Figure 5.50. Note O^+ dominates consistently within this energy range, and He^+ is generally the second most abundant ion. In all 14 orbits analyzed by Lundin et al. (1980) including storm main phase, recovery phase, and quiet periods, no example was found of a proton dominance at $L \lesssim 4$.

Thus the disagreement between the calculated equilibrium structure of radiation belt protons of Spjeldvik (1977) and the ion observations at $L \lesssim 4$ and energies $\lesssim 50$ keV is apparently due to the observed ions' being not mostly protons. It appears that the dominant ion is O^+, at least below 17 keV, an ion with a significantly longer lifetime for charge exchange with neutral hydrogen than that of protons. Composition measurements between 17 keV and several hundred keV are not yet available, although Spjeldvik's results imply that 100 keV to 1 MeV trapped ions at $L \lesssim 6.6$ are predominantly protons.

Limited observations of helium and oxygen ions within the radiation belts at energies above several hundred keV have been reported, and observed fluxes of equatorially mirroring ions have been compared with calculations of the equilibrium structure for these ions. The structure of energetic helium ions has been addressed by Spjeldvik and

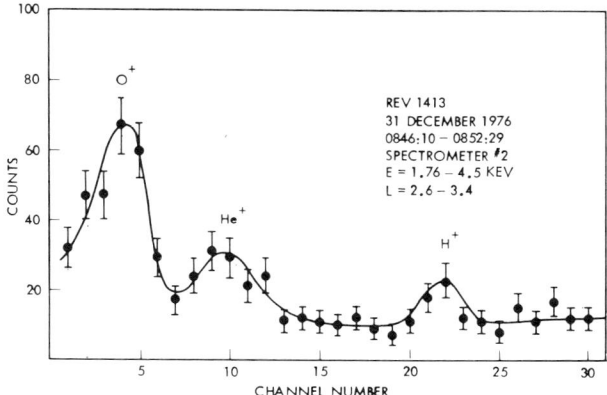

Fig. 5.49. An example of the mass spectrum from the S3−3 satellite during the recovery phase of the storm on December 29, 1976 (from Johnson et al., 1977). (© by American Geophysical Union)

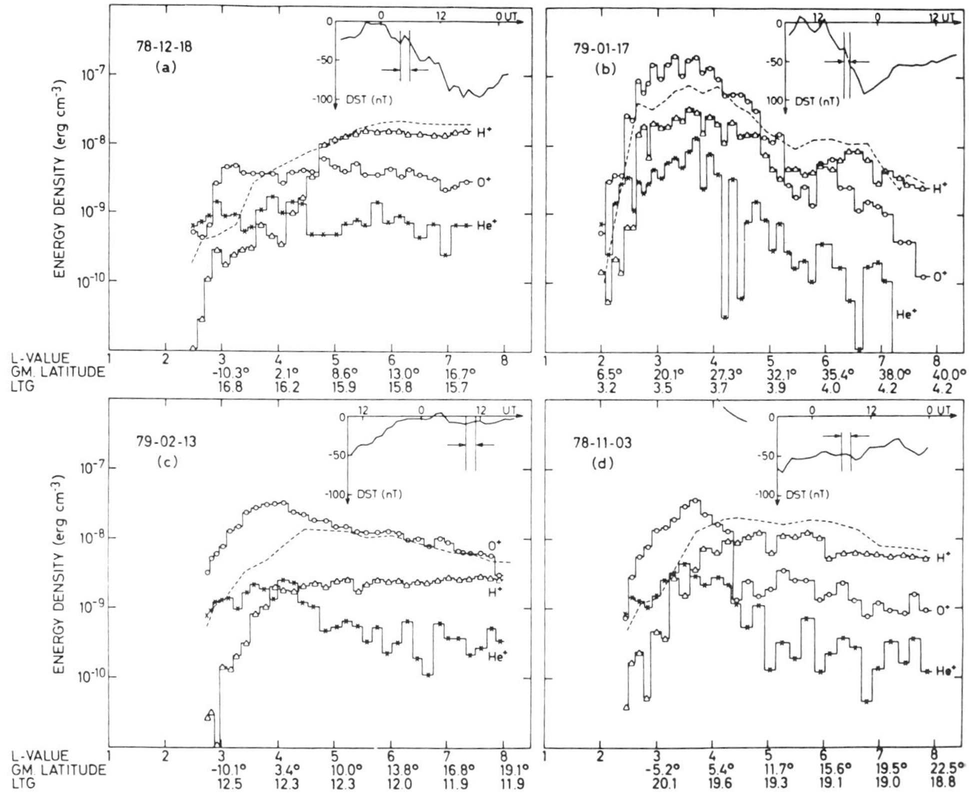

Fig. 5.50. Energy densities of H^+, He^+, and O^+ in the energy interval 0.2−17 keV vs orbital parameters for four different PROGNOZ−7 passes through the ring current. The broken curves indicate energy densities calculated from the E/q spectrometers (0.1−45 keV) under the assumption that only protons were measured. Included in each panel is a graph showing the D_{st} values during the pass (marked by the arrows) (from Lundin et al., 1980). (© by American Geophysical Union)

Fritz (1978a), who included both singly and doubly charged ions in their calculations. They needed to include three different charge transfer interactions between energetic helium ions and the ambient neutral hydrogen atoms:

$$He^{++} + H \to He^+ + H^+$$
$$He^+ + H \to He + H^+$$
$$He^+ + H \to He^{++} + H + e$$

The charge exchange cross sections used by Spjeldvik and Fritz for these interactions were based on Claflin (1970) and Brinkmann and Kramers (1930). All the other parameters employed for the helium study were the same as those employed in the proton study of Spjeldvik (1977). The outer boundary energy spectrum at $L = 6.6$ was based on ATS–6 observations (Fritz et al., 1977) above 1 MeV and low altitude observations of low energy precipitating helium ions (Johnson et al., 1975).

It should be noted that composition observations of low energy ions within the radiation belts were not available at the time of the analysis, and that the ATS–6 observations included only two energy channels (1.16–1.75 MeV and 1.75–3.62 MeV) and summed over both charge states. Thus the outer boundary condition was forced to be somewhat speculative, and this limits the accuracy of the resulting equilibrium structure calculations.

Equilibrium radial profiles of helium ion fluxes calculated by Spjeldvik and Fritz (1978a) are shown in Figure 5.51 for energies from 1 keV to 10 MeV. The solid curves are for He^{++} and the dashed curves are for He^+. Comparisons of the calculated radial profiles with Explorer 45 observations are shown in Figure 5.52 (See Fritz and Spjeldvik (1978) and Spjeldvik and Fritz (1981) for more details of the helium ion observations.) This figure gives the observed sum of He^+ and He^{++} fluxes (circles) in two energy channels, $\Delta\alpha3$ (1.6–1.74 MeV) and $\Delta\alpha4$ (1.74–3.15 MeV). Observations are shown from five consecutive geomagnetically quiet days. The lines are the results of the theoretical calculations for He^{++} ions, since the calculated He^+ ion fluxes were lower than the He^{++} fluxes at these energies. In the upper panels, theoretical results at 1.2, 1.5, and 2.0 MeV are shown and compared with the $\Delta\alpha3$ observations; in the lower panels theoretical results at 2 and 3 MeV are shown and compared with the $\Delta\alpha4$ observations. Agreement between the calculations and the observations is quite good in Figure 5.52, especially considering the uncertainties in the outer boundary condition.

Calculating an equilibrium structure for radiation belt oxygen ions requires consideration of multiply charged ions, a problem which has been addressed by Spjeldvik and Fritz (1978b). They allowed for six charge states of oxygen, O^+ through O^{6+}, the relevant charge exchange cross sections they obtained being based on data from Stebbings et al. (1960), Lo et al. (1971), MacDonald and Martin (1971), and Solov'ev et al. (1972). Many of these cross sections needed to be estimated; however, they were believed to be the best available at that time. The outer boundary energy spectrum was based on charge-insensitive observations at energies $\gtrsim 600$ keV from the synchronous orbit ATS–6 satellite (Fritz and Wilken, 1976) and on low altitude precipitation observations at energies $\lesssim 15$ keV (Sharp et al., 1976a, b). Again the applied outer boundary spectrum was required to be somewhat speculative.

Owing to the lack of information concerning the outer boundary charge state distribu-

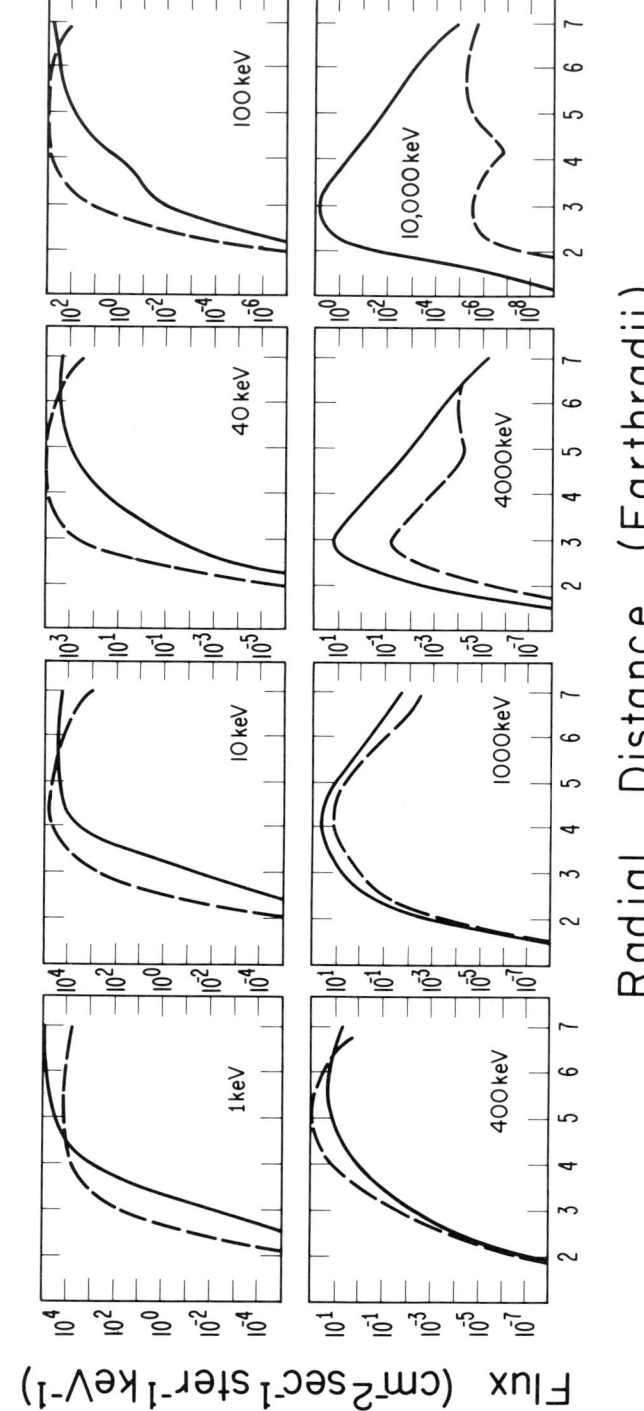

Fig. 5.51. Equilibrium radial profiles of helium ions computed for energies 1, 10, 40, 100, 400, 1000, 4000, and 10 000 keV at L-values between 1 and 7. Solid curves are doubly charged helium ions (α particles), and dashed curves are singly charged helium ions (from Spjeldvik and Fritz, 1978a). (© by American Geophysical Union)

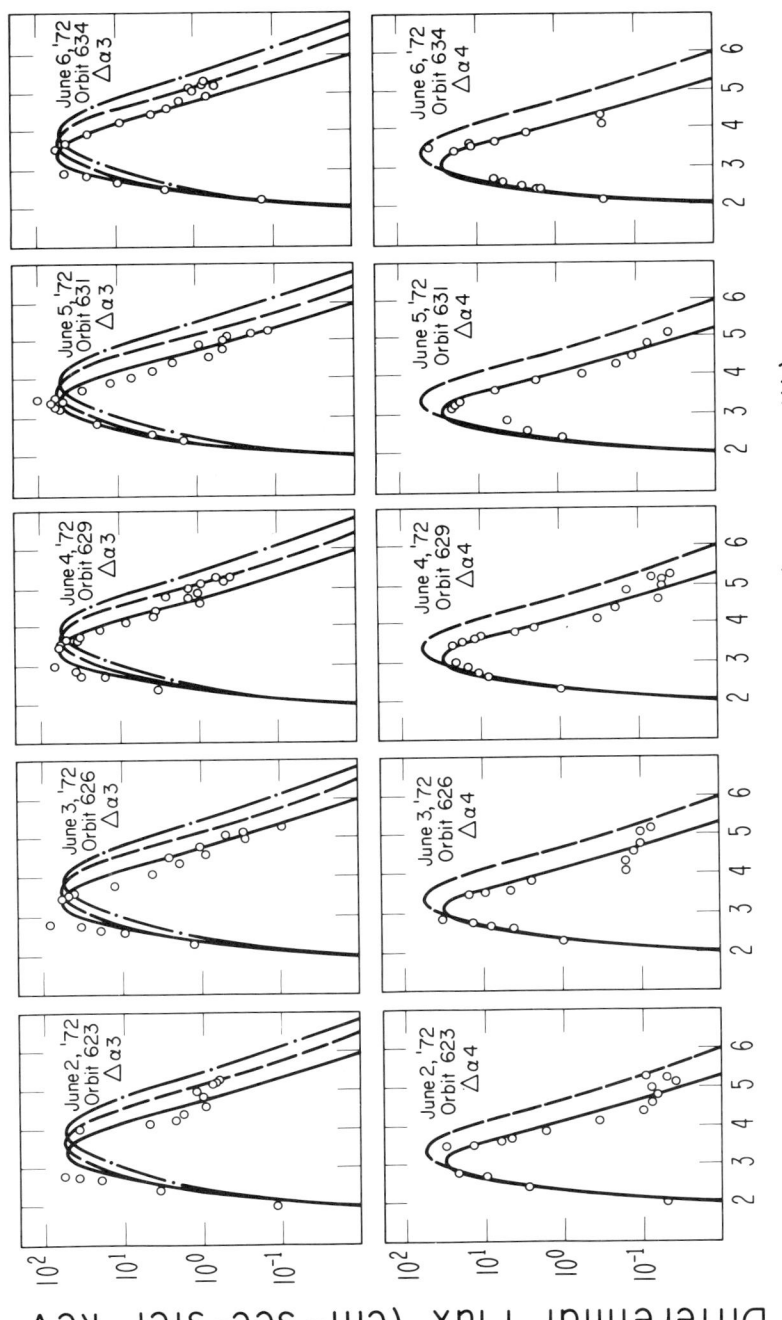

Fig. 5.52. The observed sum (circles) of He$^+$ and He^{++} fluxes in two energy channels, $\Delta\alpha 3$ (1.16–1.74 MeV) and $\Delta\alpha 4$ (1.74–3.15 MeV) from the Explorer 45 satellite. Observations are shown from five consecutive geomagnetically quiet days. The lines are the results of the theoretical calculations for He^{++} ions, the calculated He$^+$ fluxes being lower than the He^{++} fluxes at these energies. In the upper panels, theoretical results are shown at 1.2 MeV (dot-dashed), 1.5 MeV (dashed), and 2.0 MeV (solid) and compared with the $\Delta\alpha 3$ observations. In the lower panels, theoretical results are shown at 2 MeV (dashed) and 3 MeV (solid) and compared with the $\Delta\alpha 4$ observations (from Spjeldvik and Fritz, 1978a). (© by American Geophysical Union)

220   CHAPTER 5

tion, Spjeldvik and Fritz considered two extreme alternatives: a solely O^+ and a solely O^{6+} population at $L = 6.6$. A solely O^+ population would apply for ions coming directly out of the ionosphere, and solely O^{6+} population would be appropriate if the oxygen ions at $L = 6.6$ came directly from the solar wind. Figure 5.53 shows the calculated radial profiles of oxygen ions for each charge state at selected energies. Results are shown for an ionospheric source in the left-hand panels and for a solar wind source in the right-hand panels. All parameters used in this study, other than the charge exchange cross sections and the outer boundary condition, were the same as those used in the proton and helium ion equilibrium calculations.

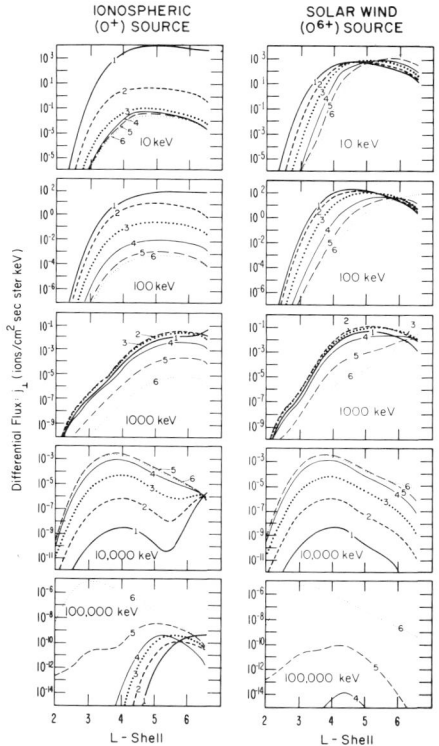

Fig. 5.53. Computed atomic oxygen ion fluxes vs L at selected energies. The six lowest positive charge states are permitted. Charge state numbers are given on the curves. The left panels show the results for a solely O^+ source at the outer boundary (ionospheric source), and the right-hand panels show results for a solely O^{6+} source at the outer boundary (solar wind source) (from Spjeldvik and Fritz, 1978b).
(© by American Geophysical Union)

An important result of Figure 5.53 is the similarity of the two sets of results well within the outer boundary, the similarity extending to higher L with increasing energy. At these radial distances, the charge state transformations generate a distribution of oxygen ions over charge states that is largely independent of the initial charge states. This gives a charge state distribution controlled by magnetospheric processes, and not by the charge state distribution of the outer zone source of oxygen ions.

All the equilibrium results we have so far discussed have used the rather low radial diffusion coefficient from magnetic fluctuations, $D_{om} = 2 \times 10^{-10} R_e^2$ day^{-1}, based on the paper of Nakada and Mead (1965). However, using synchronous orbit magnetic field observations, Lanzerotti et al. (1978) estimated that D_{om} could be significantly higher than this value, possibly approaching $10^{-8} R_e^2$ day^{-1}. From the results in Figure 5.35, we see that such an increase in the value of D_{om} could significantly affect the radial diffusion rates, particularly at energies $\gtrsim 1$ MeV. To determine the sensitivity of the equilibrium ion results to the value of D_{om}, Spjeldvik and Fritz (1978b) performed the proton, helium, and oxygen equilibrium ion calculation for values of D_{om} of 10^{-9} and $10^{-8} R_e^2$ day^{-1}. The resulting ion energy spectra at $L = 3.25$ for the different ions are shown in Figure 5.54 along with the results for $D_{om} = 2 \times 10^{-10}$. These results show that the oxygen ion fluxes at energies $\gtrsim 1$ MeV are by far the most sensitive to the value of D_{om}.

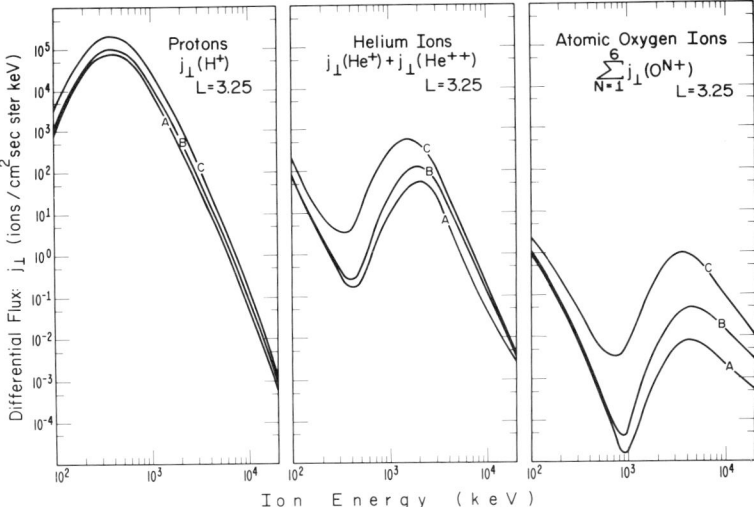

Fig. 5.54. Sensitivity of protons, helium ions, and oxygen ion calculations to variations in the radial diffusion coefficient from magnetic field fluctuations. The spectra shown for each ion species represent a flux summation over all applicable charge states at an L-shell of 3.25, where the ion fluxes generally maximize in the lower energy range. The curves labeled A, B, and C correspond to values for D_{om} of 2×10^{-10}, 10^{-9}, and $10^{-8} R_e^2$ day^{-1}, respectively (from Spjeldvik and Fritz, 1978b). (© by American Geophysical Union)

Calculated radial profiles of 4-MeV oxygen ions are shown in Figure 5.55 for $D_{om} = 10^{-8}$, 4×10^{-9}, 10^{-9}, and $2 \times 10^{-10} R_e^2$ day^{-1} and compared with 1.82–4.80 oxygen ion observations obtained from the Explorer 45 satellite by Spjeldvik and Fritz (1978b). Both the calculated and the observed fluxes are a sum over charge state. The dashed lines are for a solely O^{6+} source at $L = 6.6$, and the solid lines are for a solely O^+ source. The comparisons in this figure indicate that a value of $D_{om} \approx 4 \times 10^{-9} R_e^2$ day^{-1} might be a more reasonable value than $2 \times 10^{-10} R_e^2$ day^{-1}, though Spjeldvik and Fritz stated that the charge exchange cross sections and outer boundary energy spectrum were not sufficiently accurate to support a firm conclusion on the value of D_{om}.

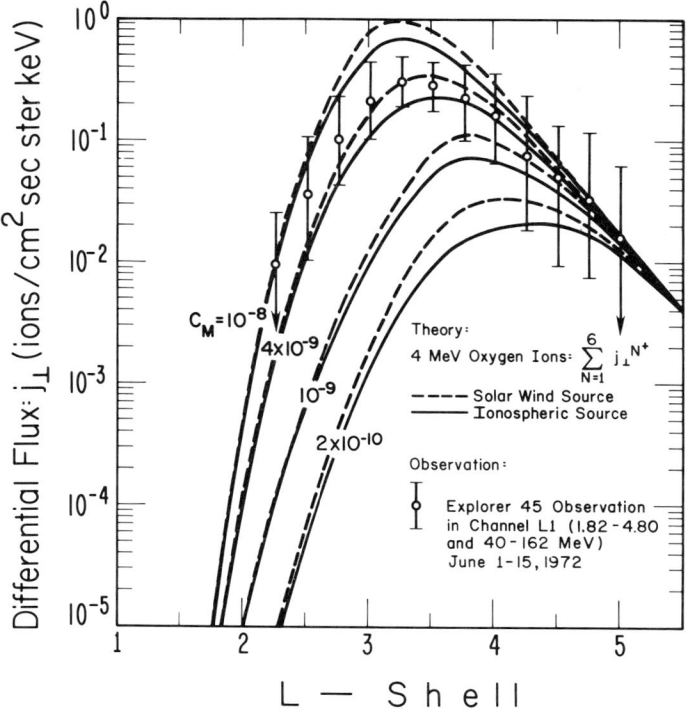

Fig. 5.55. Computed radial profiles of 4 MeV oxygen ions, summed over all charge states for $D_{om} = 10^{-8}$, 4×10^{-9}, 10^{-9}, and $10^{-8} R_e^2$ day^{-1}. Dashed lines are for a solely O^{6+} source at $L = 6.6$, and solid lines are for a solely O$^+$ source at $L = 6.6$. Data points are 1.82–4.80 oxygen ion observations from the Explorer 45 satellite. Standard deviations of the spread in the observed fluxes are indicated by vertical lines (from Spjeldvik and Fritz, 1978b). (© by American Geophysical Union)

Although the value of the coefficient D_{om} for radial diffusion from magnetic fluctuations has not yet been definitively tested, the < 1-MeV electron and proton comparisons provide a rather stringent test of the value of the coefficient D_{oe} for radial diffusion from electric potential field fluctuations. These results imply $D_{oe} \approx 1.3 \times 10^{-5} R_e^2$ day^{-1}, and Lyons and Thorne (1973) found that equilibrium calculations for electrons using a value of D_{oe} more than a factor of ~3 different from this value are in significant disagreement with radiation belt electron observations.

References

Alexander, J. K. and M. L. Kaiser, 'Terrestrial kilometric radiation, 1, Spatial structure studies', *J. Geophys. Res.* 81, 5948 (1976).

Ashour-Abdalla, M. and C. F. Kennel, 'Connective cold upper hybrid instabilities', in *Magnetospheric Particles and Fields*, B. M. McCormac (ed.), D. Reidel Publ. Co. Dordrecht, Holland, 181 (1976).

Ashour-Abdalla, M. and C. F. Kennel, 'Nonconvective and convective electron cyclotron harmonic instabilities', *J. Geophys. Res.* 83, 1531 (1978).

Ashour-Abdalla, M. and R. M. Thorne, 'The importance of electrostatic ion-cyclotron instability for quiet-time proton auroral precipitation', *Geophys. Res. Lett.* 4, 45 (1977).

Ashour-Abdalla, M. and R. M. Thorne, 'Toward a unified view of diffuse auroral precipitation', *J. Geophys. Res.* 83, 4755 (1978).

Ashour-Abdalla, M., G. Chanteur, and R. Pellat, 'A contribution to the theory of the electrostatic half-harmonic electron gyrofrequency waves in the magnetosphere', *J. Geophys. Res.* 80, 2775 (1975).

Balsiger, H., P. Eberhardt, J. Geiss, and D. T. Young, 'Magnetic storm injection of 0.9 to 16 keV/e solar and terrestrial ions in the high-altitude magnetosphere', *J. Geophys. Res.* 85, 1645 (1980).

Banks, P. M., 'Behavior of thermal plasma in the magnetosphere and topside ionosphere', in *Critical Problems of Magnetospheric Physics*, E. R. Dyer (ed.), Inter-Union Commission on Solar-Terrestrial Physics, c/o National Academy of Sciences, Washington, D.C., 157 (1972).

Benson, R. F. and W. Calvert, 'ISIS 1 observations at the source of auroral kilometric radiation', *Geophys. Res. Lett.* 6, 479 (1979).

Benson, R. F., W. Calvert, and D. M. Klumpar, 'Simultaneous wave and particle observations in the auroral kilometric radiation source region', *Geophys. Res. Lett.* 7, 959 (1980).

Bostrom, C. O., D. S. Beall, and J. C. Armstrong, 'Time history of the inner radiation zone, October 1963 to December 1968', *J. Geophys. Res.* 75, 1246 (1970).

Brinkmann, H. C. and H. C. Kramers, 'Zur Theorie der Einfangung von Elektronen duch α-Teilchen', *Kon. Akad. Wetenschapp. Amst.* 33, 973 (1930).

Brown, L. W., 'The galactic radio spectrum between 130 kHz and 2600 kHz, *Astrophys. J.* 180, 359 (1973).

Calvert, W., 'The signature of auroral kilometric radiation on ISIS 1 ionograms', *J. Geophys. Res.* 86, 76 (1981a).

Calvert, W., 'The auroral plasma cavity', *Geophys. Res. Lett.* 8, 919 (1981b).

Carpenter, D. L. and N. T. Seely, 'Cross-L plasma drifts in the outer plasmasphere: quiet time patterns and some substorm effects', *J. Geophys. Res.* 81, 2728 (1976).

Carpenter, D. L., K. Stone, J. C. Siren, and T. L. Crystal, 'Magnetospheric electric fields deduced from drifting whistler paths', *J. Geophys. Res.* 77, 2819 (1972).

Chappell, C. R., K. K. Harris, and G. W. Sharp, 'The morphology of the bulge region of the plasmasphere', *J. Geophys. Res.* 75, 3848 (1970).

Christiansen, P., M. P. Gough, G. Martelli, J. J. Block, N. Cornilleau, J. Etcheto, R. Gendrin, C. Beghin, P. Decreau, and D. Jones, 'GEOS-1 observations of electrostatic waves and their relationship with plasma parameters', *Space Sci. Rev.* 22, 383 (1978).

Claflin, E. S., 'Charge exchange cross sections for hydrogen and helium ions incident on atomic hydrogen: 1 to 1000 keV', Rep. TR-0059(6269-20)-1, Aerospace Corp., El Segundo, Calif. (1970).

Cornwall, J. M., 'Scattering of energetic trapped electrons by very-low-frequency waves', *J. Geophys. Res.* 69, 1251 (1964).

Cornwall, J. M., 'Cyclotron instabilities and electromagnetic emissions in the ultra low frequency and very low frequency ranges', *J. Geophys. Res.* 70, 61 (1965).

Cornwall, J. M., 'Diffusion processes influenced by conjugate-point wave phenomena', *Radio Sci.* 3, 740 (1968).

Cornwall, J. M., 'Radial diffusion of ionized helium and protons: a probe for magnetosphere dynamics', *J. Geophys. Res.* 77, 1756 (1972).

Cornwall, J. M., F. V. Coroniti, and R. M. Thorne, 'Turbulent loss of ring current protons', *J. Geophys. Res.* 75, 4699 (1970).

Craven, J. D., 'Temporal variations of electron intensities at low altitudes in the outer radiation zone as observed with satellite Injun 3', *J. Geophys. Res.* 71, 5643, 1966.

Croley, D. R., Jr., P. F. Mizera, and J. F. Fennell, 'Signature of a parallel electric field in ion and electron distribution functions in velocity space', *J. Geophys. Res.* 83, 2701 (1978).

Davis, L. D., Jr. and D. B. Chang, 'On the effect of geomagnetic fluctuations on trapped particles', *J. Geophys. Res.* 67, 2169 (1962).

DeForest, S. E. and C. E. McIlwain, 'Plasma clouds in the magnetosphere', *J. Geophys. Res.* 76, 3587 (1971).

Dessler, A. J. and E. N. Parker, 'Hydromagnetic theory of geomagnetic storms', *J. Geophys. Res.* **64**, 2239 (1959).

Dungey, J. W., 'Loss of Van Allen electrons due to whistlers', *Planet. Space Sci.* **11**, 591 (1963).

Dungey, J. W., 'Effects of electromagnetic perturbations on particles trapped in the radiation belts', *Space Science Rev.* **4**, 199 (1965).

Dunkel, N., B. Ficklin, L. Rorden, and R. A. Helliwell, 'Low-frequency noise observed in the distant magnetosphere with OGO 1', *J. Geophys. Res.* **75**, 1854 (1970).

Dusenbery, P. B. and R. L. Kaufmann, 'Properties of the longitudinal dielectric function: an application to the auroral plasma', *J. Geophys. Res.* **85**, 5969 (1980).

Dusenbery, P. B. and L. R. Lyons, 'General concepts on the generation of auroral kilometric radiation', *J. Geophys. Res.* **87**, 7467 (1982).

Fälthammar, C.-G., 'Effects of time-dependent electric fields on geomagnetically trapped radiation', *J. Geophys. Res.* **70**, 2503 (1965).

Fälthammar, C.-G., 'On the transport of trapped particles in the outer magnetosphere', *J. Geophys. Res.* **71**, 1487 (1966).

Fälthammar, C.-G., 'Radial diffusion by violation of the third adiabatic invariant', in *Earth's Particles and Fields*, B. M. McCormac (ed.), Reinhold, New York, 157 (1968).

Fite, W. L., R. F. Stebbings, D. G. Hummer, and R. T. Brackmann, 'Ionization and charge transfer in proton-hydrogen atom collisions', *Phys. Rev.* **119**, 663 (1960).

Fredricks, R. W., 'Plasma instabilities at $(n+1/2)f_{ce}$ and its relationship to some satellite observations', *J. Geophys. Res.* **76**, 5344 (1971).

Fredricks, R. W. and F. L. Scarf, 'Recent studies of magnetospheric electric field emissions above the electron gyrofrequency', *J. Geophys. Res.* **78**, 310 (1973).

Fritz, T. A., C. W. Arthur, P. J. Coleman, Jr., W. D. Cummings, A. J. Masley, B. H. Mauk, C. E. McIlwain, R. L. McPherron, G. A. Paulikas, K. A. Pfitzer, D. L. Reasoner, P. R. Satterblom, S. Y. Su, R. J. Walker, E. C. Whipple, Jr., and J. R. Winckler, 'Significant initial results from ATS-6', NASA Technical Paper 1101 (1977).

Fritz, T. A. and W. N. Spjeldvik, 'Observations of energetic radiation belt helium ions at the geomagnetic equator during quiet times', *J. Geophys. Res.* **83**, 2579 (1978).

Fritz, T. A. and B. Wilken, 'Substorm-generated fluxes of heavy ions at the geostationary orbit', in *Magnetospheric Particles and Fields*, B. M. McCormac (ed.), D. Reidel Publ. Co., Dordrecht, Holland, 171 (1976).

Gallagher, D. L. and D. A. Gurnett, 'Auroral kilometric radiation: time-averaged source location', *J. Geophys. Res.* **84**, 6501 (1979).

Geiss, J., H. Balsiger, P. Eberhardt, H. P. Walker, L. Weber, D. T. Young, and H. Rosenbauer, 'Dynamics of magnetospheric ion composition as observed by the GEOS mass spectrometer', *Space Sci. Rev.* **22**, 537 (1978).

Green, J. L., D. A. Gurnett, and S. D. Shawhan, 'The angular distribution of auroral kilometric radiation', *J. Geophys. Res.* **82**, 1825 (1977).

Gurnett, D. A., 'The earth as a radio source: Terrestrial kilometric radiation', *J. Geophys. Res.* **79**, 4227 (1974).

Gurnett, D. A., R. R. Anderson, F. L. Scarf, R. W. Fredricks, and E. J. Smith, 'Initial results from the ISEE−1 and −2 plasma waves investigation', *Space Sci. Rev.* **23**, 103 (1979).

Haerendel, G., 'Diffusion theory of trapped particles and the observed proton distribution', in *Earth's Particles and Fields*, B. M. McCormac (ed.), Reinhold, New York, 171 (1968).

Harris, E. G., 'Unstable plasma oscillations in a magnetic field', *Phys. Rev. Lett.* **2**, 34 (1959).

Helliwell, R. A. and J. P. Katsufrakes, 'VLF line radiation in the Earth's magnetosphere from Siple Station, Antarctica', *J. Geophys. Res.* **79**, 2511 (1974).

Hess, W. N., *The Radiation Belt and Magnetosphere*, Blaisdell Publ. Co., Waltham, Mass. (1968).

Hubbard, R. F. and T. J. Birmingham, 'Electrostatic emissions between electron gyroharmonics in the outer magnetosphere', *J. Geophys. Res.* **83**, 4837 (1978).

Hultqvist, B., W. Riedler, and H. Borg, 'Ring current protons in the upper atmosphere within the plasmapause', *Planet. Space Sci.* **24**, 783 (1976).

Imhof, W. L., R. R. Anderson, J. B. Reagan, and E. E. Gaines, 'Coordinated measurements of slot region electron precipitation by plasmaspheric wave bands', *J. Geophys. Res.* **87**, 4418 (1982).

Imhof, W. L., E. E. Gaines, and J. B. Reagan, 'Evidence for the resonance precipitation of energetic electrons from the slot region of the radiation belts', *J. Geophys. Res.* 79, 3141 (1974).

Jelly, D. and N. Brice, 'Changes in Van Allen radiation associated with polar substorms', *J. Geophys. Res.* 72, 5919 (1967).

Johnson, R. G., R. D. Sharp, and E. G. Shelley, 'Composition of the hot plasmas in the magnetosphere', in *Physics of the Hot Plasmas in the Magnetosphere*, B. Hultqvist and L. Stenflo (eds.), Plenum Press, New York, 45 (1975).

Johnson, R. G., R. D. Sharp, and E. G. Shelley, 'Observations of ions of ionospheric origin in the storm time ring current', *Geophys. Res. Lett.* 4, 403 (1977).

Joselyn, J. A. and L. R. Lyons, 'Ion cyclotron wave growth calculated from observations of the proton ring current during storm recovery', *J. Geophys. Res.* 81, 2275 (1976).

Kennel, C. F., 'Consequences of a magnetospheric plasma', *Rev. Geophys.* 1, 379 (1969).

Kennel, C. F. and F. Engelmann, 'Velocity space diffusion from weak plasma turbulence in a magnetic field', *Phys. Fluids* 9, 2377 (1966).

Kennel, C. F. and H. E. Petschek, 'Limit on stably trapped particle fluxes', *J. Geophys. Res.* 71, 1 (1966).

Kennel, C. F. and H. V. Wong, 'Resonant particle instabilities in a uniform magnetic field', *J. Plasma. Phys.* 1, 75 (1967a).

Kennel, C. F. and M. V. Wong, 'Resonantly unstable off-angle hydromagnetic waves', *J. Plasma Phys.* 1, 81 (1967b).

Kennel, C. F., F. L. Scarf, R. W. Fredricks, J. H. McGehee, and F. V. Coroniti, 'VLF electric field observations in the magnetosphere', *J. Geophys. Res.* 75, 6136 (1970).

Kimura, I., 'Effects of ions on whistler-mode ray tracing', *Radio Science* 1 (new series), 269 (1966).

Kurth, W. S., M. M. Baumback, and D. A. Gurnett, 'Direction-finding measurements of auroral kilometric radiation', *J. Geophys. Res.* 80, 2764 (1975).

Kurth, W. S., J. D. Craven, L. A. Frank, and D. A. Gurnett, 'Intense electrostatic waves near the upper hybrid resonance frequency', *J. Geophys. Res.* 84, 4145 (1979a).

Kurth, W. S., M. Ashour-Abdalla, L. A. Frank, C. F. Kennel, D. A. Gurnett, D. D. Sentman, and B. G. Burek, 'A comparison of intense electrostatic waves near f_{UHR} with linear instability theory', *Geophys. Res. Lett.* 6, 487 (1979b).

Kurth, W. S., L. A. Frank, M. Ashour-Abdalla, D. A. Gurnett, and B. G. Burek, 'Observations of a free-energy source for intense electrostatic waves', *Geophys. Res. Lett.* 7, 293 (1980).

Lanzerotti, L. J., D. C. Webb, and C. W. Arthur, 'Geomagnetic field fluctuations at synchronous orbit. 2. Radial diffusion', *J. Geophys. Res.* 83, 3866 (1978).

Lennartsson, W., R. D. Sharp, E. G. Shelley, R. G. Johnson, and H. Balsiger, 'Ion composition and energy distribution during 10 magnetic storms', *J. Geophys. Res.* 86, 4628 (1981).

Lerche, I., 'Quasilinear theory of resonant diffusion in a magneto-active relativistic plasma', *Phys. Fluids* 11, 1720 (1968).

Lincoln, J. V., 'Geomagnetic and solar data', *J. Geophys. Res.* 78, 2375 (1973).

Lo, H., H. Kurzweg, R. T. Brackman, and W. L. Fite, 'Electron capture and loss in collisions of heavy ions with atomic oxygen', *Phys. Rev. Sect. A* 4, 1462 (1971).

Lundblad, J. Å., F. Soraas, and K. Aarsnes, 'Substorm morphology of > 100 keV protons', *Planet. Space Sci.* 27, 841 (1979).

Lundin, R., L. R. Lyons, and N. Pissarenko, 'Observations of the ring current composition at $L < 4$', *Geophys. Res. Lett.* 7, 425 (1980).

Lyons, L. R., 'Comments on pitch-angle diffusion in the radiation belts', *J. Geophys. Res.* 78, 6793 (1973).

Lyons, L. R., 'General relations for particle diffusion in pitch angle and energy', *J. Plasma Phys.* 12, 45 (1974a).

Lyons, L. R., 'Pitch angle and energy diffusion coefficients from resonant interactions with ion-cyclotron and whistler waves', *J. Plasma Phys.* 12, 417 (1974b).

Lyons, L. R., 'Electron diffusion driven by magnetospheric electrostatic waves', *J. Geophys. Res.* 79, 575 (1974c).

Lyons, L. R., 'Trapped particles and waves and what can be learned from multi-satellite experiments, in *The Scientific Satellite Programme During the International Magnetospheric Study*, K. Knott and B. Battricks (eds.), D. Reidel Publ. Co., Dordrecht, Holland, 237 (1976).

Lyons, L. R. and D. S. Evans, 'The inconsistency between proton charge exchange and the observed ring current decay', *J. Geophys. Res.* **81**, 6197 (1976).

Lyons, L. R. and T. Speiser, 'Evidence for current sheet acceleration in the geomagnetic tail', *J. Geophys. Res.* **87**, 2276 (1982).

Lyons, L. R. and R. M. Thorne, 'The magnetospheric reflection of whistlers', *Planet. Space Sci.* **18**, 1753 (1970).

Lyons, L. R. and R. M. Thorne, 'Equilibrium structure of radiation belt electrons', *J. Geophys. Res.* **78**, 2142 (1973).

Lyons, L. R. and D. J. Williams, 'The quiet time structure of energetic (35–560 keV) radiation belt electrons', *J. Geophys. Res.* **80**, 943 (1975a).

Lyons, L. R. and D. J. Williams, 'The storm and post-storm evolution of energetic (35–560 keV) radiation belt electrons', *J. Geophys. Res.* **80**, 3985 (1975b).

Lyons, L. R. and D. J. Williams, 'A comment on the effects of man-made VLF waves on the radiation belts', *Geophys. Res. Lett.* **5**, 116 (1978).

Lyons, L. R., R. M. Thorne, and C. F. Kennel, 'Pitch-angle diffusion of radiation belt electrons within the plasmasphere', *J. Geophys. Res.* **77**, 3455 (1972).

MacDonald, J. R. and F. W. Martin, 'Experimental electron-transfer cross sections for collisions of oxygen ions in argon, nitrogen and helium at energies of 7–40 MeV', *Phys. Rev. Sect. A* **4**, 1965 (1971).

Mozer, F. S., 'Power spectra of the magnetospheric electric field', *J. Geophys. Res.* **76**, 3651 (1971).

Nakada, N. P. and G. D. Mead, 'Diffusion of protons in the outer radiation belt', *J. Geophys. Res.* **70**, 4777 (1965).

Newkirk, L. L. and M. Walt, 'Radial diffusion coefficient for electrons at $1.76 < L < 5$', *J. Geophys. Res.* **73**, 7231 (1968).

Omidi, N. and D. A. Gurnett, 'Growth rate calculations of auroral kilometric radiation using the relativistic resonance condition', *J. Geophys. Res.* **87**, 2377 (1982).

Owens, H. D. and L. A. Frank, 'Electron omnidirectional intensity contours in the earth's outer radiation-zone at the magnetic equator', *J. Geophys. Res.* **73**, 199 (1968).

Parady, B. K., D. D. Eberlein, J. A. Marvin, W. W. L. Taylor, and L. J. Cahill, Jr., 'Plasmaspheric hiss observations in the evening and afternoon quadrants', *J. Geophys. Res.* **80**, 2183 (1975).

Park, C. G., 'Whistler observations of the interchange of ionization between the ionosphere and the protonsphere', *J. Geophys. Res.* **75**, 4249 (1970).

Park, C. G., 'Some features of plasma distribution in the plasmasphere deduced from antarctic whistlers', *J. Geophys. Res.* **75**, 169 (1974).

Parker, E. N., 'Geomagnetic fluctuations and the form of the outer zone of the Van Allen radiation belt', *J. Geophys. Res.* **65**, 3117 (1960).

Pfitzer, K., S. Kane, and J. R. Winckler, 'The spectra and intensity of electrons in the radiation belts', *Space Res.* **6**, 702 (1966).

Pizzella, G. and L. A. Frank, 'Energy spectrums for proton (200 eV $\lesssim E \lesssim$ 1 MeV) intensities in the outer radiation zone', *J. Geophys. Res.* **76**, 88 (1971).

Roberts, C. S., 'Electron loss from the Van Allen zones due to pitch angle scattering by electromagnetic disturbances', in *Radiation Trapped in the Earth's Magnetic Field*, B. M. McCormac (ed.), D. Reidel Publ. Co., Dordrecht, Holland, 402 (1966).

Roberts, C. S., 'Pitch angle diffusion of electrons in the magnetosphere', *Rev. Geophys.* **7**, 305 (1969).

Rönmark, K., H. Borg, P. J. Christiansen, M. P. Gouph, and D. Jones, 'Banded electron cyclotron harmonic instability – a first comparison of theory and experiment', *Space Sci. Rev.* **22**, 401 (1978).

Rothwell, P. and C. Lynam, 'The plasmapause, the plasma sheet, and energetic trapped electrons in the earth's magnetosphere', *Planet. Space Sci.* **17**, 447 (1969).

Russell, C. T., R. E. Holzer, and E. J. Smith, 'Observations of ELF noise in the magnetosphere, 1. Spatial extent and frequency of occurrence', *J. Geophys. Res.* **74**, 755 (1969).

Russell, C. T. and R. M. Thorne, 'On the structure of the inner magnetosphere', *Cosmic Electrodynamics* **1**, 67 (1970).

Scarf, F. L., R. W. Fredricks, C. F. Kennel, and F. V. Coroniti, 'Satellite studies of magnetospheric substorms on August 15, 1968: OGO 5 plasma wave observations', *J. Geophys. Res.* **78**, 3119 (1973).

Schield, M. A. and L. A. Frank, 'Electron observations between the inner edge of the plasma sheet and the plasmapause', *J. Geophys. Res.* **75**, 5401 (1970).

Schulz, M., 'Particle lifetimes in strong diffusion', *Astrophys. and Space Sci.* **31**, 37 (1974).

Schulz, M. and L. J. Lanzerotti, *Particle Diffusion in the Radiation Belts*, Springer, New York (1974).

Sentman, D. P., L. A. Frank, C. F. Kennel, D. A. Gurnett, and W. S. Kurth, 'Electron distribution functions associated with electrostatic emissions in the dayside magnetosphere', *Geophys. Res. Lett.* **6**, 781 (1979).

Sharp, R. D., D. L. Carr, R. G. Johnson, and E. G. Shelley, 'Coordinated auroral-electron observations from a synchronous and a polar satellite', *J. Geophys. Res.* **76**, 7669 (1971).

Sharp, R. D., R. G. Johnson, and E. G. Shelley, 'The morphology of energetic O^+ ions during two magnetic storms: Temporal variations', *J. Geophys. Res.* **81**, 3283 (1976a).

Sharp, R. D., R. G. Johnson, and E. G. Shelley, 'The morphology of energetic O^+ ions during two magnetic storms: latitudinal variations', *J. Geophys. Res.* **81**, 3292 (1976b).

Sharp, R. D., E. G. Shelley, and R. G. Johnson, 'A search for helium ions in the recovery phase of a magnetic storm', *J. Geophys. Res.* **82**, 2361 (1977).

Shaw, R. R. and D. A. Gurnett, 'Electrostatic noise bands associated with the electron gyrofrequency and plasma frequency in the outer magnetosphere', *J. Geophys. Res.* **80**, 4259 (1975).

Shelley, E. G., R. G. Johnson, and R. D. Sharp, 'Satellite observations of energetic heavy ions during a geomagnetic storm', *J. Geophys. Res.* **77**, 6104 (1972).

Smith, P. H., N. K. Bewtra, and R. A. Hoffman, 'Inference of the ring current ion composition by means of charge exchange decay', *J. Geophys. Res.* **86**, 3470 (1981).

Solov'ev, E. S., R. N. Il'in, V. A. Oparin, I. T. Serenkov, and N. V. Fedorenko, 'Capture and loss of electrons by fast nitrogen and oxygen atoms and ions in air, nitrogen and oxygen', *Sov. Phys. Tech. Phys.* **17** (2), 267 (1972).

Spjeldvik, W. N., 'Equilibrium structure of equatorially mirroring radiation belt protons', *J. Geophys. Res.* **82**, 2801 (1977).

Spjeldvik, W. N., 'Expected charge states of energetic ions in the magnetosphere', *Space Sci. Rev.* **23**, 499 (1979).

Spjeldvik, W. N. and T. A. Fritz, 'Energetic ionized helium in the quiet time radiation belts: Theory and comparison with observation', *J. Geophys. Res.* **83**, 654 (1978a).

Spjeldvik, W. N. and T. A. Fritz, 'Theory for charge states of energetic oxygen ions in the earth's radiation belts', *J. Geophys. Res.* **83**, 1583 (1978b).

Spjeldvik, W. N. and T. A. Fritz, 'Observations of energetic helium ions in the earth's radiation belts during a sequence of geomagnetic storms', *J. Geophys. Res.* **80**, 2317 (1981).

Stebbings, R. F., W. L. Fite, and D. G. Hummer, 'Charge transfer between atomic hydrogen and N^+ and O^+', *J. Chem. Phys.* **33**, 1226 (1960).

Stix, T. H., *The Theory of Plasma Waves*, McGraw Hill, New York (1962).

Stuart, G. W., 'Satellite measured radiation., *Phys. Rev. Lett.* **2**, 417 (1959).

Theodoridis, G. C. and F. R. Paolini, 'Pitch-angle diffusion of relativistic outer belt electrons', *Ann. Geophys.* **23**, 375 (1967).

Thorne, R. M. and C. F. Kennel, 'Quasi-trapped VLF propagation in the outer magnetosphere', *J. Geophys. Res.* **72**, 857 (1967).

Thorne, R. M. and B. T. Tsurutani, Power line radiation: Can it significantly affect the earth's radiation belts?', *Science* **204**, 839 (1979).

Thorne, R. M. and B. T. Tsurutani, 'Comment on "Sunday decreases in magnetospheric VLF wave activity"', *J. Geophys. Res.* **86**, 1639 (1981).

Thorne, R. M., E. J. Smith, R. K. Burton, and R. E. Holzer, 'Plasmaspheric hiss', *J. Geophys. Res.* **78**, 1581 (1973).

Thorne, R. M., S. R. Church, and D. J. Gorney, 'On the origin of plasmaspheric hiss: The importance of wave propagation and the plasmapause', *J. Geophys. Res.* **84**, 5241 (1979).

Tinsley, B. A., 'Evidence that the recovery phase ring current consists of helium ions', *J. Geophys. Res.* **81**, 6193 (1976).

Tverskoy, B. A., 'Dynamics of the radiation belts of the earth. 2', *Geomagn. Aeron.* **4**, 351 (1964).

Tverskoy, B. A., 'Transport and acceleration of charged particles in the earth's magnetosphere', *Geomagn. Aeron.* **5**, 617 (1965).

Tverskoy, B. A., 'Stability of the earth's radiation belts', *Geomag. Aeron.* **7**, 177 (1967).

Vampola, A. L., 'Natural variations in the geomagnetically trapped electron population', Report No. TR-0059(6260-20)-17, Aerospace Corp., Los Angeles, Calif. (1971).

Vernov, S. N., E. V. Gorchakov, S. N. Kuznetsov, Yu. I. Logachev, E. N. Sosnovets, and V. G. Stolpovsky, 'Particle fluxes in the outer geomagnetic field', *Rev. Geophys. Space. Phys.* **7**, 257 (1969).

Walt, M., 'Radial diffusion of trapped particles', in *Particles and Fields in the Magnetosphere*, B. M. McCormac (ed.), D. Reidel Publ. Co., Dordrecht, Holland, 410 (1970).

Walt, M. and W. M. MacDonald, 'Diffusion of electrons in the Van Allen radiation belt, 1. Treatment of particles with mirror points at high altitude', *J. Geophys. Res.* **67**, 5013 (1962).

Wentworth, R. C., W. M. MacDonald, and S. F. Singer, 'Lifetimes of trapped radiation belt particles determined by Coulomb scattering', *Phys. Fluids* **2**, 499 (1959a).

Wentworth, R. C., W. M. MacDonald, and S. F. Singer, Lifetimes of trapped radiation belt particles, *Bull. Am. Phys. Soc. Ser. 2* **4**, 7 (1959).

West, H. I., Jr., R. M. Buck, and J. R. Walton, 'Electron pitch angle distributions throughout the magnetosphere as observed on OGO5', *J. Geophys. Res.* **78**, 1064 (1973).

Whalen, B. A. and I. B. McDiarmid, 'Pitch angle diffusion of low-energy auroral electrons', *J. Geophys. Res.* **78**, 1608 (1973).

Williams, D. J., 'Phase space variations of near equatorially mirroring ring current ions', *J. Geophys. Res.* **86**, 189 (1981).

Williams, D. J. and L. R. Lyons, 'The proton ring current and its interaction with the plasmapause: Storm recovery phase', *J. Geophys. Res.* **79**, 4195 (1974a).

Williams, D. J. and L. R. Lyons, 'Further aspects of the proton ring current interaction with the plasmapause: Main and recovery phases', *J. Geophys. Res.* **79**, 4791 (1974b).

Williams, D. J. and A. M. Smith, 'Daytime trapped electron intensities at high latitudes at 1100 km', *J. Geophys. Res.* **70**, 541 (1965).

Williams, D. J., J. F. Arens, and L. J. Lanzerotti, 'Observations of trapped electrons at low and high altitudes', *J. Geophys. Res.* **73**, 5673 (1968).

Williams, D. J., T. A. Fritz, and A. Konradi, 'Observations of proton spectra ($1.0 \leq E_p \leq 300$ keV) and fluxes at the plasmapause', *J. Geophys. Res.* **78**, 4751 (1973).

Williams, D. J., G. Hernandez, and L. R. Lyons, 'Simultaneous observations of the proton ring current and stable auroral red arcs', *J. Geophys. Res.* **81**, 608 (1976).

Wu, C. S. and L. C. Lee, 'A theory of terrestrial kilometric radiation', *Astrophys. J.* **230**, 621 (1979).

Wu, C. S., H. K. Wong, D. J. Gorney, and L. C. Lee, 'Generation of the auroral kilometric radiation', *J. Geophys. Res.* **87**, 4476 (1982).

Young, T. S. T., J. D. Callen, and J. E. McCune, 'High-frequency electrostatic waves in the magnetosphere', *J. Geophys. Res.* **78**, 1082 (1973).

INDEX

Adiabatic invariants 3, 6, 7, 29, 30, 46, 47, 80, 196
    First 45, 80, 81
    Integral 14, 29
    Second 13, 45
    Third 20
Alfven layer 113
Argus nuclear detonations 28–30
Aurora 3, 61, 99
    Diffuse 3, 98, 99, 193–195
    Discrete 3, 99ff, 145, 146
    Oval (zone) 2, 62, 112, 116
    Polar cap 112
Auroral electrojet 62

B, L coordinate system 29–32
Backscattered electrons 101, 102
Betatron acceleration 3, 4, 48
Bounce motion between mirror points 7, 11–14
    Frequency of 13, 25
    Period of 13, 24, 26
Bow shock 2

Charge exchange 3, 51–53, 206, 207, 211–217
Cold plasma waves 142, 143
    Cutoff frequency for right-hand waves 146, 147, 150, 151, 156
    Dispersion relation 142
    Linear growth rate for 142, 143
    Lower hybrid frequency 159
    Quasi linear diffusion equation for 143
    Upper hybrid frequency 146, 147
Collisions
    Auroral electrons 106
    Ionosphere 60
    Proton-neutral 1
    Radiation belt electrons 176, 177, 199–201, 205
    Radiation belt ions 200, 207
Corotation 2, 4, 50, 75, 76, 79, 80
Current sheet energization 86ff, 189
Current sheet particle motion 87–91
Currents
    Ionosphere 60–63, 113, 114
    Magnetic field aligned 56, 104, 106, 109–120, 123, 125

Magnetopause 30
Magnetosphere tail 2, 30, 86, 98
Trapped particle 44

Dessler-Parker relation 42
Differential flux, relation to phase space distribution function 20, 21, 201
Diffusion in velocity space 133–135, 139–141, 172, 173, 200
    Pitch angle 40, 50–53, 101, 111, 112, 156, 157, 173–178, 182–186, 191, 193–195, 199
        Strong 166, 167, 183, 193
        Weak 167
        Within the radiation belts, general 163–168
    Quasi linear diffusion equation 139–141
Double layers 112
Drift motion of particles 14, 15, 18–20
    Around earth 2, 7, 36–40, 80
    Cold plasma 75–79
    Curvature 2, 16
    Electric field 3, 15, 19, 20, 48, 50, 56, 60, 61, 75, 92
    Energetic plasmma 79–86
    Frequency of 18, 25, 197
    Gradient 2, 15–17
    Inertial 17
    Period of 19, 20, 25, 26
    Polarization 18
    Velocity of 18, 19, 25

Electric fields 40, 99
    Convection (cross-tail) 3, 48, 50, 56ff, 86–89, 92, 98, 112
    Corotation 75, 76, 79
    Divergence 108, 112, 113, 116, 120, 123, 125
    Ionosphere 59–64, 108, 112–125
    Parallel 3, 4, 49, 56, 100–125, 148
Energy
    Flux of precipitating auroral electrons 100–108, 111, 115–122, 124, 125
    Storage 2, 3
    Transfer in tail current sheet 86, 87, 98, 99
Euler potentials 30

INDEX

Fermi acceleration 3, 4, 48
Field line interconnection (merging, reconnection) 4

Geomagnetic storms 32, 34, 41, 44–47, 50, 168–171, 182–190, 211
Guidling center approximation 6–8, 56, 87
Gyromotion 6–10
    Frequency of 10, 24
    Period of 10, 24, 26
    Radius of 9, 24

Hall conductivity 60
Hall current 60, 61, 113
Height-integrated ionospheric current 60, 61
Helium ions 3, 52, 212–219, 221
"Inverted-V" event 108, 112, 116, 117, 125

Ionosphere
    Collisions 60
    Currents 60–64
        S_D 62, 63
        S_q 63
        S_q^p 63
    Electric field 59–64, 112
    Electrons 2, 109
    Ions 2, 49, 109
    Particle deposition from magnetosphere 52
    Particle source for magnetosphere 3, 48, 49
    Plasma within magnetosphere 77, 78

Jupiter 3, 4

L-shell splitting 36, 37
L-valve 29, 30
    Of plasmapause 79
Limit of resonant diffusion 133, 134
Lioville's Theorem 20, 40, 46, 66
Loss cone 3, 13, 39, 66, 92, 98, 104, 110, 111, 148, 157, 164, 167, 182, 183
Low latitude boundry layer 2

Magnetic field
    Arc length along field lines 8
    Curvature 16, 17, 40
    Dipole 7, 8, 11
    Earth 1–3, 7, 8, 23, 28–40, 45, 56–59, 67, 68, 73, 74, 112
    Earth's dipole moment 8
    Field lines 8
    Ground response to ionospheric currents 60, 62–64
    Interplanetary 56–59, 64, 67, 68, 73, 74
Magnetic moment 7, 9, 10

Magnetopause 1, 2, 30, 40, 77, 87
Magnetosheath 2, 40, 93, 112, 117
Magnetosphere
    General 1, 3, 4, 74, 75
Magnetospheric tail 2, 3, 30, 40, 75, 79, 86–89, 92, 98, 99
Magnetotail boundary layer 2, 92
Mercury 4
Mirror points 12–14

Neutral sheet 2

Omnidirectional flux 22
Oxygen ions 3, 47, 52, 213–217, 220–222

Particle acceleration, general 4
Particle precipitation
    Auroral
        Ions 98, 189, 195
        Electrons 3, 99ff, 147, 193–195
    Lifetime for radiation belt particles 165–168, 178, 180, 182
    Oxygen ions 213, 214
    Polar caps 66–73
    Radiation belt electrons 178
    Radiation belt ions 189, 213, 214
Particle trapping 2, 28–32, 50
    Associated currents 41–43, 50
Particle trapping boundary 32, 34, 40
Pedersen conductivity 60, 114–119, 124
Pedersen current 60, 114
Phase space distribution function 20, 201
    Relation to differential flux 20, 201
Pitch angle 10–13
Plasma mantle 2, 92–95
Plasma sheet 2, 3, 48, 50, 92, 95–98, 117
Plasmapause 52, 53, 77–79, 157–163, 182, 189, 202
Plasmasphere 2, 48, 52, 157–163, 182, 188, 189, 207
Polar caps 56–59, 61, 62, 64–75, 112
    Potential difference across 59, 74, 75, 77, 78
Pseudo-trapping 36, 38, 40

R, λ coordinate system 30
Radial diffusion 195ff
    From electric potential field fluctuations 196–199, 201, 208, 222
    Of electrons 201–205
    From magnetic field fluctuations 196–199, 208, 221, 222
    Of ions 207–209, 217–222
Radiation belts 2, 3, 157, 159, 199
    Electron loss in slot region 172–180

INDEX

Energetic oxygen distribution 214, 215, 217, 220, 221
Energetic electron distribution 168–171, 175–178, 202–205, 222
Energetic ion (ring current ion) distribution 182–187, 189, 208–215
Energetic helium distribution 214–217, 221
Energetic proton distribution 208, 212, 214, 215
Lifetime from precipitation into atmosphere 165–168, 178, 182, 200, 201
Pitch angle diffusion within, general 163–168
Pitch angle distributions, theory 166–168
Reasonance between waves and particles 133–139, 141, 163
  Cyclotron 133–136, 138, 139, 148–152, 157, 158, 172–178, 182, 186, 192, 193, 195, 199
    Characteristic energy for 157–159, 182
  Landau 133–136, 138, 148, 172–175
  Parallel particle energy for 142, 157, 182
Ring current 2, 3, 30, 41–53, 182, 189, 211, 214, 215
Riometer measurements 66, 67

Saturn 4
Secondary electrons 101, 102
Single wave characteristics 134–139, 141, 148, 150
Solar particles 66–73
Solar wind 1, 2, 48, 57, 63, 87
Sub-auroral red arcs 52, 53

Waves
  Associated with accelerated auroral electrons 101, 111, 112

Auroral kilometric radiation 3, 145ff
  Association with discrete aurora and density depletions 145–147
  Dispersion relation for 147, 150, 155
  Generation of 147–156
  Group velocity of 156
Electrostatic
  Dispersion relation for 191, 192
  Generation of 192, 193, 195
  Growth rate for 191, 192
  Interaction with trapped electrons 3, 193–195
  Interaction with trapped ions 195
  Observations of 191
Growth and damping, general 133, 135–138
Ion-cyclotron
  Dispersion relation for 181
  Generation of 157, 180, 182, 186, 189
  Group velocity 180
  Interaction with radiation belt ions 3, 52, 53, 180–189, 209, 212
  Observations of 186
Linear growth rate for resonant interactions 138, 139, 142, 143
Total energy 138
Whistlers, lightning generated 157
Whistler-mode
  Dispersion relation for 157
  Generation of 157–163
  Interaction with radiation belt electrons 172–180, 199–201, 205
  Limit to stably trapped electron flux 159
  Observations of 159, 163
  Propagation of 159–162
  Reflection of 159, 160
Whistler ducts 79

RAYMOND H. FOGLER LIBRARY
DATE DUE

BOOKS ARE SUBJECT TO
RECALL AFTER TWO WEEKS